OPTIMIZATION METHODS IN FINANCE

Optimization models are playing an increasingly important role in financial decisions. This is the first textbook devoted to explaining how recent advances in optimization models, methods and software can be applied to solve problems in computational finance ranging from asset allocation to risk management, from option pricing to model calibration more efficiently and more accurately. Chapters discussing the theory and efficient solution methods for all major classes of optimization problems alternate with chapters illustrating their use in modeling problems of mathematical finance.

The reader is guided through the solution of asset/liability cash flow matching using linear programming techniques, which are also used to explain asset pricing and arbitrage. Volatility estimation is discussed using nonlinear optimization models. Quadratic programming formulations are provided for portfolio optimization problems based on a mean-variance model, for returns-based style analysis and for risk-neutral density estimation. Conic optimization techniques are introduced for modeling volatility constraints in asset management and for approximating covariance matrices. For constructing an index fund, the authors use an integer programming model. Option pricing is presented in the context of dynamic programming and so is the problem of structuring asset backed securities. Stochastic programming is applied to asset/liability management, and in this context the notion of Conditional Value at Risk is described. The final chapters are devoted to robust optimization models in finance.

The book is based on Master's courses in financial engineering and comes with worked examples, exercises and case studies. It will be welcomed by applied mathematicians, operational researchers and others who work in mathematical and computational finance and who are seeking a text for self-learning or for use with courses.

GERARD CORNUEJOLS is an IBM University Professor of Operations Research at the Tepper School of Business, Carnegie Mellon University

REHA TÜTÜNCÜ is a Vice President in the Quantitative Resources Group at Goldman Sachs Asset Management, New York

OPTIMIZATION METHODS IN FINANCE

GERARD CORNUEJOLS

Carnegie Mellon University

REHA TÜTÜNCÜ

Goldman Sachs Asset Management

CAMBRIDGE
UNIVERSITY PRESS

CAMBRIDGE UNIVERSITY PRESS
Cambridge, New York, Melbourne, Madrid, Cape Town, Singapore, São Paulo, Delhi

Cambridge University Press
The Edinburgh Building, Cambridge CB2 8RU, UK

Published in the United States of America by Cambridge University Press, New York

www.cambridge.org
Information on this title: www.cambridge.org/9780521861700

© Gerard Cornuejols and Reha Tütüncü 2007

First published 2007
Reprinted 2009

Printed in the United Kingdom at the University Press, Cambridge

A catalog record for this publication is available from the British Library

ISBN-13   978-0-521-86170-0 hardback

To Julie
and
to Paz

Contents

Foreword

The use of sophisticated mathematical tools in modern finance is now commonplace. Researchers and practitioners routinely run simulations or solve differential equations to price securities, estimate risks, or determine hedging strategies. Some of the most important tools employed in these computations are optimization algorithms. Many computational finance problems ranging from asset allocation to risk management, from option pricing to model calibration, can be solved by optimization techniques. This book is devoted to explaining how to solve such problems efficiently and accurately using recent advances in optimization models, methods, and software.

Optimization is a mature branch of applied mathematics. Typical optimization problems have the objective of allocating limited resources to alternative activities in order to maximize the total benefit obtained from these activities. Through decades of intensive and innovative research, fast and reliable algorithms and software have become available for many classes of optimization problems. Consequently, optimization is now being used as an effective management and decision-support tool in many industries, including the financial industry.

This book discusses several classes of optimization problems encountered in financial models, including linear, quadratic, integer, dynamic, stochastic, conic, and robust programming. For each problem class, after introducing the relevant theory (optimality conditions, duality, etc.) and efficient solution methods, we discuss several problems of mathematical finance that can be modeled within this problem class. The reader is guided through the solution of asset/liability cash-flow matching using linear programming techniques, which are also used to explain asset pricing and arbitrage. Volatility estimation is discussed using nonlinear optimization models. Quadratic programming formulations are provided for portfolio optimization problems based on a mean-variance model for returns-based style analysis and for risk-neutral density estimation. Conic optimization techniques are introduced for modeling volatility constraints in asset management and for approximating covariance matrices. For constructing an index fund, we use an integer programming model. Option pricing is presented in the context of dynamic programming and so is the problem of structuring asset-backed securities. Stochastic programming is applied to asset/liability management, and in this context the notion of Conditional Value at Risk is described. Robust optimization models for portfolio selection and option pricing are also discussed.

This book is intended as a textbook for Master's programs in financial engineering, finance, or computational finance. In addition, the structure of chapters, alternating between optimization methods and financial models that employ these methods, allows the use of this book as a primary or secondary text in upper level undergraduate or introductory graduate courses in operations research, management science, and applied mathematics.

Optimization algorithms are sophisticated tools and the relationship between their inputs and outputs is sometimes opaque. To maximize the value one gets from these tools and to understand how they work, users often need a significant amount of guidance and practical experience with them. This book aims to provide this guidance and serve as a reference tool for the finance practitioners who use or want to use optimization techniques.

This book has its origins in courses taught at Carnegie Mellon University in the Masters program in Computational Finance and in the MBA program at the Tepper School of Business (Gérard Cornuéjols), and at the Tokyo Institute of Technology, Japan, and the University of Coimbra, Portugal (Reha Tütüncü). We thank the attendants of these courses for their feedback and for many stimulating discussions. We would also like to thank the colleagues who provided the initial impetus for this project or collaborated with us on various research projects that are reflected in the book, especially Rick Green, Raphael Hauser, John Hooker, Mark Koenig, Masakazu Kojima, Vijay Krishnamurthy, Yanjun Li, Ana Margarida Monteiro, Mustafa Pınar, Sanjay Srivastava, Michael Trick, and Luís Vicente. Various drafts of this book were experimented with in class by Javier Peña, François Margot, Miguel Lejeune, Miroslav Karamanov, and Kathie Cameron, and we thank them for their comments. Initial drafts of this book were completed when the second author was on the faculty of the Department of Mathematical Sciences at Carnegie Mellon University; he gratefully acknowledges their financial support.

1

Introduction

Optimization is a branch of applied mathematics that derives its importance both from the wide variety of its applications and from the availability of efficient algorithms. Mathematically, it refers to the minimization (or maximization) of a given *objective function* of several *decision variables* that satisfy functional *constraints*. A typical optimization model addresses the allocation of scarce resources among possible alternative uses in order to maximize an objective function such as total profit.

Decision variables, the objective function, and constraints are three essential elements of any optimization problem. Problems that lack constraints are called *unconstrained optimization* problems, while others are often referred to as *constrained optimization* problems. Problems with no objective functions are called *feasibility* problems. Some problems may have multiple objective functions. These problems are often addressed by reducing them to a single-objective optimization problem or a sequence of such problems.

If the decision variables in an optimization problem are restricted to integers, or to a discrete set of possibilities, we have an *integer* or *discrete optimization* problem. If there are no such restrictions on the variables, the problem is a *continuous optimization* problem. Of course, some problems may have a mixture of discrete and continuous variables. We continue with a list of problem classes that we will encounter in this book.

1.1 Optimization problems

We start with a generic description of an optimization problem. Given a function $f(x) : I\!R^n \rightarrow I\!R$ and a set $S \subset I\!R^n$, the problem of finding an $x^* \in I\!R^n$ that solves

$$\begin{aligned} &\min_x f(x) \\ &\text{s.t.} \quad x \in S \end{aligned} \qquad (1.1)$$

is called an optimization problem. We refer to f as the *objective function* and to S as the *feasible region*. If S is empty, the problem is called *infeasible*. If it is possible to find a sequence $x^k \in S$ such that $f(x^k) \to -\infty$ as $k \to +\infty$, then the problem is *unbounded*. If the problem is neither infeasible nor unbounded, then it is often possible to find a solution $x^* \in S$ that satisfies

$$f(x^*) \le f(x), \ \forall x \in S.$$

Such an x^* is called a *global minimizer* of the problem (1.1). If

$$f(x^*) < f(x), \ \forall x \in S, \ x \ne x^*,$$

then x^* is a *strict global minimizer*. In other instances, we may only find an $x^* \in S$ that satisfies

$$f(x^*) \le f(x), \ \forall x \in S \cap B_{x^*}(\varepsilon)$$

for some $\varepsilon > 0$, where $B_{x^*}(\varepsilon)$ is the open ball with radius ε centered at x^*, i.e.,

$$B_{x^*}(\varepsilon) = \{x : \|x - x^*\| < \varepsilon\}.$$

Such an x^* is called a *local minimizer* of the problem (1.1). A *strict local minimizer* is defined similarly.

In most cases, the feasible set S is described explicitly using functional constraints (equalities and inequalities). For example, S may be given as

$$S := \{x : g_i(x) = 0, i \in \mathcal{E} \ \text{and} \ g_i(x) \ge 0, i \in \mathcal{I}\},$$

where $\mathcal{E}$ and $\mathcal{I}$ are the index sets for equality and inequality constraints. Then, our generic optimization problem takes the following form:

$$
\begin{aligned}
\min_x \ & f(x) \\
& g_i(x) = 0, \quad i \in \mathcal{E} \\
& g_i(x) \ge 0, \quad i \in \mathcal{I}.
\end{aligned}
\tag{1.2}
$$

Many factors affect whether optimization problems can be solved efficiently. For example, the number n of decision variables, and the total number of constraints $|\mathcal{E}| + |\mathcal{I}|$, are generally good predictors of how difficult it will be to solve a given optimization problem. Other factors are related to the properties of the functions f and g_i that define the problem. Problems with a linear objective function and linear constraints are easier, as are problems with convex objective functions and convex feasible sets. For this reason, instead of general purpose optimization algorithms, researchers have developed different algorithms for problems with special characteristics. We list the main types of optimization problems we will encounter. A more complete list can be found, for example, on the *Optimization Tree* available from www-fp.mcs.anl.gov/otc/Guide/OptWeb/.

1.1.1 Linear and nonlinear programming

One of the most common and easiest optimization problems is *linear optimization* or *linear programming* (LP). This is the problem of optimizing a linear objective function subject to linear equality and inequality constraints. It corresponds to the case where the functions f and g_i in (1.2) are all linear. If either f or one of the functions g_i is not linear, then the resulting problem is a *nonlinear programming* (NLP) problem.

The standard form of the LP is given below:

$$\min_x c^{\mathrm{T}}x$$
$$Ax = b \qquad\qquad (1.3)$$
$$x \geq 0,$$

where $A \in IR^{m \times n}$, $b \in IR^m$, $c \in IR^n$ are given, and $x \in IR^n$ is the variable vector to be determined. In this book, a k-vector is also viewed as a $k \times 1$ matrix. For an $m \times n$ matrix M, the notation M^{T} denotes the *transpose* matrix, namely the $n \times m$ matrix with entries $M_{ij}^{\mathrm{T}} = M_{ji}$. As an example, in the above formulation c^{T} is a $1 \times n$ matrix and $c^{\mathrm{T}}x$ is the 1×1 matrix with entry $\sum_{j=1}^{n} c_j x_j$. The objective in (1.3) is to minimize the linear function $\sum_{j=1}^{n} c_j x_j$.

As with (1.1), the problem (1.3) is said to be *feasible* if its constraints are consistent (i.e., they define a nonempty region) and it is called *unbounded* if there exists a sequence of feasible vectors $\{x^k\}$ such that $c^{\mathrm{T}}x^k \to -\infty$. When (1.3) is feasible but not unbounded it has an *optimal solution*, i.e., a vector x that satisfies the constraints and minimizes the objective value among all feasible vectors. Similar definitions apply to nonlinear programming problems.

The best known and most successful methods for solving LPs are the simplex and interior-point methods. NLPs can be solved using gradient search techniques as well as approaches based on Newton's method such as interior-point and sequential quadratic programming methods.

1.1.2 Quadratic programming

A more general optimization problem is the *quadratic optimization* or the *quadratic programming* (QP) problem, where the objective function is now a quadratic function of the variables. The standard form QP is defined as follows:

$$\min_x \tfrac{1}{2}x^{\mathrm{T}}Qx + c^{\mathrm{T}}x$$
$$Ax = b \qquad\qquad (1.4)$$
$$x \geq 0,$$

where $A \in IR^{m \times n}$, $b \in IR^m$, $c \in IR^n$, $Q \in IR^{n \times n}$ are given, and $x \in IR^n$. Since $x^{\mathrm{T}}Qx = \tfrac{1}{2}x^{\mathrm{T}}(Q + Q^{\mathrm{T}})x$, one can assume without loss of generality that Q is *symmetric*, i.e., $Q_{ij} = Q_{ji}$.

The objective function of the problem (1.4) is a convex function of x when Q is a *positive semidefinite matrix*, i.e., when $y^T Q y \geq 0$ for all y (see Appendix A for a discussion on convex functions). This condition is equivalent to Q having only nonnegative eigenvalues. When this condition is satisfied, the QP problem is a convex optimization problem and can be solved in *polynomial time* using interior-point methods. Here we are referring to a classical notion used to measure computational complexity. Polynomial time algorithms are efficient in the sense that they always find an optimal solution in an amount of time that is guaranteed to be at most a polynomial function of the input size.

1.1.3 Conic optimization

Another generalization of (1.3) is obtained when the nonnegativity constraints $x \geq 0$ are replaced by general conic inclusion constraints. This is called a *conic optimization* (CO) problem. For this purpose, we consider a closed convex cone C (see Appendix B for a brief discussion on cones) in a finite-dimensional vector space X and the following conic optimization problem:

$$
\begin{aligned}
\min_x \ & c^T x \\
& Ax = b \\
& x \in C.
\end{aligned} \tag{1.5}
$$

When $X = I\!R^n$ and $C = I\!R^n_+$, this problem is the standard form LP. However, much more general nonlinear optimization problems can also be formulated in this way. Furthermore, some of the most efficient and robust algorithmic machinery developed for linear optimization problems can be modified to solve these general optimization problems. Two important subclasses of conic optimization problems we will address are: (i) second-order cone optimization, and (ii) semidefinite optimization. These correspond to the cases when C is the second-order cone:

$$
C_q := \left\{ x = (x_1, x_2, \ldots, x_n) \in I\!R^n : x_1^2 \geq x_2^2 + \cdots + x_n^2, \, x_1 \geq 0 \right\},
$$

and the cone of symmetric positive semidefinite matrices:

$$
C_s := \left\{ X = \begin{bmatrix} x_{11} & \cdots & x_{1n} \\ \vdots & \ddots & \vdots \\ x_{n1} & \cdots & x_{nn} \end{bmatrix} \in I\!R^{n \times n} : X = X^T, \, X \text{ is positive semidefinite} \right\}.
$$

When we work with the cone of positive semidefinite matrices, the standard inner products used in $c^T x$ and Ax in (1.5) are replaced by an appropriate inner product for the space of n-dimensional square matrices.

1.1.4 Integer programming

Integer programs are optimization problems that require some or all of the variables to take integer values. This restriction on the variables often makes the problems very hard to solve. Therefore we will focus on *integer linear programs*, which have a linear objective function and linear constraints. A *pure* integer linear program (ILP) is given by:

$$
\begin{aligned}
\min_x \ & c^\mathrm{T} x \\
& Ax \geq b \\
& x \geq 0 \ \text{and integral},
\end{aligned}
\tag{1.6}
$$

where $A \in I\!R^{m \times n}$, $b \in I\!R^m$, $c \in I\!R^n$ are given, and $x \in I\!N^n$ is the variable vector to be determined.

An important case occurs when the variables x_j represent binary decision variables, that is, $x \in \{0, 1\}^n$. The problem is then called a *0–1 linear program*.

When there are both continuous variables and integer constrained variables, the problem is called a *mixed integer linear program* (MILP):

$$
\begin{aligned}
\min_x \ & c^\mathrm{T} x \\
& Ax \geq \ b \\
& x \geq \ 0 \\
& x_j \in I\!N \ \text{for} \ j = 1, \ldots, p.
\end{aligned}
\tag{1.7}
$$

where A, b, c are given data and the integer p (with $1 \leq p < n$) is also part of the input.

1.1.5 Dynamic programming

Dynamic programming refers to a computational method involving recurrence relations. This technique was developed by Richard Bellman in the early 1950s. It arose from studying programming problems in which changes over time were important, thus the name "dynamic programming." However, the technique can also be applied when time is not a relevant factor in the problem. The idea is to divide the problem into "stages" in order to perform the optimization recursively. It is possible to incorporate stochastic elements into the recursion.

1.2 Optimization with data uncertainty

In all the problem classes discussed so far (except dynamic programming), we made the implicit assumption that the *data* of the problem, namely the parameters such as Q, A, b and c in QP, are all known. This is not always the case. Often, the problem parameters correspond to quantities that will only be realized in the future, or cannot be known exactly at the time the problem must be formulated and solved. Such situations are especially common in models involving financial quantities, such as returns on investments, risks, etc. We will discuss two fundamentally different approaches that address optimization with data uncertainty. *Stochastic programming* is an approach used when the data uncertainty is random and can be explained by some probability distribution. *Robust optimization* is used when one wants a solution that behaves well in all possible realizations of the uncertain data. These two alternative approaches are not problem classes (as in LP, QP, etc.) but rather modeling techniques for addressing data uncertainty.

1.2.1 Stochastic programming

The term *stochastic programming* refers to an optimization problem in which some problem data are random. The underlying optimization problem might be a linear program, an integer program, or a nonlinear program. An important case is that of *stochastic linear programs.*

A stochastic program *with recourse* arises when some of the decisions (recourse actions) can be taken after the outcomes of some (or all) random events have become known. For example, a *two-stage stochastic linear program with recourse* can be written as follows:

$$
\begin{aligned}
\max_x \quad & a^\mathrm{T}x + E[\max_{y(\omega)} c(\omega)^\mathrm{T} y(\omega)] \\
& Ax && = b \\
& B(\omega)x + && C(\omega)y(\omega) = d(\omega) \\
& x \geq 0, && y(\omega) \geq 0,
\end{aligned}
\tag{1.8}
$$

where the first-stage decisions are represented by vector x and the second-stage decisions by vector $y(\omega)$, which depend on the realization of a random event ω. A and b define deterministic constraints on the first-stage decisions x, whereas $B(\omega)$, $C(\omega)$, and $d(\omega)$ define stochastic linear constraints linking the recourse decisions $y(\omega)$ to the first-stage decisions. The objective function contains a deterministic term $a^\mathrm{T}x$ and the expectation of the second-stage objective $c(\omega)^\mathrm{T} y(\omega)$ taken over all realizations of the random event ω.

Note that, once the first-stage decisions x have been made and the random event ω has been realized, one can compute the optimal second-stage decisions by solving the following linear program:

$$
\begin{aligned}
f(x, \omega) = \max \ & c(\omega)^\mathrm{T} y(\omega) \\
& C(\omega)y(\omega) = d(\omega) - B(\omega)x \\
& y(\omega) \geq 0.
\end{aligned}
\tag{1.9}
$$

Let $f(x) = E[f(x, \omega)]$ denote the expected value of the optimal value of this problem. Then, the two-stage stochastic linear program becomes

$$\max a^T x + f(x)$$
$$Ax = b \tag{1.10}$$
$$x \geq 0.$$

Thus, if the (possibly nonlinear) function $f(x)$ is known, the problem reduces to a nonlinear programming problem. When the data $c(\omega)$, $B(\omega)$, $C(\omega)$, and $d(\omega)$ are described by finite distributions, one can show that f is piecewise linear and concave. When the data are described by probability densities that are absolutely continuous and have finite second moments, one can show that f is differentiable and concave. In both cases, we have a convex optimization problem with linear constraints for which specialized algorithms are available.

1.2.2 Robust optimization

Robust optimization refers to the modeling of optimization problems with data uncertainty to obtain a solution that is guaranteed to be "good" for all possible realizations of the uncertain parameters. In this sense, this approach departs from the randomness assumption used in stochastic optimization for uncertain parameters and gives the same importance to all possible realizations. Uncertainty in the parameters is described through *uncertainty sets* that contain all (or most) possible values that can be realized by the uncertain parameters.

There are different definitions and interpretations of robustness and the resulting models differ accordingly. One important concept is *constraint robustness*, often called *model robustness* in the literature. This refers to solutions that remain *feasible* for all possible values of the uncertain inputs. This type of solution is required in several engineering applications. Here is an example adapted from Ben-Tal and Nemirovski [8]. Consider a multi-phase engineering process (a chemical distillation process, for example) and a related process optimization problem that includes balance constraints (materials entering a phase of the process cannot exceed what is used in that phase plus what is left over for the next phase). The quantities of the end products of a particular phase may depend on external, uncontrollable factors and are therefore uncertain. However, no matter what the values of these uncontrollable factors are, the balance constraints *must* be satisfied. Therefore, the solution must be constraint robust with respect to the uncertainties of the problem. A mathematical model for finding constraint-robust solutions will be described. First, consider an optimization problem of the form:

$$\min_x \quad f(x)$$
$$G(x, p) \in K. \tag{1.11}$$

Here, x are the decision variables, f is the (certain) objective function, G and K are the structural elements of the constraints that are assumed to be certain and p are the uncertain parameters of the problem. Consider an uncertainty set $\mathcal{U}$ that contains all possible values of the uncertain parameters p. Then, a constraint-robust optimal solution can be found by solving the following problem:

$$\min_x \quad f(x)$$
$$G(x, p) \in K, \quad \forall p \in \mathcal{U}. \tag{1.12}$$

A related concept is *objective robustness*, which occurs when uncertain parameters appear in the objective function. This is often referred to as solution robustness in the literature. Such robust solutions must remain close to optimal for all possible realizations of the uncertain parameters. Next, consider an optimization problem of the form:

$$\min_x f(x, p)$$
$$x \in S. \tag{1.13}$$

Here, S is the (certain) feasible set and f is the objective function that depends on uncertain parameters p. Assume as above that $\mathcal{U}$ is the uncertainty set that contains all possible values of the uncertain parameters p. Then, an objective-robust solution is obtained by solving:

$$\min_{x \in S} \max_{p \in \mathcal{U}} f(x, p). \tag{1.14}$$

Note that objective robustness is a special case of constraint robustness. Indeed, by introducing a new variable t (to be minimized) into (1.13) and imposing the constraint $f(x, p) \leq t$, we get an equivalent problem to (1.13). The constraint-robust formulation of the resulting problem is equivalent to (1.14).

Constraint robustness and objective robustness are concepts that arise in conservative decision making and are not always appropriate for optimization problems with data uncertainty.

1.3 Financial mathematics

Modern finance has become increasingly technical, requiring the use of sophisticated mathematical tools in both research and practice. Many find the roots of this trend in the portfolio selection models and methods described by Markowitz [54] in the 1950s and the option pricing formulas developed by Black, Scholes, and Merton [15, 55] in the late 1960s and early 1970s. For the enormous effect these works produced on modern financial practice, Markowitz was awarded the Nobel prize in Economics in 1990, while Scholes and Merton won the Nobel prize in Economics in 1997.

Below, we introduce topics in finance that are especially suited for mathematical analysis and involve sophisticated tools from mathematical sciences.

1.3.1 Portfolio selection and asset allocation

The theory of optimal selection of portfolios was developed by Harry Markowitz in the 1950s. His work formalized the diversification principle in portfolio selection and, as mentioned above, earned him the 1990 Nobel prize for Economics. Here we give a brief description of the model and relate it to QPs.

Consider an investor who has a certain amount of money to be invested in a number of different securities (stocks, bonds, etc.) with random returns. For each security $i = 1, \ldots, n$, estimates of its expected return μ_i and variance σ_i^2 are given. Furthermore, for any two securities i and j, their correlation coefficient ρ_{ij} is also assumed to be known. If we represent the proportion of the total funds invested in security i by x_i, one can compute the expected return and the variance of the resulting portfolio $x = (x_1, \ldots, x_n)$ as follows:

$$E[x] = x_1\mu_1 + \cdots + x_n\mu_n = \mu^{\mathrm{T}}x,$$

and

$$\mathrm{Var}[x] = \sum_{i,j} \rho_{ij}\sigma_i\sigma_j x_i x_j = x^{\mathrm{T}}Qx,$$

where $\rho_{ii} \equiv 1$, $Q_{ij} = \rho_{ij}\sigma_i\sigma_j$, and $\mu = (\mu_1, \ldots, \mu_n)$.

The portfolio vector x must satisfy $\sum_i x_i = 1$ and there may or may not be additional feasibility constraints. A feasible portfolio x is called *efficient* if it has the maximal expected return among all portfolios with the same variance, or, alternatively, if it has the minimum variance among all portfolios that have at least a certain expected return. The collection of efficient portfolios form the *efficient frontier* of the portfolio universe.

Markowitz' *portfolio optimization problem*, also called the *mean-variance optimization* (MVO) problem, can be formulated in three different but equivalent ways. One formulation results in the problem of finding a minimum variance portfolio of the securities 1 to n that yields at least a target value R of expected return. Mathematically, this formulation produces a convex quadratic programming problem:

$$\begin{aligned}
\min_x\ & x^{\mathrm{T}}Qx \\
& e^{\mathrm{T}}x = 1 \\
& \mu^{\mathrm{T}}x \geq R \\
& x \geq 0,
\end{aligned} \tag{1.15}$$

where e is an n-dimensional vector with all components equal to 1. The first constraint indicates that the proportions x_i should sum to 1. The second constraint indicates that the expected return is no less than the target value and, as we discussed above, the objective function corresponds to the total variance of the portfolio. Nonnegativity constraints on x_i are introduced to rule out short sales (selling a security that you do not have). Note that the matrix Q is positive semidefinite since x^TQx, the variance of the portfolio, must be nonnegative for every portfolio (feasible or not) x.

As an alternative to problem (1.15), we may choose to maximize the expected return of a portfolio while limiting the variance of its return. Or, we can maximize a risk-adjusted expected return, which is defined as the expected return minus a multiple of the variance. These two formulations are essentially equivalent to (1.15), as we will see in Chapter 8.

The model (1.15) is rather versatile. For example, if short sales are permitted on some or all of the securities, then this can be incorporated into the model simply by removing the nonnegativity constraint on the corresponding variables. If regulations or investor preferences limit the amount of investment in a subset of the securities, the model can be augmented with a linear constraint to reflect such a limit. In principle, any linear constraint can be added to the model without making it significantly harder to solve.

Asset allocation problems have the same mathematical structure as portfolio selection problems. In these problems the objective is not to choose a portfolio of stocks (or other securities) but to determine the optimal investment among a set of asset classes. Examples of asset classes are large capitalization stocks, small capitalization stocks, foreign stocks, government bonds, corporate bonds, etc. There are many mutual funds focusing on specific asset classes and one can therefore conveniently invest in these asset classes by purchasing the relevant mutual funds. After estimating the expected returns, variances, and covariances for different asset classes, one can formulate a QP identical to (1.15) and obtain efficient portfolios of these asset classes.

A different strategy for portfolio selection is to try to mirror the movements of a broad market population using a significantly smaller number of securities. Such a portfolio is called an index fund. No effort is made to identify mispriced securities. The assumption is that the market is efficient and therefore no superior risk-adjusted returns can be achieved by stock picking strategies since the stock prices reflect all the information available in the marketplace. Whereas actively managed funds incur transaction costs that reduce their overall performance, index funds are not actively traded and incur low management fees. They are typical of a passive management strategy. How do investment companies construct index funds? There are numerous ways of doing this. One way is to solve a clustering problem where similar stocks have one representative in the index fund. This naturally leads to an integer programming formulation.

1.3.2 Pricing and hedging of options

We first start with a description of some of the well-known financial options. A *European call option* is a contract with the following conditions:

- At a prescribed time in the future, known as the *expiration date*, the *holder* of the option has the right, but not the obligation, to
- purchase a prescribed asset, known as the *underlying*, for a
- prescribed amount, known as the *strike price* or *exercise price*.

A *European put option* is similar, except that it confers the right to sell the underlying asset (instead of buying it for a call option). An *American option* is like a European option, but it can be exercised any time before the expiration date.

Since the payoff from an option depends on the value of the underlying security, its price is also related to the current value and expected behavior of this underlying security. To find the fair value of an option, we need to solve a *pricing* problem. When there is a good model for the stochastic behavior of the underlying security, the option pricing problem can be solved using sophisticated mathematical techniques.

Option pricing problems are often solved using the following strategy. We try to determine a portfolio of assets with known prices which, if updated properly through time, will produce the same payoff as the option. Since the portfolio and the option will have the same eventual payoffs, we conclude that they must have the same value today (otherwise, there is *arbitrage*) and we can therefore obtain the price of the option. A portfolio of other assets that produces the same payoff as a given financial instrument is called a *replicating portfolio* (or a *hedge*) for that instrument. Finding the right portfolio, of course, is not always easy and leads to a *replication* (or *hedging*) problem.

Let us consider a simple example to illustrate these ideas. Let us assume that one share of stock XYZ is currently valued at $40. The price of XYZ a month from today is random with two possible states. In the "up" state (denoted by u) the price will double, and in the "down" state (denoted by d) the price will halve. Assume that up and down states have equal probabilities.

$$S_0 = \$40 \begin{cases} 80 = S_1(u) \\ 20 = S_1(d) \end{cases}$$

Today, we purchase a European call option to buy one share of XYZ stock for $50 a month from today. What is the fair price of this option?

Let us assume that we can borrow or lend money with no interest between today and next month, and that we can buy or sell any amount of the XYZ stock without any commissions, etc. These are part of the "frictionless market" assumptions we will address later. Further assume that XYZ will not pay any dividends within the next month.

To solve the option pricing problem, we consider the following hedging problem: can we form a portfolio of the underlying stock (bought or sold) and cash (borrowed or lent) today, such that the payoff from the portfolio at the expiration date of the option will match the payoff of the option? Note that the option payoff will be $30 if the price of the stock goes up and $0 if it goes down. Assume this portfolio has Δ shares of XYZ and $\$B$ cash. This portfolio would be worth $40\Delta + B$ today. Next month, payoffs for this portfolio will be:

$$P_0 = 40\Delta + B \begin{cases} 80\Delta + B = P_1(u) \\ 20\Delta + B = P_1(d) \end{cases}$$

Let us choose Δ and B such that

$$80\Delta + B = 30,$$
$$20\Delta + B = 0,$$

so that the portfolio replicates the payoff of the option at the expiration date. This gives $\Delta = 1/2$ and $B = -10$, which is the *hedge* we were looking for. This portfolio is worth $P_0 = 40\Delta + B = \$10$ today, therefore, the fair price of the option must also be $10.

1.3.3 Risk management

Risk is inherent in most economic activities. This is especially true of financial activities where results of decisions made today may have many possible different outcomes depending on future events. Since companies cannot usually insure themselves completely against risk, they have to manage it. This is a hard task even with the support of advanced mathematical techniques. Poor risk management led to several spectacular failures in the financial industry during the 1990s (e.g., Barings Bank, Long Term Capital Management, Orange County).

A coherent approach to risk management requires quantitative risk measures that adequately reflect the vulnerabilities of a company. Examples of risk measures include portfolio variance as in the Markowitz MVO model, the Value-at-Risk (VaR) and the expected shortfall (also known as conditional Value-at-Risk, or CVaR). Furthermore, risk control techniques need to be developed and implemented to adapt to rapid changes in the values of these risk measures. Government regulators already mandate that financial institutions control their holdings in certain ways and place margin requirements for "risky" positions.

Optimization problems encountered in financial risk management often take the following form. Optimize a performance measure (such as expected investment return) subject to the usual operating constraints and the constraint that a particular risk measure for the company's financial holdings does not exceed a prescribed amount. Mathematically, we may have the following problem:

$$\begin{aligned} \max_x \quad & \mu^\mathrm{T} x \\ & \mathrm{RM}[x] \leq \gamma \\ & e^\mathrm{T} x = 1 \\ & x \geq 0. \end{aligned} \qquad (1.16)$$

As in the Markowitz MVO model, x_i represent the proportion of the total funds invested in security i. The objective is to maximize the expected portfolio return and μ is the expected return vector for the different securities. $\mathrm{RM}[x]$ denotes the value of a particular risk measure for portfolio x and γ is the prescribed upper limit on this measure. Since $\mathrm{RM}[x]$ is generally a nonlinear function of x, (1.16) is a nonlinear programming problem. Alternatively, we can minimize the risk measure while constraining the expected return of the portfolio to achieve or exceed a given target value R. This will produce a problem very similar to (1.15).

1.3.4 Asset/liability management

How should a financial institution manage its assets and liabilities? A static mean-variance optimization model, such as the one we discussed for asset allocation, fails to incorporate the dynamic nature of asset management and multiple liabilities with different maturities faced by financial institutions. Furthermore, it penalizes returns both above and below the mean. A multi-period model that emphasizes the need to meet liabilities in each period for a finite (or possibly infinite) horizon is often required. Since liabilities and asset returns usually have random components, their optimal management requires tools of "Optimization under Uncertainty" and, most notably, stochastic programming approaches.

Let L_t be the liability of the company in period t for $t = 1, \ldots, T$. Here, we assume that the liabilities L_t are random with known distributions. A typical problem to solve in asset/liability management is to determine which assets (and in what quantities) the company should hold in each period to maximize its expected wealth at the end of period T. We can further assume that the asset classes the company can choose from have random returns (again, with known distributions) denoted by R_{it} for asset class i in period t. Since the company can make the holding decisions for each period after observing the asset returns and liabilities in the previous periods, the resulting problem can be cast as a stochastic program with recourse:

$$\max_x \qquad\qquad E\left[\sum_i x_{i,T}\right]$$
$$\sum_i (1 + R_{it})x_{i,t-1} - \sum_i x_{i,t} = L_t, \ t = 1, \ldots, T, \qquad (1.17)$$
$$x_{i,t} \geq 0 \quad \forall i, t.$$

The objective function represents the expected total wealth at the end of the last period. The constraints indicate that the surplus left after liability L_t is covered will be invested as follows: $x_{i,t}$ invested in asset class i. In this formulation, $x_{i,0}$ are the fixed and possibly nonzero initial positions in the different asset classes.

2

Linear programming: theory and algorithms

2.1 The linear programming problem

One of the most common and fundamental optimization problems is the linear optimization, or *linear programming* (LP) problem. LP is the problem of optimizing a linear objective function subject to linear equality and inequality constraints. A generic linear optimization problem has the following form:

$$
\min_x \ c^T x
$$
$$
a_i^T x = b_i, \ i \in \mathcal{E},
$$
$$
a_i^T x \geq b_i, \ i \in \mathcal{I},
$$

(2.1)

where $\mathcal{E}$ and $\mathcal{I}$ are the index sets for equality and inequality constraints, respectively. Linear programming is arguably the best known and the most frequently solved optimization problem. It owes its fame mostly to its great success; real-world problems coming from as diverse disciplines as sociology, finance, transportation, economics, production planning, and airline crew scheduling have been formulated and successfully solved as LPs.

For algorithmic purposes, it is often desirable to have the problems structured in a particular way. Since the development of the simplex method for LPs the following form has been a popular standard and is called the *standard form LP*:

$$
\min_x \ c^T x
$$
$$
Ax = b
$$
$$
x \geq 0.
$$

(2.2)

Here $A \in I\!R^{m \times n}$, $b \in I\!R^m$, $c \in I\!R^n$ are given, and $x \in I\!R^n$ is the variable vector to be determined as the solution of the problem.

The standard form is not restrictive: inequalities other than nonnegativity constraints can be rewritten as equalities after the introduction of a so-called *slack* or

15

surplus variable that is restricted to be nonnegative. For example,

$$\begin{aligned}
\min \quad &-x_1 - x_2 \\
&2x_1 + x_2 \leq 12 \\
&x_1 + 2x_2 \leq 9 \\
&x_1 \geq 0, \ x_2 \geq 0,
\end{aligned} \tag{2.3}$$

can be rewritten as

$$\begin{aligned}
\min \quad &-x_1 - \ x_2 \\
&2x_1 + \ x_2 + x_3 \qquad\quad = 12 \\
&x_1 + 2x_2 \qquad + x_4 = 9 \\
&x_1 \geq 0, \ x_2 \geq 0, \ x_3 \geq 0, \ x_4 \geq 0.
\end{aligned} \tag{2.4}$$

Variables that are unrestricted in sign can be expressed as the difference of two new nonnegative variables. Maximization problems can be written as minimization problems by multiplying the objective function by a negative constant. Simple transformations are available to rewrite any given LP in the standard form above. Therefore, in the rest of our theoretical and algorithmic discussion we assume that the LP is in the standard form.

Exercise 2.1   Write the following linear program in standard form.

$$\begin{aligned}
\min \quad & x_2 \\
&x_1 + x_2 \geq 1, \\
&x_1 - x_2 \leq 0, \\
&x_1, x_2 \text{ unrestricted in sign.}
\end{aligned}$$

Answer: After writing $x_i = y_i - z_i$, $i = 1, 2$ with $y_i \geq 0$ and $z_i \geq 0$ and introducing surplus variable s_1 for the first constraint and slack variable s_2 for the second constraint we obtain:

$$\begin{aligned}
\min \quad & y_2 - z_2 \\
&y_1 - z_1 + y_2 - z_2 - s_1 \qquad = 1 \\
&y_1 - z_1 - y_2 + z_2 \qquad + s_2 = 0 \\
&y_1 \geq 0, \ z_1 \geq 0, \ y_2 \geq 0, \ z_2 \geq 0, \ s_1 \geq 0, \ s_2 \geq 0.
\end{aligned}$$

Exercise 2.2   Write the following linear program in standard form.

$$\begin{aligned}
\max \ \ &4x_1 + x_2 - x_3 \\
&x_1 \qquad\ \ + 3x_3 \leq 6 \\
&3x_1 + x_2 + 3x_3 \geq 9 \\
&x_1 \geq 0, \ x_2 \geq 0, \ x_3 \text{ unrestricted in sign.}
\end{aligned}$$

Recall the following definitions from the Chapter 1: the LP (2.2) is said to be *feasible* if its constraints are consistent and it is called *unbounded* if there exists a sequence of feasible vectors $\{x^k\}$ such that $c^T x^k \to -\infty$. When we talk about a *solution* (without any qualifiers) to (2.2) we mean any candidate vector $x \in I\!R^n$. A *feasible solution* is one that satisfies the constraints, and an *optimal solution* is a vector x that satisfies the constraints and minimizes the objective value among all feasible vectors. When LP is feasible but not unbounded it has an optimal solution.

Exercise 2.3

(i) Write a two-variable linear program that is unbounded.
(ii) Write a two-variable linear program that is infeasible.

Exercise 2.4   Draw the feasible region of the following two-variable linear program.

$$\max\ 2x_1 - x_2$$
$$x_1 + x_2 \geq 1$$
$$x_1 - x_2 \leq 0$$
$$3x_1 + x_2 \leq 6$$
$$x_1 \geq 0,\ x_2 \geq 0.$$

Determine the optimal solution to this problem by inspection.

The most important questions we will address in this chapter are the following: how do we recognize an optimal solution and how do we find such solutions? One of the most important tools in optimization to answer these questions is the notion of a dual problem associated with the LP problem (2.2). We describe the dual problem in the next section.

2.2 Duality

Consider the standard form LP in (2.4) above. Here are a few alternative feasible solutions:

$$(x_1, x_2, x_3, x_4) = \left(0, \frac{9}{2}, \frac{15}{2}, 0\right) \qquad \text{Objective value} = -\frac{9}{2}$$
$$(x_1, x_2, x_3, x_4) = (6, 0, 0, 3) \qquad \text{Objective value} = -6$$
$$(x_1, x_2, x_3, x_4) = (5, 2, 0, 0) \qquad \text{Objective value} = -7.$$

Since we are minimizing, the last solution is the best among the three feasible solutions we found, but is it the optimal solution? We can make such a claim if we can, somehow, show that there is no feasible solution with a smaller objective value.

Note that the constraints provide some bounds on the value of the objective function. For example, for any feasible solution, we must have

$$-x_1 - x_2 \geq -2x_1 - x_2 - x_3 = -12$$

using the first constraint of the problem. The inequality above must hold for all feasible solutions since the x_i's are all nonnegative and the coefficient of each variable on the LHS is at least as large as the coefficient of the corresponding variable on the RHS. We can do better using the second constraint:

$$-x_1 - x_2 \geq -x_1 - 2x_2 - x_4 = -9$$

and even better by adding a negative third of each constraint:

$$-x_1 - x_2 \geq -x_1 - x_2 - \frac{1}{3}x_3 - \frac{1}{3}x_4$$

$$= -\frac{1}{3}(2x_1 + x_2 + x_3) - \frac{1}{3}(x_1 + 2x_2 + x_4) = -\frac{1}{3}(12 + 9) = -7.$$

This last inequality indicates that, for any feasible solution, the objective function value cannot be smaller than -7. Since we already found a feasible solution achieving this bound, we conclude that this solution, namely $(x_1, x_2, x_3, x_4) = (5, 2, 0, 0)$, must be an optimal solution of the problem.

This process illustrates the following strategy: if we find a feasible solution to the LP problem, and a bound on the optimal value of the problem such that the bound and the objective value of the feasible solution coincide, then we can conclude that our feasible solution is an optimal solution. We will comment on this strategy shortly. Before that, though, we formalize our approach for finding a bound on the optimal objective value.

Our strategy was to find a linear combination of the constraints, say with multipliers y_1 and y_2 for the first and second constraint respectively, such that the combined coefficient of **each** variable forms a lower bound on the objective coefficient of that variable. Namely, we tried to choose multipliers y_1 and y_2 associated with constraints 1 and 2 such that

$$y_1(2x_1 + x_2 + x_3) + y_2(x_1 + 2x_2 + x_4)$$
$$= (2y_1 + y_2)x_1 + (y_1 + 2y_2)x_2 + y_1x_3 + y_2x_4$$

provides a lower bound on the optimal objective value. Since the x_i's must be nonnegative, the expression above would necessarily give a lower bound if the coefficient of each x_i is less than or equal to the corresponding objective function coefficient, or if:

$$2y_1 + y_2 \leq -1$$
$$y_1 + 2y_2 \leq -1$$
$$y_1 \qquad\quad \leq 0$$
$$\qquad y_2 \leq 0.$$

Note that the objective coefficients of x_3 and x_4 are zero. Naturally, to obtain the largest possible lower bound, we would like to find y_1 and y_2 that achieve the maximum combination of the right-hand-side values:

$$\max 12y_1 + 9y_2.$$

This process results in a linear programming problem that is strongly related to the LP we are solving:

$$
\begin{aligned}
\max \; & 12y_1 + 9y_2 \\
& 2y_1 + y_2 \le -1 \\
& y_1 + 2y_2 \le -1 \\
& y_1 \le 0 \\
& y_2 \le 0.
\end{aligned}
\tag{2.5}
$$

This problem is called the *dual* of the original problem we considered. The original LP in (2.2) is often called the *primal* problem. For a generic primal LP problem in standard form (2.2) the corresponding dual problem can be written as follows:

$$
\begin{aligned}
\max_y \; & b^\mathrm{T} y \\
& A^\mathrm{T} y \le c,
\end{aligned}
\tag{2.6}
$$

where $y \in I\!R^m$. Rewriting this problem with explicit *dual slacks*, we obtain the standard form dual linear programming problem:

$$
\begin{aligned}
\max_{y,s} \; & b^\mathrm{T} y \\
& A^\mathrm{T} y + s = c \\
& s \ge 0,
\end{aligned}
\tag{2.7}
$$

where $s \in I\!R^n$.

Exercise 2.5 Consider the following LP:

$$
\begin{aligned}
\min \; & 2x_1 + 3x_2 \\
& x_1 + x_2 \ge 5 \\
& x_1 \ge 1 \\
& x_2 \ge 2.
\end{aligned}
$$

Prove that $x^* = (3, 2)$ is the optimal solution by showing that the objective value of any feasible solution is at least 12.

Next, we make some observations about the relationship between solutions of the primal and dual LPs. The objective value of any primal feasible solution is at least as large as the objective value of any feasible dual solution. This fact is known as the *weak duality theorem*:

Theorem 2.1 (Weak duality theorem) *Let x be any feasible solution to the primal LP (2.2) and y be any feasible solution to the dual LP (2.6). Then*

$$c^{\mathrm{T}}x \geq b^{\mathrm{T}}y.$$

Proof: Since $x \geq 0$ and $c - A^{\mathrm{T}}y \geq 0$, the inner product of these two vectors must be nonnegative:

$$(c - A^{\mathrm{T}}y)^{\mathrm{T}}x = c^{\mathrm{T}}x - y^{\mathrm{T}}Ax = c^{\mathrm{T}}x - y^{\mathrm{T}}b \geq 0.$$

$\square$

The quantity $c^{\mathrm{T}}x - y^{\mathrm{T}}b$ is often called the *duality gap*. The following three results are immediate consequences of the weak duality theorem.

Corollary 2.1 *If the primal LP is unbounded, then the dual LP must be infeasible.*

Corollary 2.2 *If the dual LP is unbounded, then the primal LP must be infeasible.*

Corollary 2.3 *If x is feasible for the primal LP, y is feasible for the dual LP, and $c^{\mathrm{T}}x = b^{\mathrm{T}}y$, then x must be optimal for the primal LP and y must be optimal for the dual LP.*

Exercise 2.6 Show that the dual of the linear program

$$\min_x c^{\mathrm{T}}x$$
$$Ax \geq b$$
$$x \geq 0$$

is the linear program

$$\max_y b^{\mathrm{T}}y$$
$$A^{\mathrm{T}}y \leq c$$
$$y \geq 0.$$

Exercise 2.7 We say that two linear programming problems are equivalent if one can be obtained from the other by (i) multiplying the objective function by -1 and changing it from min to max, or max to min, and/or (ii) multiplying some or all constraints by -1. For example, $\min\{c^{\mathrm{T}}x : Ax \geq b\}$ and $\max\{-c^{\mathrm{T}}x : -Ax \leq -b\}$ are equivalent problems. Find a linear program which is equivalent to its own dual.

Exercise 2.8 Give an example of a linear program such that it and its dual are both infeasible.

Exercise 2.9   For the following pair of primal–dual problems, determine whether the listed solutions are optimal.

$$
\begin{array}{ll}
\min 2x_1 + 3x_2 & \max -30y_1 + 10y_2 \\
\quad 2x_1 + 3x_2 \le 30 & \quad -2y_1 + \ \ y_2 + y_3 \le 2 \\
\quad \ x_1 + 2x_2 \ge 10 & \quad -3y_1 + 2y_2 - y_3 \le 3 \\
\quad \ x_1 - \ \ x_2 \ge 0 & \qquad y_1, \quad y_2, \quad y_3 \ge 0. \\
\quad \ x_1, \quad \ x_2 \ge 0 &
\end{array}
$$

(i) $x_1 = 10$, $x_2 = 10/3$; $y_1 = 0$, $y_2 = 1$, $y_3 = 1$.
(ii) $x_1 = 20$, $x_2 = 10$; $y_1 = -1$, $y_2 = 4$, $y_3 = 0$.
(iii) $x_1 = 10/3$, $x_2 = 10/3$; $y_1 = 0$, $y_2 = 5/3$, $y_3 = 1/3$.

2.3 Optimality conditions

Corollary 2.3 in the previous section identified a sufficient condition for optimality of a primal–dual pair of feasible solutions, namely that their objective values coincide. One natural question to ask is whether this is a necessary condition. The answer is yes, as we illustrate next.

Theorem 2.2 (Strong duality theorem)   *If the primal (dual) problem has an optimal solution x (y), then the dual (primal) has an optimal solution y (x) such that $c^\mathsf{T} x = b^\mathsf{T} y$.*

The reader can find a proof of this result in most standard linear programming textbooks (see Chvátal [21] for example). A consequence of the strong duality theorem is that if both the primal LP problem and the dual LP have feasible solutions then they both have optimal solutions and for any primal optimal solution x and dual optimal solution y we have that $c^\mathsf{T} x = b^\mathsf{T} y$.

The strong duality theorem provides us with conditions to identify optimal solutions (called *optimality conditions*): $x \in I\!R^n$ is an optimal solution of (2.2) if and only if:

1. x is primal feasible: $Ax = b$, $x \ge 0$, and there exists a $y \in I\!R^m$ such that
2. y is dual feasible: $A^\mathsf{T} y \le c$; and
3. there is no duality gap: $c^\mathsf{T} x = b^\mathsf{T} y$.

Further analyzing the last condition above, we can obtain an alternative set of optimality conditions. Recall from the proof of the weak duality theorem that $c^\mathsf{T} x - b^\mathsf{T} y = (c - A^\mathsf{T} y)^\mathsf{T} x \ge 0$ for any feasible primal–dual pair of solutions, since it is given as an inner product of two nonnegative vectors. This inner product is 0 ($c^\mathsf{T} x = b^\mathsf{T} y$) if and only if the following statement holds: for each $i = 1, \ldots, n$, either x_i or $(c - A^T y)_i = s_i$ is zero. This equivalence is easy to see. All the terms in the summation on the RHS of the following equation are nonnegative:

$$0 = (c - A^T y)^T x = \sum_{i=1}^{n} (c - A^T y)_i x_i$$

Since the sum is zero, each term must be zero. Thus we have found an alternative set of optimality conditions: $x \in I\!R^n$ is an optimal solution of (2.2) if and only if:

1. x is primal feasible: $Ax = b$, $x \geq 0$, and there exists a $y \in I\!R^m$ such that
2. y is dual feasible: $s := c - A^T y \geq 0$; and
3. there is complementary slackness: for each $i = 1, \ldots, n$ we have $x_i s_i = 0$.

Exercise 2.10   Consider the linear program

$$\min 5x_1 + 12x_2 + 4x_3$$
$$x_1 + 2x_2 + x_3 = 10$$
$$2x_1 - x_2 + 3x_3 = 8$$
$$x_1 \geq 0, \ x_2 \geq 0, \ x_3 \geq 0.$$

You are given the information that x_2 and x_3 are positive in the optimal solution. Use the complementary slackness conditions to find the optimal dual solution.

Exercise 2.11   Consider the following linear programming problem:

$$\max 6x_1 + 5x_2 + 4x_3 + 5x_4 + 6x_5$$
$$x_1 + x_2 + x_3 + x_4 + x_5 \leq 3$$
$$5x_1 + 4x_2 + 3x_3 + 2x_4 + x_5 \leq 14$$
$$x_1 \geq 0, \ x_2 \geq 0, \ x_3 \geq 0, \ x_4 \geq 0, \ x_5 \geq 0.$$

Solve this problem using the following strategy:

(i) Find the dual of the above LP. The dual has only two variables. Solve the dual by inspection after drawing a graph of the feasible set.
(ii) Now using the optimal solution to the dual problem, and complementary slackness conditions, determine which primal constraints are active, and which primal variables must be zero at an optimal solution. Using this information determine the optimal solution to the primal problem.

Exercise 2.12   Using the optimality conditions for

$$\min_x c^T x$$
$$Ax = b$$
$$x \geq 0,$$

deduce that the optimality conditions for

$$\max_x \ c^T x$$
$$Ax \le b$$
$$x \ge 0$$

are $Ax \le b$, $x \ge 0$ and there exists y such that $A^T y \ge c$, $y \ge 0$, $c^T x = b^T y$.

Exercise 2.13 Consider the following investment problem over T years, where the objective is to maximize the value of the investments in year T. We assume a perfect capital market with the same annual lending and borrowing rate $r > 0$ each year. We also assume that exogenous investment funds b_t are available in year t, for $t = 1, \ldots, T$. Let n be the number of possible investments. We assume that each investment can be undertaken fractionally (between 0 and 1). Let a_{tj} denote the cash flow associated with investment j in year t. Let c_j be the value of investment j in year T (including all cash flows subsequent to year T discounted at the interest rate r).

The linear program that maximizes the value of the investments in year T is the following. Denote by x_j the fraction of investment j undertaken, and let y_t be the amount borrowed (if negative) or lent (if positive) in year t.

$$\max \quad \sum_{j=1}^n c_j x_j + y_T$$
$$-\sum_{j=1}^n a_{1j} x_j + y_1 \le b_1$$
$$-\sum_{j=1}^n a_{tj} x_j - (1+r)y_{t-1} + y_t \le b_t \qquad \text{for } t = 2, \ldots, T,$$
$$0 \le x_j \le 1 \qquad \text{for } j = 1, \ldots, n.$$

(i) Write the dual of the above linear program.
(ii) Solve the dual linear program found in (i). [Hint: Note that some of the dual variables can be computed by backward substitution.]
(iii) Write the complementary slackness conditions.
(iv) Deduce that the first T constraints in the primal linear program hold as equalities.
(v) Use the complementary slackness conditions to show that the solution obtained by setting $x_j = 1$ if $c_j + \sum_{t=1}^T (1+r)^{T-t} a_{tj} > 0$, and $x_j = 0$ otherwise, is an optimal solution.
(vi) Conclude that the above investment problem always has an optimal solution where each investment is either undertaken completely or not at all.

2.4 The simplex method

The best known and most successful methods for solving LPs are *interior-point methods* (IPMs) and the *simplex method*. We discuss the simplex method here and postpone our discussion of IPMs till we study quadratic programming problems, as IPMs are also applicable to quadratic programs and other more general classes of optimization problems.

We introduce the essential elements of the simplex method using a simple bond portfolio selection problem.

Example 2.1   *A bond portfolio manager has $100,000 to allocate to two differ-ent bonds: one corporate and one government bond. The corporate bond has a yield of 4%, a maturity of 3 years and an A rating from a rating agency that is translated into a numerical rating of 2 for computational purposes. In contrast, the government bond has a yield of 3%, a maturity of 4 years and rating of Aaa with the corresponding numerical rating of 1 (lower numerical ratings correspond to higher quality bonds). The portfolio manager would like to allocate funds so that the average rating for the portfolio is no worse than Aa (numerical equivalent 1.5) and average maturity of the portfolio is at most 3.6 years. Any amount not invested in the two bonds will be kept in a cash account that is assumed to earn no interest for simplicity and does not contribute to the average rating or maturity computations.[1] How should the manager allocate funds between these two bonds to achieve the objective of maximizing the yield from this investment?*

Letting variables x_1 and x_2 denote the allocation of funds to the corporate and government bond respectively (in thousands of dollars) we obtain the following formulation for the portfolio manager's problem:

$$\max \quad Z = 4x_1 + 3x_2$$
subject to:
$$x_1 + x_2 \leq 100$$
$$\frac{2x_1 + x_2}{100} \leq 1.5$$
$$\frac{3x_1 + 4x_2}{100} \leq 3.6$$
$$x_1, x_2 \geq \quad 0.$$

We first multiply the second and third inequalities by 100 to avoid fractions. After we add slack variables to each of the functional inequality constraints we obtain a representation of the problem in the standard form, suitable for the simplex method.[2] For example, letting x_3 denote the amount we keep as cash, we can rewrite the first constraint as $x_1 + x_2 + x_3 = 100$ with the additional condition of $x_3 \geq 0$.

---

[1] In other words, we are assuming a quality rating of 0 – "perfect" quality, and maturity of 0 years for cash.
[2] This representation is not exactly in the standard form since the objective is maximization rather than minimiza-tion. However, any maximization problem can be transformed into a minimization problem by multiplying the objective function by -1. Here, we avoid such a transformation to leave the objective function in its natural form – it should be straightforward to adapt the steps of the algorithm in the following discussion to address minimization problems.

Continuing with this strategy we obtain the following formulation:

$$\max \ Z = 4x_1 + 3x_2$$
subject to:

$$
\begin{aligned}
x_1 + x_2 + x_3 \qquad\qquad &= 100 \\
2x_1 + x_2 \qquad + x_4 \qquad &= 150 \\
3x_1 + 4x_2 \qquad\qquad + x_5 &= 360 \\
x_1 \geq 0, \ x_2 \geq 0, \ x_3 \geq 0, \ x_4 \geq 0, \ x_5 \geq 0.
\end{aligned}
$$

(2.8)

2.4.1 Basic solutions

Let us consider a general LP problem in the following form:

$$\max cx \tag{2.9}$$

$$Ax \leq b \tag{2.10}$$

$$x \geq 0, \tag{2.11}$$

where A is an $m \times n$ matrix, b is an m-dimensional column vector and c is an n-dimensional row vector. The n-dimensional column vector x represents the variables of the problem. (In the bond portfolio example we have $m = 3$ and $n = 2$.) Here is how we can represent these vectors and matrices:

$$
A = \begin{bmatrix}
a_{11} & a_{12} & \cdots & a_{1n} \\
a_{21} & a_{22} & \cdots & a_{2n} \\
\vdots & \vdots & \ddots & \vdots \\
a_{m1} & a_{m2} & \cdots & a_{mn}
\end{bmatrix}, \quad
b = \begin{bmatrix}
b_1 \\
b_2 \\
\vdots \\
b_m
\end{bmatrix}, \quad
c = \begin{bmatrix} c_1 & c_2 & \cdots & c_n \end{bmatrix},
$$

$$
x = \begin{bmatrix}
x_1 \\
x_2 \\
\vdots \\
x_n
\end{bmatrix}, \quad
0 = \begin{bmatrix}
0 \\
0 \\
\vdots \\
0
\end{bmatrix}.
$$

Next, we add slack variables to each of the functional constraints to get the augmented form of the problem. Let x_s denote the vector of slack variables:

$$
x_s = \begin{bmatrix}
x_{n+1} \\
x_{n+2} \\
\vdots \\
x_{n+m}
\end{bmatrix}
$$

and let I denote the $m \times m$ identity matrix. Now, the constraints in the augmented form can be written as

$$[A, I]\begin{bmatrix} x \\ x_s \end{bmatrix} = b, \quad \begin{bmatrix} x \\ x_s \end{bmatrix} \geq 0. \tag{2.12}$$

There are many potential solutions to system (2.12). Let us focus on the equation $[A, I]\begin{bmatrix} x \\ x_s \end{bmatrix} = b$. By choosing $x = 0$ and $x_s = b$, we immediately satisfy this equation – but not necessarily all the inequalities. More generally, we can consider partitions of the augmented matrix $[A, I]$:[3]

$$[A, I] \equiv [B, N],$$

where B is an $m \times m$ square matrix that consists of linearly independent columns of $[A, I]$. Such a B matrix is called a *basis matrix* and this partition is called a *basis partition*. If we partition the variable vector $\begin{bmatrix} x \\ x_s \end{bmatrix}$ in the same way:

$$\begin{bmatrix} x \\ x_s \end{bmatrix} \equiv \begin{bmatrix} x_B \\ x_N \end{bmatrix},$$

we can rewrite the equality constraints in (2.12) as

$$[B, N]\begin{bmatrix} x_B \\ x_N \end{bmatrix} = Bx_B + Nx_N = b,$$

or, by multiplying both sides by B^{-1} from the left,

$$x_B + B^{-1}Nx_N = B^{-1}b.$$

By our construction, the following three systems of equations are equivalent in the sense that any solution to one is a solution for the other two:

$$[A, I]\begin{bmatrix} x \\ x_s \end{bmatrix} = b,$$
$$Bx_B + Nx_N = b$$
$$x_B + B^{-1}Nx_N = B^{-1}b.$$

Indeed, the second and third linear systems are just other representations of the first one in terms of the matrix B. As we observed above, an obvious solution to the last system (and, therefore, to the other two) is $x_N = 0$, $x_B = B^{-1}b$. In fact, for any fixed values of the components of x_N we can obtain a solution by simply setting

$$x_B = B^{-1}b - B^{-1}Nx_N. \tag{2.13}$$

---

[3] Here, we are using the notation $U \equiv V$ to indicate that the matrix V is obtained from the matrix U by permuting its columns. Similarly, for the column vectors u and v, $u \equiv v$ means that v is obtained from u by permuting its elements.

One can think of x_N as the *independent* variables that we can choose freely, and, once they are chosen, the *dependent* variables x_B are determined uniquely. We call a solution of the systems above a *basic solution* if it is of the form

$$x_N = 0, \; x_B = B^{-1}b.$$

If, in addition, $x_B = B^{-1}b \geq 0$, the solution $x_B = B^{-1}b, x_N = 0$ is a *basic feasible solution* of the LP problem above. The variables x_B are called the *basic variables*, while x_N are the *nonbasic variables*. Geometrically, basic feasible solutions correspond to extreme points of the feasible set $\{x : Ax \leq b, x \geq 0\}$. Extreme points of a set are those that cannot be written as a convex combination of two other points in the set.

The objective function $Z = cx$ can be represented similarly using the basis partition. Let $c = [c_B, \; c_N]$ represent the partition of the objective vector. Now, we have the following sequence of equivalent representations of the objective function equation:

$$Z = cx \Leftrightarrow Z - cx = 0,$$

$$Z - [c_B, \; c_N]\begin{bmatrix} x_B \\ x_N \end{bmatrix} = 0,$$

$$Z - c_B x_B - c_N x_N = 0.$$

Now substituting $x_B = B^{-1}b - B^{-1}Nx_N$ from (2.13) we obtain

$$Z - c_B(B^{-1}b - B^{-1}Nx_N) - c_N x_N = 0$$

$$Z - (c_N - c_B B^{-1}N)x_N = c_B B^{-1}b.$$

Note that the last equation does not contain the basic variables. This representation allows us to determine the *net* effect on the objective function of changing a nonbasic variable. This is an essential property used by the simplex method as we discuss in the following subsection. The vector of objective function coefficients $c_N - c_B B^{-1}N$ corresponding to the nonbasic variables is often called the vector of *reduced costs* since they contain the cost coefficients c_N "reduced" by the cross effects of the basic variables given by $c_B B^{-1}N$.

Exercise 2.14 Consider the following linear programming problem:

$$\begin{aligned}
\max \; & 4x_1 + 3x_2 \\
& 3x_1 + x_2 \leq 9 \\
& 3x_1 + 2x_2 \leq 10 \\
& x_1 + x_2 \leq 4 \\
& x_1 \geq 0, \; x_2 \geq 0.
\end{aligned}$$

Transform this problem into the standard form. How many basic solutions does the standard form problem have? What are the basic feasible solutions and what are the extreme points of the feasible region?

Exercise 2.15   A plant can manufacture five products P_1, P_2, P_3, P_4, and P_5. The plant consists of two work areas: the job shop area A_1 and the assembly area A_2. The time required to process one unit of product P_j in work area A_i is p_{ij} (in hours), for $i = 1, 2$ and $j = 1, \ldots, 5$. The weekly capacity of work area A_i is C_i (in hours). The company can sell all it produces of product P_j at a profit of s_j, for $i = 1, \ldots, 5$.

The plant manager thought of writing a linear program to maximize profits, but never actually did for the following reason: from past experience, he observed that the plant operates best when at most two products are manufactured at a time. He believes that if he uses linear programming, the optimal solution will consist of producing all five products and therefore it will not be of much use to him. Do you agree with him? Explain, based on your knowledge of linear programming.

Answer: The linear program has two constraints (one for each of the work areas). Therefore, at most two variables are positive in a basic solution. In particular, this is the case for an optimal basic solution. So the plant manager is mistaken in his beliefs. There is always an optimal solution of the linear program in which at most two products are manufactured.

2.4.2 Simplex iterations

A key result of linear programming theory is that when a linear programming problem has an optimal solution, it **must** have an optimal solution that is an extreme point. The significance of this result lies in the fact that when we are looking for a solution of a linear programming problem we can focus on the objective value of extreme point solutions only. There are only a finite number of them, so this reduces our search space from an infinite space to a finite one.

The simplex method solves a linear programming problem by moving from one extreme point to an adjacent extreme point. Since, as we discussed in the previous section, extreme points of the feasible set correspond to basic feasible solutions (BFSs), algebraically this is achieved by moving from one BFS to another. We describe this strategy in detail in this section.

The process we mentioned in the previous paragraph must start from an initial BFS. How does one find such a point? While finding a basic solution is almost trivial, finding feasible basic solutions can be difficult. Fortunately, for problems of the form (2.9), such as the bond portfolio optimization problem (2.8) there is a simple strategy. Choosing

$$B = \begin{bmatrix} 1 & 0 & 0 \\ 0 & 1 & 0 \\ 0 & 0 & 1 \end{bmatrix}, \quad x_B = \begin{bmatrix} x_3 \\ x_4 \\ x_5 \end{bmatrix}, \quad N = \begin{bmatrix} 1 & 1 \\ 2 & 1 \\ 5 & 10 \end{bmatrix}, \quad x_N = \begin{bmatrix} x_1 \\ x_2 \end{bmatrix},$$

we get an initial basic feasible solution with $x_B = B^{-1}b = [100, 150, 360]^{\mathrm{T}}$. The objective value for this BFS is $4 \cdot 0 + 3 \cdot 0 = 0$.

Once we obtain a BFS, we first need to determine whether this solution is optimal or whether there is a way to improve the objective value. Recall that the basic variables are uniquely determined once we choose to set the nonbasic variables to a specific value, namely zero. So, the only way to obtain alternative solutions is to modify the values of the nonbasic variables. We observe that both the nonbasic variables x_1 and x_2 would improve the objective value if they were introduced into the basis. Why? The initial basic feasible solution has $x_1 = x_2 = 0$ and we can get other feasible solutions by increasing the value of one of these two variables. To preserve the feasibility of the equality constraints, this will require adjusting the values of the basic variables $x_3, x_4,$ and x_5. But since all three are strictly positive in the initial basic feasible solution, it is possible to make x_1 strictly positive without violating any of the constraint, including the nonnegativity requirements.

None of the variables x_3, x_4, x_5 appear in the objective row. Thus, we only have to look at the coefficient of the nonbasic variable we would increase to see what effect this would have on the objective value. The rate of improvement in the objective value for x_1 is 4 and for x_2 this rate is only 3. While a different method may choose to increase both of these variables simultaneously, the simplex method requires that only one nonbasic variable is modified at a time. This requirement is the algebraic equivalent of the geometric condition of moving from one extreme point to an *adjacent* extreme point. Between x_1 and x_2, we choose the variable x_1 to enter the basis since it has a faster rate of improvement.

The basis holds as many variables as there are equality constraints in the standard form formulation of the problem. Since x_1 is to enter the basis, one of $x_3, x_4,$ and x_5 must leave the basis. Since nonbasic variables have value zero in a basic solution, we need to determine how much to increase x_1 so that one of the current basic variables becomes zero and can be designated as nonbasic. The important issue here is to maintain the nonnegativity of all basic variables. Because each basic variable only appears in one row, this is an easy task. As we increase x_1, all current basic variables will decrease since x_1 has positive coefficients in each row.[4] We

---

[4] If x_1 had a zero coefficient in a particular row, then increasing it would not effect the basic variable in that row. If x_1 had a negative coefficient in a row, then as x_1 was being increased the basic variable of that row would need to be increased to maintain the equality in that row; but then we would not worry about that basic variable becoming negative.

guarantee the nonnegativity of the basic variables of the next iteration by using the ratio test. We observe that

$$\text{increasing } x_1 \text{ beyond } 100/1 = 100 \quad \Rightarrow \quad x_3 < 0,$$
$$\text{increasing } x_1 \text{ beyond } 150/2 = 75 \quad \Rightarrow \quad x_4 < 0,$$
$$\text{increasing } x_1 \text{ beyond } 360/3 = 120 \quad \Rightarrow \quad x_5 < 0,$$

so we should not increase x_1 more than $\min\{100, 75, 120\} = 75$. On the other hand, if we increase x_1 by exactly 75, x_4 will become zero. The variable x_4 is said to *leave the basis*. It has now become a nonbasic variable.

Now we have a new basis: $\{x_3, x_1, x_5\}$. For this basis we have the following basic feasible solution:

$$B = \begin{bmatrix} 1 & 1 & 0 \\ 0 & 2 & 0 \\ 0 & 3 & 1 \end{bmatrix}, \quad x_B = \begin{bmatrix} x_3 \\ x_1 \\ x_5 \end{bmatrix} = B^{-1}b = \begin{bmatrix} 1 & -1/2 & 0 \\ 0 & 1/2 & 0 \\ 0 & -3/2 & 1 \end{bmatrix} \begin{bmatrix} 100 \\ 150 \\ 360 \end{bmatrix} = \begin{bmatrix} 25 \\ 75 \\ 135 \end{bmatrix},$$

$$N = \begin{bmatrix} 1 & 0 \\ 1 & 1 \\ 4 & 0 \end{bmatrix}, \quad x_N = \begin{bmatrix} x_2 \\ x_4 \end{bmatrix} = \begin{bmatrix} 0 \\ 0 \end{bmatrix}.$$

After finding a new feasible solution, we always ask the question "Is this the optimal solution, or can we still improve it?" Answering that question was easy when we started, because none of the basic variables were in the objective function. Now that we have introduced x_1 into the basis, the situation is more complicated. If we now decide to increase x_2, the objective row coefficient of x_2 does not tell us how much the objective value changes per unit change in x_2, because changing x_2 requires that we also change x_1, a basic variable that appears in the objective row. It may happen that increasing x_2 by 1 unit does not increase the objective value by 3 units, because x_1 may need to be decreased, pulling down the objective function. It could even happen that increasing x_2 actually decreases the objective value even though x_2 has a positive coefficient in the objective function. So, what do we do? We could still do what we did with the initial basic solution **if** x_1 did not appear in the objective row and the rows where it is not the basic variable. To achieve this, all we need to do is to use the row where x_1 is the basic variable (in this case the second row) to solve for x_1 in terms of the nonbasic variables and then substitute this expression for x_1 in the objective row and other equations. So, the second equation

$$2x_1 + x_2 + x_4 = 150$$

would give us:

$$x_1 = 75 - \frac{1}{2}x_2 - \frac{1}{2}x_4.$$

Substituting this value in the objective function we get:

$$Z = 4x_1 + 3x_2 = 4\left(75 - \frac{1}{2}x_2 - \frac{1}{2}x_4\right) + 3x_2 = 300 + x_2 - 2x_4.$$

Continuing the substitution we get the following representation of the original bond portfolio problem:

max Z

subject to:

$$Z \; -x_2 \; + 2x_4 \qquad\qquad\qquad = 300$$
$$\tfrac{1}{2}x_2 \; - \; \tfrac{1}{2}x_4 + x_3 \qquad\qquad = 25$$
$$\tfrac{1}{2}x_2 \; + \; \tfrac{1}{2}x_4 \qquad + x_1 \qquad = 75$$
$$\tfrac{5}{2}x_2 \; - \; \tfrac{3}{2}x_4 \qquad\qquad + x_5 = 135$$
$$x_2 \geq 0, \; x_4 \geq 0, \; x_3 \geq 0, \; x_1 \geq 0, \; x_5 \geq 0.$$

This representation looks exactly like the initial system. Once again, the objective row is free of basic variables and basic variables only appear in the row where they are basic, with a coefficient of 1. Therefore, we now can tell how a change in a nonbasic variables would effect the objective function: increasing x_2 by 1 unit will increase the objective function by 1 unit (not 3!) and increasing x_4 by 1 unit will decrease the objective function by 2 units.

Now that we have represented the problem in a form identical to the original, we can repeat what we did before, until we find a representation that gives the optimal solution. If we repeat the steps of the simplex method, we find that x_2 will be introduced into the basis next, and the leaving variable will be x_3. If we solve for x_1 using the first equation and substitute for it in the remaining ones, we get the following representation:

max Z

subject to:

$$Z + \quad 2x_3 + x_4 \qquad\qquad\qquad = 350$$
$$2x_3 - x_4 + x_2 \qquad\qquad = 50$$
$$-x_3 + x_4 \qquad + x_1 \qquad = 50$$
$$-5x_3 + x_4 \qquad\qquad + x_5 = \quad 10$$
$$x_3 \geq 0, \; x_4 \geq 0, \; x_2 \geq 0, \; x_1 \geq 0, \; x_5 \geq 0.$$

The basis and the basic solution that corresponds to the system above is:

$$B = \begin{bmatrix} 1 & 1 & 0 \\ 1 & 2 & 0 \\ 4 & 3 & 1 \end{bmatrix}, \quad x_B = \begin{bmatrix} x_2 \\ x_1 \\ x_5 \end{bmatrix} = B^{-1}b = \begin{bmatrix} 2 & -1 & 0 \\ -1 & 1 & 0 \\ -5 & 1 & 1 \end{bmatrix} \begin{bmatrix} 100 \\ 150 \\ 360 \end{bmatrix} = \begin{bmatrix} 50 \\ 50 \\ 10 \end{bmatrix},$$

$$N = \begin{bmatrix} 1 & 0 \\ 0 & 1 \\ 0 & 0 \end{bmatrix}, \quad x_N = \begin{bmatrix} x_3 \\ x_4 \end{bmatrix} = \begin{bmatrix} 0 \\ 0 \end{bmatrix}.$$

At this point we can conclude that this basic solution is the optimal solution. Let us try to understand why. From the objective function row of our final representation of the problem we have that, for any feasible solution $x = (x_1, x_2, x_3, x_4, x_5)$, the objective function Z satisfies

$$Z + 2x_3 + x_4 = 350.$$

Since $x_3 \geq 0$ and $x_4 \geq 0$ is also required, this implies that in every feasible solution

$$Z \leq 350.$$

But we just found a basic feasible solution with value 350. So this is the optimal solution.

More generally, recall that for any BFS $x = (x_B, x_N)$, the objective value Z satisfies

$$Z - (c_N - c_B B^{-1} N) x_N = c_B B^{-1} b.$$

If for a BFS $x_B = B^{-1} b \geq 0$, $x_N = 0$, we have

$$c_N - c_B B^{-1} N \leq 0,$$

then this solution is an optimal solution since it has objective value $Z = c_B B^{-1} b$ whereas, for all other solutions, $x_N \geq 0$ implies that $Z \leq c_B B^{-1} b$.

Exercise 2.16   What is the solution to the following linear programming problem?

$$\max \ Z = c_1 x_1 + c_2 x_2 + \cdots + c_n x_n$$
$$\text{s.t. } a_1 x_1 + a_2 x_2 + \cdots + a_n x_n \leq b,$$
$$0 \leq x_i \leq u_i \ (i = 1, 2, \ldots, n).$$

Assume that all the data elements (c_i, a_i, and u_i) are strictly positive and the coefficients are arranged such that:

$$\frac{c_1}{a_1} \geq \frac{c_2}{a_2} \geq \cdots \geq \frac{c_n}{a_n}.$$

Write the problem in standard form and apply the simplex method to it. What will be the steps of the simplex method when applied to this problem, i.e., in what order will the variables enter and leave the basis?

2.4.3 The tableau form of the simplex method

In most linear programming textbooks, the simplex method is described using *tableaus* that summarize the information in the different representations of the problem we saw above. Since the reader will likely encounter simplex tableaus elsewhere, we include a brief discussion for the purpose of completeness. To study the tableau form of the simplex method, we recall the bond portfolio example of the previous subsection. We begin by rewriting the objective row as

$$Z - 4x_1 - 3x_2 = 0$$

and represent this system using the following tableau:

Basic var.	x_1	x_2	x_3	x_4	x_5	
Z	-4	-3	0	0	0	0
x_3	1	1	1	0	0	100
x_4	2^*	1	0	1	0	150
x_5	3	4	0	0	1	360

This tableau is often called the *simplex tableau*. The columns labeled by each variable contain the coefficients of that variable in each equation, including the objective row equation. The leftmost column is used to keep track of the basic variable in each row. The arrows and the asterisk will be explained below.

Step 0 *Form the initial tableau.*

Once we have formed this tableau we look for an *entering variable*, i.e., a variable that has a negative coefficient in the objective row and will improve the objective function if it is introduced into the basis. In this case, two of the variables, namely x_1 and x_2, have negative objective row coefficients. Since x_1 has the most negative coefficient we will pick that one (this is indicated by the arrow pointing down on x_1), but in principle any variable with a negative coefficient in the objective row can be chosen to enter the basis.

Step 1 *Find a variable with a negative coefficient in the first row (the objective row). If all variables have nonnegative coefficients in the objective row, STOP, the current tableau is optimal.*

After we choose x_1 as the entering variable, we need to determine a *leaving variable*. The leaving variable is found by performing a *ratio test*. In the ratio test, one looks at the column that corresponds to the entering variable, and for each *positive* entry in that column computes the ratio of that positive number to the right-hand-side value in that row. The minimum of these ratios tells us how much we can increase our entering variable without making any of the other variables negative. The basic variable in the row that gives the minimum ratio becomes the leaving variable. In the tableau above the column for the entering variable, the column for the right-hand-side values, and the ratios of corresponding entries are

$$x_1 \qquad \text{RHS} \qquad \text{ratio}$$

$$\begin{bmatrix} 1 \\ 2 \\ 5 \end{bmatrix}, \quad \begin{bmatrix} 100 \\ 150 \\ 360 \end{bmatrix}, \quad \begin{matrix} 100/1 \\ 150/2 \\ 360/3 \end{matrix}, \quad \min \left\{ \frac{100}{1}, \frac{150^*}{2}, \frac{360}{3} \right\} = 75,$$

and therefore x_4, the basic variable in the second row, is chosen as the leaving variable, as indicated by the left-pointing arrow in the tableau.

One important issue here is that we only look at the positive entries in the column when we perform the ratio test. Notice that if some of these entries were negative, then increasing the entering variable would only increase the basic variable in those rows, and would not force them to be negative, therefore we need not worry about those entries. Now, if all of the entries in a column for an entering variable turn out to be zero or negative, then we conclude that the problem must be *unbounded*; we can increase the entering variable (and the objective value) indefinitely, the equalities can be balanced by *increasing* the basic variables appropriately, and none of the nonnegativity constraints will be violated.

Step 2 *Consider the column picked in Step 1. For each positive entry in this column, calculate the ratio of the right-hand-side value to that entry. Find the row that gives the minimum such ratio and choose the basic variable in that row as the leaving variable. If all the entries in the column are zero or negative, STOP, the problem is unbounded.*

Before proceeding to the next iteration, we need to update the tableau to reflect the changes in the set of basic variables. For this purpose, we choose a *pivot element*, which is the entry in the tableau that lies in the intersection of the column for the entering variable (the *pivot column*), and the row for the leaving variable (the *pivot row*). In the tableau above, the pivot element is the number 2, marked with an asterisk. The next job is *pivoting*. When we pivot, we aim to get the number 1 in the position of the pivot element (which can be achieved by dividing the entries in the pivot row by the pivot element), and zeros elsewhere in the pivot column (which can be achieved by adding suitable multiples of the pivot row to the other rows, including the objective row). All these operations are row operations on the matrix that consists of the numbers in the tableau, and what we are doing is essentially Gaussian elimination on the pivot column. Pivoting on the tableau above yields:

	Basic var.	x_1	x_2	x_3	x_4	x_5	
	Z	0	-1	0	2	0	300
$\Leftarrow$	x_3	0	$1/2^*$	1	$-1/2$	0	25
	x_1	1	$1/2$	0	$1/2$	0	75
	x_5	0	$5/2$	0	$-3/2$	1	135

Step 3 *Find the entry (the pivot element) in the intersection of the column picked in Step 1 (the pivot column) and the row picked in Step 2 (the pivot row). Pivot on that entry, i.e., divide all the entries in the pivot row by the pivot element, add appropriate multiples of the pivot row to the others in order to get zeros in other components of the pivot column. Go to Step 1.*

If we repeat the steps of the simplex method, this time working with the new tableau, we first identify x_2 as the only candidate to enter the basis. Next, we do the ratio test:

$$\min\left\{\frac{25^*}{1/2}, \frac{75}{1/2}, \frac{135}{5/2}\right\} = 50,$$

so x_3 leaves the basis. Now, one more pivot produces the optimal tableau:

Basic var.	x_1	x_2	x_3	x_4	x_5	
Z	0	0	2	1	0	350
x_2	0	1	2	-1	0	50
x_1	1	0	-1	1	0	50
x_5	0	0	-5	1	1	10

This solution is optimal since all the coefficients in the objective row are nonnegative.

Exercise 2.17 Solve the following linear program by the simplex method.

$$\begin{aligned}
\max \quad & 4x_1 + x_2 - x_3 \\
& x_1 \qquad\quad + 3x_3 \leq 6 \\
& 3x_1 + x_2 + 3x_3 \leq 9 \\
& x_1 \geq 0, \ x_2 \geq 0, \ x_3 \geq 0.
\end{aligned}$$

Answer:

	x_1	x_2	x_3	s_1	s_2	
Z	-4	-1	1	0	0	0
s_1	1	0	3	1	0	6
s_2	3	1	3	0	1	9
Z	0	1/3	5	0	4/3	12
s_1	0	$-1/3$	2	1	$-1/3$	3
x_1	1	1/3	1	0	1/3	3

The optimal solution is $x_1 = 3$, $x_2 = x_3 = 0$.

Exercise 2.18 Solve the following linear program by the simplex method.

$$\begin{aligned}
\max\ 4x_1 + x_2 - \ x_3 \\
x_1 \qquad\quad + 3x_3 \le 6 \\
3x_1 + x_2 + 3x_3 \le 9 \\
x_1 + x_2 - \ x_3 \le 2 \\
x_1 \ge 0,\ x_2 \ge 0,\ x_3 \ge 0.
\end{aligned}$$

Exercise 2.19 Suppose the following tableau was obtained in the course of solving a linear program with nonnegative variables x_1, x_2, x_3, and two inequalities. The objective function is maximized and slack variables x_4 and x_5 were added.

Basic var.	x_1	x_2	x_3	x_4	x_5	
Z	0	a	b	0	4	82
x_4	0	-2	2	1	3	c
x_1	1	-1	3	0	-5	3

Give conditions on a, b and c that are required for the following statements to be true:

(i) The current basic solution is a basic feasible solution. Assume that the condition found in (i) holds in the rest of the exercise.
(ii) The current basic solution is optimal.
(iii) The linear program is unbounded (for this question, assume that $b > 0$).
(iv) The current basic solution is optimal and there are alternate optimal solutions (for this question, assume that $a > 0$).

2.4.4 Graphical interpretation

Figure 2.1 shows the feasible region for Example 2.1. The five inequality constraints define a convex pentagon. The five corner points of this pentagon (the black dots on the figure) are the basic feasible solutions: each such solution satisfies two of the constraints with equality.

Which are the solutions explored by the simplex method? The simplex method starts from the basic feasible solution ($x_1 = 0$, $x_2 = 0$) (in this solution, x_1 and x_2 are the nonbasic variables. The basic variables $x_3 = 100$, $x_4 = 150$ and $x_5 = 360$ correspond to the slack in the constraints that are not satisfied with equality). The first iteration of the simplex method makes x_1 basic by increasing it along an edge of the feasible region until some other constraint is satisfied with equality. This leads to the new basic feasible solution ($x_1 = 75$, $x_2 = 0$) (in this solution, x_2 and x_4 are nonbasic, which means that the constraints $x_2 \ge 0$ and $2x_1 + x_2 \le 150$ are

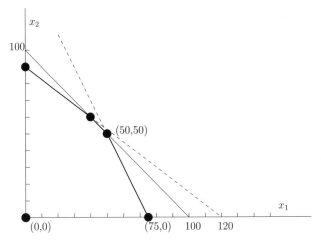

Figure 2.1 Graphical interpretation of the simplex iterations

satisfied with equality). The second iteration makes x_2 basic while keeping x_4 nonbasic. This corresponds to moving along the edge $2x_1 + x_2 = 150$. The value x_2 is increased until another constraint becomes satisfied with equality. The new solution is $x_1 = 50$ and $x_2 = 50$. No further movement from this point can increase the objective, so this is the optimal solution.

Exercise 2.20   Solve the linear program of Exercise 2.14 by the simplex method. Give a graphical interpretation of the simplex iterations.

Exercise 2.21   Find basic solutions of Example 2.1 that are not feasible. Identify these solutions in Figure 2.1.

2.4.5 The dual simplex method

The previous sections describe the *primal* simplex method, which moves from a basic feasible solution to another until all the reduced costs are nonpositive. There are certain applications where the *dual simplex method* is faster. In contrast to the primal simplex method, this method keeps the reduced costs nonpositive and moves from a basic (infeasible) solution to another until a basic feasible solution is reached.

We illustrate the dual simplex method with an example. Consider Example 2.1 with the following additional constraint:

$$6x_1 + 5x_2 \le 500$$

The feasible set resulting from this additional constraint is shown in Figure 2.2, where the bold line represents the boundary of the new constraint. Adding a slack

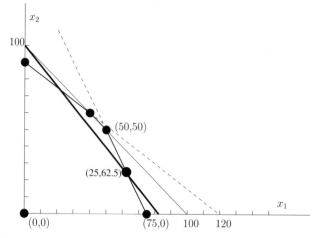

Figure 2.2 Graphical interpretation of the dual simplex iteration

variable x_6, we get $6x_1 + 5x_2 + x_6 = 500$. To initialize the dual simplex method, we can start from any basic solution with nonpositive reduced costs. For example, we can start from the optimal solution that we found in Section 2.4.3, without the additional constraint, and make x_6 basic. This gives the following tableau:

Basic var.	x_1	x_2	x_3	x_4	x_5	x_6	
Z	0	0	2	1	0	0	350
x_2	0	1	2	−1	0	0	50
x_1	1	0	−1	1	0	0	50
x_5	0	0	−5	1	1	0	10
x_6	6	5	0	0	0	1	500

Actually, this tableau is not yet in the correct format. Indeed, x_1 and x_2 are basic and therefore their columns in the tableau should be unit vectors. To restore this property, it suffices to eliminate the 6 and 5 in the x_6 row by subtracting appropriate multiples of the x_1 and x_2 rows. This now gives the tableau in the correct format:

Basic var.	x_1	x_2	x_3	x_4	x_5	x_6	
Z	0	0	2	1	0	0	350
x_2	0	1	2	−1	0	0	50
x_1	1	0	−1	1	0	0	50
x_5	0	0	−5	1	1	0	10
x_6	0	0	−4	−1	0	1	−50

Observe that the basic variable x_6 has a negative value in this representation and therefore the basic solution is not feasible. This is confirmed in Figure 2.2 by the fact that the point (50, 50) corresponding to the current basic solution is on the wrong side of the new constraint boundary. Now we are ready to apply the dual simplex algorithm. Note that the current basic solution $x_1 = 50, x_2 = 50$, $x_3 = x_4 = 0, x_5 = 10, x_6 = -50$ is infeasible since x_6 is negative. We will pivot to make it nonnegative. As a result, variable x_6 will *leave the basis*. The pivot element will be one of the negative entries in the x_6 row, namely -4 or -1. Which one should we choose in order to keep all the reduced costs nonnegative? The minimum ratio between $2/|-4|$ and $1/|-1|$ determines the variable that *enters the basis*. Here the minimum is $1/2$, which means that x_3 enters the basis. After pivoting on -4, the tableau becomes:

Basic var.	x_1	x_2	x_3	x_4	x_5	x_6	
Z	0	0	0	0.5	0	0.5	325
x_2	0	1	0	-1.5	0	0.5	25
x_1	1	0	0	1.25	0	-0.25	62.5
x_5	0	0	0	2.25	1	-1.25	72.5
x_3	0	0	1	0.25	0	-0.25	12.5

The corresponding basic solution is $x_1 = 62.5, x_2 = 25, x_3 = 12.5, x_4 = 0, x_5 = 72.5, x_6 = 0$. Since it is feasible and all reduced costs are nonpositive, this is the optimum solution. If there had still been negative basic variables in the solution, we would have continued pivoting using the rules outlined above: the variable that leaves the basis is one with a negative value, the pivot element is negative, and the variable that enters the basis is chosen by the minimum ratio rule.

Exercise 2.22 Solve the following linear program by the dual simplex method, starting from the solution found in Exercise 2.17 .

$$
\begin{array}{ll}
\max & 4x_1 + x_2 - x_3 \\
& x_1 \quad\;\; + 3x_3 \le 6 \\
& 3x_1 + x_2 + 3x_3 \le 9 \\
& x_1 + x_2 - x_3 \le 2 \\
& x_1 \ge 0, \; x_2 \ge 0, \; x_3 \ge 0.
\end{array}
$$

2.4.6 Alternatives to the simplex method

Performing a pivot of the simplex method is extremely fast on today's computers, even for problems with thousands of variables and hundreds of constraints. This explains the success of the simplex method. For large problems, however, the number of iterations also tends to be large. At the time of writing, LPs with tens of thousands of constraints and 100 000 or more variables are generally considered large problems. Such models are not uncommon in financial applications and can often be handled by the simplex method.

Although the simplex method demonstrates satisfactory performance for the solution of most practical problems, it has the disadvantage that, in the worst case, the amount of computing time (the so-called *worst-case complexity*) can grow exponentially with the size of the problem. Here *size* refers to the space required to write all the data in binary. If all the numbers are bounded (say between 10^{-6} and 10^{6}), a good proxy for the size of a linear program is the number of variables times the number of constraints. One of the important concepts in the theoretical study of optimization algorithms is the concept of *polynomial-time algorithms*. This refers to an algorithm whose running time can be bounded by a polynomial function of the input size for all instances of the problem class that it is intended for. After it was discovered in the 1970s that the worst-case complexity of the simplex method is exponential (and, therefore, that the simplex method is not a polynomial-time algorithm) there was an effort to identify alternative methods for linear programming with polynomial-time complexity. The first such method, called the *ellipsoid method*, was originally developed by Yudin and Nemirovski [84] in the mid 1970s for convex nonlinear optimization problems. In 1979, Khachiyan [44] proved that the ellipsoid method is a polynomial-time algorithm for linear programming. But the more exciting and enduring development was the announcement by Karmarkar in 1984 that an *interior-point method* (IPM) can solve LPs in polynomial time. What distinguished Karmarkar's [43] IPM from the ellipsoid method was that, in addition to having this desirable theoretical property, it could solve some real-world LPs much faster than the simplex method. These methods use a different strategy to reach the optimum, generating iterates in the interior of the feasible region rather than at its extreme points. Each iteration is fairly expensive, compared to simplex iterations, but the number of iterations needed does not depend much on the size of the problem and is often less than 50. As a result, interior-point methods can be faster than the simplex method for large-scale problems. Most state-of-the-art linear programming packages (Cplex, Xpress, OSL, etc.) provide the option to solve linear programs by either method.

We present interior-point methods in Chapter 7, in the context of solving quadratic programs.

3

LP models: asset/liability cash-flow matching

3.1 Short-term financing

Corporations routinely face the problem of financing short-term cash commitments. Linear programming can help in figuring out an optimal combination of financial instruments to meet these commitments. To illustrate this, consider the following problem. For simplicity of exposition, we keep the example very small.

A company has the following short-term financing problem:

Month	Jan	Feb	Mar	Apr	May	Jun
Net cash flow	−150	−100	200	−200	50	300

Net cash flow requirements are given in thousands of dollars. The company has the following sources of funds:

- a line of credit of up to $100k at an interest rate of 1% per month;
- in any one of the first three months, it can issue 90-day commercial paper bearing a total interest of 2% for the three-month period;
- excess funds can be invested at an interest rate of 0.3% per month.

There are many questions that the company might want to answer. What interest payments will the company need to make between January and June? Is it economical to use the line of credit in some of the months? If so, when? How much? Linear programming gives us a mechanism for answering these questions quickly and easily. It also allows to answer some "what if" questions about changes in the data without having to resolve the problem. What if the net cash flow in January was −$200k (instead of −$150k)? What if the limit on the credit line was increased from $100k to $200k? What if the negative net cash flow in January was due to the purchase of a machine worth $150k and the vendor allowed part or all of the payment on this machine to be made in June at an interest rate of 3% for the five-month period? The answers to these questions are readily available when this problem is formulated and solved as a linear program.

There are three steps in applying linear programming: modeling, solving, and interpreting.

3.1.1 Modeling

We begin by modeling the above short-term financing problem. That is, we write it in the language of linear programming. There are rules about what one can and cannot do within linear programming. These rules are in place to make certain that the remaining steps of the process (solving and interpreting) can be successful.

Key to a linear program are the *decision variables*, *objective*, and *constraints*.

Decision variables

The decision variables represent (unknown) decisions to be made. This is in contrast to *problem data*, which are values that are either given or can be simply calculated from what is given. For the short-term financing problem, there are several possible choices of decision variables. We will use the following decision variables: the amount x_i drawn from the line of credit in month i, the amount y_i of commercial paper issued in month i, the excess funds z_i in month i and the company's wealth v in June. Note that, alternatively, one could use the decision variables x_i and y_i only, since excess funds and company's wealth can be deduced from these variables.

Objective

Every linear program has an objective. This objective is to be either minimized or maximized. This objective has to be *linear* in the decision variables, which means it must be the sum of constants times decision variables. $3x_1 - 10x_2$ is a linear function. $x_1 x_2$ is not a linear function. In this case, our objective is simply to maximize v.

Constraints

Every linear program also has constraints limiting feasible decisions. Here we have three types of constraints: (i) cash inflow = cash outflow for each month, (ii) upper bounds on x_i, and (iii) nonnegativity of the decision variables x_i, y_i and z_i.

For example, in January ($i = 1$), there is a cash requirement of $150k. To meet this requirement, the company can draw an amount x_1 from its line of credit and issue an amount y_1 of commercial paper. Considering the possibility of excess funds z_1 (possibly 0), the cash-flow balance equation is as follows:

$$x_1 + y_1 - z_1 = 150.$$

Next, in February ($i = 2$), there is a cash requirement of $100k. In addition, principal plus interest of $1.01x_1$ is due on the line of credit and $1.003z_1$ is received on the invested excess funds. To meet the requirement in February, the company can draw an amount x_2 from its line of credit and issue an amount y_2 of commercial paper. So, the cash-flow balance equation for February is as follows:

$$x_2 + y_2 - 1.01x_1 + 1.003z_1 - z_2 = 100.$$

Similarly, for March we get the following equation:

$$x_3 + y_3 - 1.01x_2 + 1.003z_2 - z_3 = -200.$$

For the months of April, May, and June, issuing commercial paper is no longer an option, so we will not have variables y_4, y_5, and y_6 in the formulation. Furthermore, any commercial paper issued between January and March requires a payment with 2% interest three months later. Thus, we have the following additional equations:

$$
\begin{aligned}
x_4 - 1.02y_1 - 1.01x_3 + 1.003z_3 - z_4 &= 200 \\
x_5 - 1.02y_2 - 1.01x_4 + 1.003z_4 - z_5 &= {-50} \\
- 1.02y_3 - 1.01x_5 + 1.003z_5 - v &= -300.
\end{aligned}
$$

Note that x_i is the balance on the credit line in month i, not the incremental borrowing in month i. Similarly, z_i represents the overall excess funds in month i. This choice of variables is quite convenient when it comes to writing down the upper bound and nonnegativity constraints.

$$
\begin{aligned}
0 &\le x_i \le 100 \\
y_i &\ge 0 \\
z_i &\ge 0.
\end{aligned}
$$

Final Model

This gives us the complete model of this problem:

$$
\begin{aligned}
\max \quad & v \\
x_1 + y_1 - z_1 &= 150 \\
x_2 + y_2 - 1.01x_1 + 1.003z_1 - z_2 &= 100 \\
x_3 + y_3 - 1.01x_2 + 1.003z_2 - z_3 &= -200 \\
x_4 - 1.02y_1 - 1.01x_3 + 1.003z_3 - z_4 &= 200 \\
x_5 - 1.02y_2 - 1.01x_4 + 1.003z_4 - z_5 &= {-50} \\
- 1.02y_3 - 1.01x_5 + 1.003z_5 - v &= -300 \\
x_1 &\le 100 \\
x_2 &\le 100 \\
x_3 &\le 100 \\
x_4 &\le 100 \\
x_5 &\le 100 \\
x_i, y_i, z_i &\ge 0.
\end{aligned}
$$

Formulating a problem as a linear program means going through the above process of clearly defining the decision variables, objective, and constraints.

Exercise 3.1   How would the formulation of the short-term financing problem above change if the commercial papers issued had a two-month maturity instead of three?

Exercise 3.2   A company will face the following cash requirements in the next eight quarters (positive entries represent cash needs while negative entries represent cash surpluses):

Q1	Q2	Q3	Q4	Q5	Q6	Q7	Q8
100	500	100	−600	−500	200	600	−900

The company has three borrowing possibilities:

• A two-year loan available at the beginning of Q1, with an interest rate of 1% per quarter.
• The other two borrowing opportunities are available at the beginning of every quarter: a six-month loan with an interest rate of 1.8% per quarter, and a quarterly loan at an interest rate of 2.5% for the quarter.

Any surplus can be invested at an interest rate of 0.5% per quarter.
    Formulate a linear program that maximizes the wealth of the company at the beginning of Q9.

Exercise 3.3   A home buyer in France can combine several mortgage loans to finance the purchase of a house. Given borrowing needs B and a horizon of T months for paying back the loans, the home buyer would like to minimize the total cost (or equivalently, the monthly payment p made during each of the next T months). Regulations impose limits on the amount that can be borrowed from certain sources. There are n different loan opportunities available. Loan i has a fixed interest rate r_i, a length $T_i \leq T$, and a maximum amount borrowed b_i. The monthly payment on loan i is not required to be the same every month, but a minimum payment m_i is required each month. However, the total monthly payment p over all loans is constant. Formulate a linear program that finds a combination of loans that minimizes the home buyer's cost of borrowing. [Hint: In addition to variables x_{ti} for the payment on loan i in month t, it may be useful to introduce a variable for the amount of outstanding principal on loan i in month t.]

3.1.2 Solving the model with SOLVER

Special computer programs can be used to find solutions to linear programming models. The most widely available program is undoubtedly SOLVER, included in the Excel spreadsheet program. Here are other suggestions:

- MATLAB has a linear programming solver that can be accessed with the command linprog. Type help linprog to find out details.
- Even if one does not have access to any linear programming software, it is possible to solve linear programs (and other optimization problems) using the website www-neos.mcs. anl.gov/neos/. This is the website for the Network Enabled Optimization Server. Using the JAVA submission tool on this site, one can submit a linear programming problem (in some standard format) and have a remote computer solve the problem using one of the several solver options. The solution is then transmitted to the submitting person by e-mail.
- A good open-source LP code written in C is CLP, available from the following website at the time of writing: www.coin-or.org/.

SOLVER, while not a state-of-the-art code is a reasonably robust, easy-to-use tool for linear programming. SOLVER uses standard spreadsheets together with an interface to define variables, objective, and constraints.

We briefly outline how to create a SOLVER spreadsheet:

- Start with a spreadsheet that has all of the data entered in some reasonably neat way.

    In the short-term financing example, the spreadsheet might contain the cash flows, interest rates, and credit limit.
- The model will be created in a separate part of the spreadsheet. Identify one cell with each decision variable. SOLVER will eventually put the optimal values in these cells.

    In the short-term financing example, we could associate cells $B$2 to $B$6 with variables x_1 to x_5 respectively, cells $C$2 to $C$4 with the y_i variables, cells $D$2 to $D$6 with the z_i variables, and, finally, $E$2 with the variable v.
- A separate cell represents the objective. Enter a formula that represents the objective.

    For the short-term financing example, we might assign cell $B$8 to the objective function. Then, in cell $B$8, we enter the function =$E$2.

    This formula must be a linear formula, so, in general, it must be of the form: c1*x1 +c2*x2+···, where the cells c1, c2 and so on contain constant values and the cells x1, x2 and so on contain the decision variables.
- We then choose a cell to represent the left-hand side of each constraint (again a linear function) and another cell to represent the right-hand side (a constant).

    In the short-term financing example, let us choose cells $B$10 to $B$15 for the amounts generated through financing, for each month, and cells $D$10 to $D$15 for the cash requirements. For example, cell $B$10 would contain the function =$C$2+$B$2-$D$2 and cell $D$10 the value 150. Similarly, rows 16 to 20 could be used to write the credit limit constraints.

    Helpful hint: Excel has a function sumproduct() that is designed for linear programs. sumproduct(A1:A10,B1:B10) is identical to A1*B1+A2*B2+A3*B3+···+A10*B10. This function can save much time. All that is needed is that the length of the first range be the same as the length of the second range (so one can be horizontal and the other vertical).

- We then select `Solver` under the `Tools` menu. This gives a form to fill out to define the linear program.
- In the `Set Target Cell` box, select the objective cell. Choose `Max` or `Min` depending on whether you want to maximize or minimize the objective.
- In the `By Changing Cells` box, type the range (or ranges) containing the variable cells. In our short-term financing example, this would be `$B$2:$B$6,$C$2:$C$4, $D$2:$D$6,$E$2`.
- Next we add the constraints. Press the `Add` button to add constraints. The dialog box has three parts: the left-hand side, the type of constraint, and the right-hand side. The box associated with the left-hand side is called `Cell Reference`. Type the appropriate cell (`$B$10` for the first constraint in the short-term financing example). In the second box select the type of constraint (= in our example), and in the third box, called `Constraint:`, type the cell containing the right-hand side (`$D$10` in our example). Then press `Add`. Repeat the process for the second constraint. Continue until all constraints are added. On the final constraint, press `OK`.

  Helpful hint: It is possible to include ranges of constraints, as long as they all have the same type. `$B$10:$B$15<=$D$10:$D$15` means `$B$10<=$D$10, $B$11<= $D$11...$B$15<=$D$15`. Similarly `$B$2:$B$6>=0` means each individual cell from `$B$2` to `$B$6` must be greater than or equal to 0.

- Push the `Options` button and check the `Assume Linear Model` in the resulting dialog box. This tells Excel to use the simplex method rather than a nonlinear programming routine. This is important, because the simplex method is more efficient and reliable. This also gives you sensitivity ranges, which are not available for nonlinear models.

  Note that if you want your variables to assume nonnegative values only you need to specify this in the options box (alternatively, you can add nonnegativity constraints in the previous step, in your constraints). Click on `OK`.

- Push the `Solve` button. In the resulting dialog box, select `Answer` and `Sensitivity`. This will put the answer and sensitivity analysis in two new sheets. Ask Excel to `Keep Solver Solution`, and your worksheet will be updated so that the optimal values are in the variable cells.

Exercise 3.4 Solve the linear program formulated in Exercise 3.2 with your favorite software package.

3.1.3 Interpreting the output of SOLVER

If we were to solve the short-term financing problem above using SOLVER, the solution given in the `Answer` report would look as follows:

Target Cell (Max)

Cell	Name	Original Value	Final Value
$B$8	Objective	0	92.49694915

Adjustable Cells

Cell	Name	Original Value	Final Value
$B$2	$x1$	0	0
$B$3	$x2$	0	50.98039216
$B$4	$x3$	0	0
$B$5	$x4$	0	0
$B$6	$x5$	0	0
$C$2	$y1$	0	150
$C$3	$y2$	0	49.01960784
$C$4	$y3$	0	203.4343636
$D$2	$z1$	0	0
$D$3	$z2$	0	0
$D$4	$z3$	0	351.9441675
$D$5	$z4$	0	0
$D$6	$z5$	0	0
$E$2	v	0	92.49694915

Constraints

Cell	Name	Cell Value	Formula	Slack
$B$10	January	150	$B$10 = $D$10	0
$B$11	February	100	$B$11 = $D$11	0
$B$12	March	−200	$B$12 = $D$12	0
$B$13	April	200	$B$13 = $D$13	0
$B$14	May	−50	$B$14 = $D$14	0
$B$15	June	−300	$B$15 = $D$15	0
$B$16	$x1$limit	0	$B$16 <= $D$16	100
$B$17	$x2$limit	50.98039216	$B$17 <= $D$17	49.01960784
$B$18	$x3$limit	0	$B$18 <= $D$18	100
$B$19	$x4$limit	0	$B$19 <= $D$19	100
$B$20	$x5$limit	0	$B$20 <= $D$20	100

This report is fairly easy to read: the company's wealth v in June will be $92 497. This is reported in Final Value of the Target Cell (recall that our units are in $1000). To achieve this, the company will issue $150 000 in commercial paper in January, $49 020 in February and $203 434 in March. In addition, it will draw $50 980 from its line of credit in February. Excess cash of $351 944 in March will be invested for just one month. All this is reported in the `Adjustable Cells` section of the report. For this particular application, the `Constraints` section of the report does not contain anything useful. On the other hand, very useful information can be found in the sensitivity report. This will be discussed in Section 3.3.

Exercise 3.5   Formulate and solve the variation of the short-term financing problem you developed in Exercise 3.1 using SOLVER. Interpret the solution.

Exercise 3.6   Recall Example 2.1. Solve the problem using your favorite linear programming solver. Compare the output provided by the solver to the solution we obtained in Chapter 2.

3.1.4 Modeling languages

Linear programs can be formulated using modeling languages such as AMPL, GAMS, MOSEL, or OPL. The need for these modeling languages arises because the Excel spreadsheet format becomes inadequate when the size of the linear program increases. A modeling language lets people use common notation and familiar concepts to formulate optimization models and examine solutions. Most importantly, large problems can be formulated in a compact way. Once the problem has been formulated using a modeling language, it can be solved using any number of solvers. A user can switch between solvers with a single command and select options that may improve solver performance. The short-term financing model would be formulated as follows (all variables are assumed to be nonnegative unless otherwise specified):

```
DATA
LET T=6 be the number of months to plan for
L(t) = Liability in month t=1,...,T
ratex = monthly interest rate on line of credit
ratey = 3-month interest rate on commercial paper
ratez = monthly interest rate on excess funds
VARIABLES
x(t) = Amount drawn from line of credit in month t
y(t) = Amount of commercial paper issued in month t
z(t) = Excess funds in month t
OBJECTIVE (Maximize wealth in June)
Max z(6)
CONSTRAINTS
```

```
Month(t=1:T): x(t)-(1+ratex)*x(t-1)+y(t)
  -(1+ratey)*y(t-3)-z(t)+(1+ratez)*z(t-1) = L(t)
Month(t=1:T-1): x(t) < 100
Boundary conditions on x: x(0)=x(6) = 0
Boundary conditions on y: y(-2)=y(-1)=y(0)=y(4)=y(5)=
  y(6) = 0
Boundary conditions on z: z(0) = 0
END
```

Exercise 3.7 Formulate the linear program of Exercise 3.3 with one of the modeling languages AMPL, GAMS, MOSEL, or OPL.

3.1.5 Features of linear programs

Hidden in each linear programming formulation are a number of assumptions. The usefulness of an LP model is directly related to how closely reality matches up with these assumptions.

The first two assumptions are due to the linear form of the objective and constraint functions. The contribution to the objective of any decision variable is proportional to the value of the decision variable. Similarly, the contribution of each variable to the left-hand side of each constraint is proportional to the value of the variable. This is the *proportionality assumption*.

Furthermore, the contribution of a variable to the objective and constraints is independent of the values of the other variables. This is the *additivity assumption*. When additivity or proportionality assumptions are not satisfied, a *nonlinear programming* model may be more appropriate. We discuss such models in Chapters 5 and 6.

The next assumption is the *divisibility assumption*: is it possible to take any fraction of any variable? A fractional production quantity may be worrisome if we are producing a small number of battleships or be innocuous if we are producing millions of paperclips. If the divisibility assumption is important and does not hold, then a technique called *integer programming* rather than linear programming is required. This technique takes orders of magnitude more time to find solutions but may be necessary to create realistic solutions. We discuss integer programming models and methods in Chapters 11 and 12.

The final assumption is the *certainty assumption*: linear programming allows for no uncertainty about the input parameters such as the cash-flow requirements or interest rates we used in the short-term financing model. Problems with uncertain parameters can be addressed using *stochastic programming* or *robust optimization* approaches. We discuss such models in Chapters 16 through 20.

It is very rare that a problem will meet all of the assumptions exactly. That does not negate the usefulness of a model. A model can still give useful managerial insight even if reality differs slightly from the rigorous requirements of the model.

Exercise 3.8   Give an example of an optimization problem where the proportionality assumption is not satisfied.

Exercise 3.9   Give an example of an optimization problem where the additivity assumption is not satisfied.

Exercise 3.10   Consider the LP model we develop for the cash-flow matching problem in Section 3.2. Which of the linear programming assumptions used for this formulation is the least realistic one? Why?

3.2 Dedication

Dedication or *cash-flow matching* is a technique used to fund known liabilities in the future. The intent is to form a portfolio of assets whose cash inflows will exactly offset the cash outflows of the liabilities. The liabilities will therefore be paid off, as they become due, without the need to sell or buy assets in the future. The portfolio is formed today and held until all liabilities are paid off. Dedicated portfolios usually only consist of risk-free non-callable bonds since the portfolio future cash inflows need to be known when the portfolio is constructed. This eliminates interest rate risk completely. It is used by some municipalities and small pension funds. For example, municipalities sometimes want to fund liabilities stemming from bonds they have issued. These pre-refunded municipal bonds can be taken off the books of the municipality. This may allow them to evade restrictive covenants in the bonds that have been pre-refunded and perhaps allow them to issue further debt.

It should be noted, however, that dedicated portfolios cost typically from 3% to 7% more in dollar terms than do "immunized" portfolios that are constructed based on matching present value, duration, and convexity of the assets and liabilities. The *present value* of the liability stream L_t for $t = 1, \ldots, T$ is $P = \sum_{t=1}^{T} L_t/(1 + r_t)^t$, where r_t denotes the risk-free rate in year t. Its *duration* is $D = (1/P)\sum_{t=1}^{T} t L_t/(1 + r_t)^t$ and its *convexity* is $C = (1/P)\sum_{t=1}^{T} t(t + 1)L_t/(1 + r_t)^{t+2}$. Intuitively, duration is the average (discounted) time at which the liabilities occur, whereas convexity, a bit like variance, indicates how concentrated the cash flows are over time. For a portfolio that consists only of risk-free bonds, the present value P^* of the portfolio future cash inflows can be computed using the same risk-free rate r_t (this would not be the case for a portfolio containing risky bonds). Similarly for the duration D^* and convexity C^* of the portfolio future cash inflows. An "immunized" portfolio can be constructed based on matching $P^* = P$, $D^* = D$ and $C^* = C$. Portfolios that are constructed by matching these three factors are immunized against parallel shifts in the yield curve, but there may still be a great deal of exposure and vulnerability to other types of shifts, and they need to be actively managed, which can be costly. By contrast, dedicated portfolios do not need to be managed after they are constructed.

When municipalities use cash-flow matching, the standard custom is to call a few investment banks, send them the liability schedule, and request bids. The municipality then buys its securities from the bank that offers the lowest price for a successful cash-flow match.

Assume that a bank receives the following liability schedule:

Year 1	Year 2	Year 3	Year 4	Year 5	Year 6	Year 7	Year 8
12 000	18 000	20 000	20 000	16 000	15 000	12 000	10 000

The bonds available for purchase today (Year 0) are given in the next table. All bonds have a face value of $100. The coupon figure is annual. For example, Bond 5 costs $98 today, and it pays back $4 in Year 1, $4 in Year 2, $4 in Year 3, and $104 in Year 4. All these bonds are widely available and can be purchased in any quantities at the stated price.

Bond	1	2	3	4	5	6	7	8	9	10
Price	102	99	101	98	98	104	100	101	102	94
Coupon	5	3.5	5	3.5	4	9	6	8	9	7
Maturity year	1	2	2	3	4	5	5	6	7	8

We would like to formulate and solve a linear program to find the least-cost portfolio of bonds to purchase today, to meet the obligations of the municipality over the next eight years. To eliminate the possibility of any reinvestment risk, we assume a 0% reinvestment rate.

Using a modeling language, the formulation might look as follows.

```
DATA
LET T=8 be the number of years to plan for.
LET N=10 be the number of bonds available for purchase
  today.
L(t)  = Liability in year t=1,...,T
P(i)  = Price of bond i, i=1,...,N
C(i)  = Annual coupon for bond i, i=1,...,N
```

```
M(i) = Maturity year of bond i, i=1,...,N
VARIABLES
x(i) = Amount of bond i in the portfolio, i=1,...,N
z(t) = Surplus at the end of year t, for t=0,...,T
OBJECTIVE (Minimize cost)
Min z(0) + SUM(i=1:N) P(i)*x(i)
CONSTRAINTS Year(t=1:T):
SUM(i=1:N | M(i)>t-1) C(i)*x(i) + SUM(i=1:N | M(i)=t)
   100*x(i)-z(t)+z(t-1) = L(t)
END
```

Exercise 3.11   Solve the dedication linear program above using an LP software package and verify that we can optimally meet the municipality's liabilities for $93 944 with the following portfolio: 62 Bond1, 125 Bond3, 152 Bond4, 157 Bond5, 123 Bond6, 124 Bond8, 104 Bond9, and 93 Bond10.

Exercise 3.12   A small pension fund has the following liabilities (in million dollars):

Year 1	Year 2	Year 3	Year 4	Year 5	Year 6	Year 7	Year 8	Year 9
24	26	28	28	26	29	32	33	34

It would like to construct a dedicated bond portfolio. The bonds available for purchase are the following:

Bond	1	2	3	4	5	6	7	8
Price	102.44	99.95	100.02	102.66	87.90	85.43	83.42	103.82
Coupon	5.625	4.75	4.25	5.25	0.00	0.00	0.00	5.75
Maturity year	1	2	2	3	3	4	5	5

Bond	9	10	11	12	13	14	15	16
Price	110.29	108.85	109.95	107.36	104.62	99.07	103.78	64.66
Coupon	6.875	6.5	6.625	6.125	5.625	4.75	5.5	0.00
Maturity year	6	6	7	7	8	8	9	9

Formulate an LP that minimizes the cost of the dedicated portfolio, assuming a 2% reinvestment rate. Solve the LP using your favorite software package.

3.3 Sensitivity analysis for linear programming

The optimal solution to a linear programming model is the most important output of LP solvers, but it is not the only useful information they generate. Most linear programming packages produce a tremendous amount of *sensitivity information*, or information about what happens when data values are changed.

Recall that, in order to formulate a problem as a linear program, we had to invoke a *certainty assumption*: we had to know what value the data took on, and we made decisions based on that data. Often this assumption is somewhat dubious: the data might be unknown, or guessed at, or otherwise inaccurate. How can we determine the effect on the optimal decisions if the values change? Clearly some numbers in the data are more important than others. Can we find the "important" numbers? Can we determine the effect of estimation errors?

Linear programming offers extensive capabilities for addressing these questions. We give examples of how to interpret the SOLVER output. To access the information in SOLVER, one can simply ask for the sensitivity report after optimizing. Rather than giving rules for reading the reports, we show how to answer a set of questions from the output.

3.3.1 Short-term financing

The SOLVER sensitivity report looks as follows:

Adjustable Cells

Cell	Name	Final Value	Reduced Cost	Objective Coefficient	Allowable Increase	Allowable Decrease
$B$2	$x1$	0	−0.0032	0	0.0032	$1E + 30$
$B$3	$x2$	50.98	0	0	0.0032	0
$B$4	$x3$	0	−0.0071	0	0.0071	$1E + 30$
$B$5	$x4$	0	−0.0032	0	0.0032	$1E + 30$
$B$6	$x5$	0	0	0	0	$1E + 30$
$C$2	$y1$	150	0	0	0.0040	0.0032
$C$3	$y2$	49.02	0	0	0	0.0032
$C$4	$y3$	203.43	0	0	0.0071	0
$D$2	$z1$	0	−0.0040	0	0.0040	$1E + 30$
$D$3	$z2$	0	−0.0071	0	0.0071	$1E + 30$
$D$4	$z3$	351.94	0	0	0.0039	0.0032
$D$5	$z4$	0	−0.0039	0	0.0039	$1E + 30$
$D$6	$z5$	0	−0.007	0	0.007	$1E + 30$
$E$2	v	92.50	0	1	$1E + 30$	1

Constraints

Cell	Name	Final Value	Shadow Price	Constraint R.H. Side	Allowable Increase	Allowable Decrease
$B$10	January	150	−1.0373	150	89.17	150
$B$11	February	100	−1.030	100	49.020	50.980
$B$12	March	−200	−1.020	−200	90.683	203.434
$B$13	April	200	−1.017	200	90.955	204.044
$B$14	May	−50	−1.010	−50	50	52
$B$15	June	−300	−1	−300	92.497	$1E + 30$
$B$16	$x1$	0	0	100	$1E + 30$	100
$B$17	$x2$	50.98	0	100	$1E + 30$	49.020
$B$18	$x3$	0	0	100	$1E + 30$	100
$B$19	$x4$	0	0	100	$1E + 30$	100
$B$20	$x5$	0	0	100	$1E + 30$	100

The key columns for sensitivity analysis are the Reduced Cost and Shadow Price columns in SOLVER. The *shadow price u* of a constraint C has the following interpretation:

> If the right-hand side of the constraint C changes by an amount Δ, the optimal objective value changes by $u\Delta$, as long as the amount of change Δ is within the allowable range.

For a linear program, the shadow price u is an exact figure, as long as the amount of change Δ is within the allowable range given in the last two columns of the SOLVER output. When the change Δ falls outside this range, the rate of change in the optimal objective value changes and the shadow price u cannot be used. When this occurs, one has to resolve the linear program using the new data.

Next, we consider several examples of sensitivity questions and demonstrate how they can be answered using shadow prices and reduced costs.

- For example, assume that the net cash flow in January was −$200k (instead of −150). By how much would the company's wealth decrease at the end of June?

  The answer is in the shadow price of the January constraint, $u = -1.0373$. The RHS of the January constraint would go from 150 to 200, an increase of $\Delta = 50$, which is within the allowable increase (89.17). (Recall that these figures are in thousand dollars.) So the company's wealth in June would decrease by 1.0373 * 50 000 = $51 865.

- Now assume that the net cash flow in March was $250k (instead of 200). By how much would the company's wealth increase at the end of June?

Again, the change $\Delta = -50$ is within the allowable decrease (203.434), so we can use the shadow price $u = -1.02$ to calculate the change in objective value. The increase is $(-1.02) * (-50) = \$51\,000$.

- Assume that the credit limit was increased from \$100k to \$200k. By how much would the company's wealth increase at the end of June?

  In each month, the change $\Delta = 100$ is within the allowable increase ($+\infty$) and the shadow price for the credit limit constraint is $u = 0$. So there is no effect on the company's wealth in June. Note that non-binding constraints – such as the credit limit constraint for months January through May – always have zero shadow price.

- Assume that the negative net cash flow in January is due to the purchase of a machine worth \$150\,000. The vendor allows the payment to be made in June at an interest rate of 3% for the five-month period. Would the company's wealth increase or decrease by using this option? What if the interest rate for the 5-month period was 4%?

  The shadow price of the January constraint is -1.0373. This means that reducing cash requirements in January by \$1 increases the wealth in June by \$1.0373. In other words, the break-even interest rate for the five-month period is 3.73%. So, if the vendor charges 3%, we should accept, but if he/she charges 4% we should not. Note that the analysis is valid since the amount $\Delta = -150$ is within the allowable decrease.

- Now, let us consider the reduced costs. The basic variables always have a zero reduced cost. The nonbasic variables (which by definition take the value 0) have a nonpositive reduced cost and, frequently, their reduced cost is strictly negative. There are two useful interpretations of the reduced cost c, for a nonbasic variable x.

  First, assume that x is set to a positive value Δ instead of its optimal value 0. Then, the objective value is changed by $c\Delta$. For example, what would be the effect of financing part of the January cash needs through the line of credit? The answer is in the reduced cost of variable x_1. Because this reduced cost -0.0032 is strictly negative, the objective function would decrease. Specifically, each dollar financed through the line of credit in January would result in a decrease of \$0.0032 in the company's wealth v in June.

  The second interpretation of c is that its magnitude $|c|$ is the minimum amount by which the objective coefficient of x must be increased in order for the variable x to become positive in an optimal solution. For example, consider the variable x_1 again. Its value is zero in the current optimal solution, with objective function v. However, if we changed the objective to $v + 0.0032x_1$, it would now be optimal to use the line of credit in January. In other words, the reduced cost on x_1 can be viewed as the minimum rebate that the bank would have to offer (payable in June) to make it attractive to use the line of credit in January.

Exercise 3.13 Recall Example 2.1. Determine the shadow price and reduced cost information for this problem using an LP software package. How would the solution change if the average maturity of the portfolio is required to be 3.3 instead of 3.6?

Exercise 3.14 Generate the sensitivity report for Exercise 3.2 with your favorite LP solver.

(i) Suppose the cash requirement in Q2 is 300 (instead of 500). How would this affect the wealth in Q9?
(ii) Suppose the cash requirement in Q2 is 100 (instead of 500). Can the sensitivity report be used to determine the wealth in Q9?
(iii) One of the company's suppliers may allow deferred payments of $50 from Q3 to Q4. What would be the value of this?

Exercise 3.15 *Workforce planning*: Consider a restaurant that is open seven days a week. Based on past experience, the number of workers needed on a particular day is given as follows:

Day	Mon	Tue	Wed	Thu	Fri	Sat	Sun
Number	14	13	15	16	19	18	11

Every worker works five consecutive days, and then takes two days off, repeating this pattern indefinitely. How can we minimize the number of workers that staff the restaurant?

Answer: Let the days be numbers 1 through 7 and let x_i be the number of workers who begin their five-day shift on day i. The linear programming formulation is as follows:

$$\min \qquad \sum_i x_i$$

s.t.

$$x_1 + x_4 + x_5 + x_6 + x_7 \geq 14$$
$$x_1 + x_2 + x_5 + x_6 + x_7 \geq 13$$
$$x_1 + x_2 + x_3 + x_6 + x_7 \geq 15$$
$$x_1 + x_2 + x_3 + x_4 + x_7 \geq 16$$
$$x_1 + x_2 + x_3 + x_4 + x_5 \geq 19$$
$$x_2 + x_3 + x_4 + x_5 + x_6 \geq 18$$
$$x_3 + x_4 + x_5 + x_6 + x_7 \geq 11$$
$$x_i \geq 0 \text{ (for all } i\text{)}.$$

Sensitivity analysis

The following table gives the sensitivity report for the solution of the workforce planning problem:

Adjustable Cells

Cell	Name	Final Value	Reduced Cost	Objective Coefficient	Allowable Increase	Allowable Decrease
$B$14	Shift1	4	0	1	0.5	1
$B$15	Shift2	7	0	1	0	0.333333
$B$16	Shift3	1	0	1	0.5	0
$B$17	Shift4	4	0	1	0.5	0
$B$18	Shift5	3	0	1	0	0.333333
$B$19	Shift6	3	0	1	0.5	1
$B$20	Shift7	0	0.333333	1	$1E+30$	0.333333

Constraints

Cell	Name	Final Value	Shadow Price	Constraint R.H. Side	Allowable Increase	Allowable Decrease
$B$24	Monday	14	0.333333	14	1.5	6
$B$25	Tuesday	17	0	13	4	$1E+30$
$B$26	Wednesday	15	0.333333	15	6	3
$B$27	Thursday	16	0	16	3	4
$B$28	Friday	19	0.333333	19	4.5	3
$B$29	Saturday	18	0.333333	18	1.5	6
$B$30	Sunday	11	0	11	4	1

Answer each of the following questions independently of the others.

(i) What is the current total number of workers needed to staff the restaurant?

(ii) Due to a special offer, demand on Thursdays increases. As a result, 18 workers are needed instead of 16. What is the effect on the total number of workers needed to staff the restaurant?

(iii) Assume that demand on Mondays decreases, so that 11 workers are needed instead of 14. What is the effect on the total number of workers needed to staff the restaurant?

(iv) Every worker in the restaurant is paid $1000 per month. Therefore, the objective function in the formulation can be viewed as total wage expenses (in thousand dollars). Workers have complained that Shift 4 is the least desirable shift. Management is considering increasing the wages of workers on Shift 4 to $1100. Would this change the optimal solution? What would be the effect on total wage expenses?

(v) Shift 2, on the other hand, is very desirable (Sundays off while on duty Fridays and Saturdays, which are the best days for tips). Management is considering

reducing the wages of workers on Shift 2 to $900 per month. Would this change
the optimal solution? What would be the effect on total wage expenses?

(vi) Management is considering introducing a new shift with the days off on Tues-
days and Sundays. Because these days are not consecutive, the wages will be
$1200 per month. Will this increase or reduce the total wage expenses?

3.3.2 *Dedication*

We end this section with the sensitivity report of the dedication problem formulated
in Section 3.2.

Adjustable Cells

Cell	Name	Final Value	Reduced Cost	Objective Coefficient	Allowable Increase	Allowable Decrease
$B$5	$x1$	62.13612744	0	102	3	5.590909091
$B$6	$x2$	0	0.830612245	99	$1E+30$	0.830612245
$B$7	$x3$	125.2429338	0	101	0.842650104	3.311081442
$B$8	$x4$	151.5050805	0	98	3.37414966	4.712358277
$B$9	$x5$	156.8077583	0	98	4.917243419	17.2316607
$B$10	$x6$	123.0800686	0	104	9.035524153	3.74817022
$B$11	$x7$	0	8.786840002	100	$1E+30$	8.786840002
$B$12	$x8$	124.1572748	0	101	3.988878399	8.655456271
$B$13	$x9$	104.0898568	0	102	9.456887408	0.860545483
$B$14	$x10$	93.45794393	0	94	0.900020046	1E+30
$H$4	$z0$	0	0.028571429	1	$1E+30$	0.028571429
$H$5	$z1$	0	0.055782313	0	$1E+30$	0.055782313
$H$6	$z2$	0	0.03260048	0	$1E+30$	0.03260048
$H$7	$z3$	0	0.047281187	0	$1E+30$	0.047281187
$H$8	$z4$	0	0.179369792	0	$1E+30$	0.179369792
$H$9	$z5$	0	0.036934059	0	$1E+30$	0.036934059
$H$10	$z6$	0	0.086760435	0	$1E+30$	0.086760435
$H$11	$z7$	0	0.008411402	0	$1E+30$	0.008411402
$H$12	$z8$	0	0.524288903	0	$1E+30$	0.524288903

Constraints

Cell	Name	Final Value	Shadow Price	Constraint R.H. Side	Allowable Increase	Allowable Decrease
$B$19	year1	12000	0.971428571	12000	1E+30	6524.293381
$B$20	year2	18000	0.915646259	18000	137010.161	13150.50805
$B$21	year3	20000	0.883045779	20000	202579.3095	15680.77583
$B$22	year4	20000	0.835764592	20000	184347.1716	16308.00686
$B$23	year5	16000	0.6563948	16000	89305.96314	13415.72748
$B$24	year6	15000	0.619460741	15000	108506.7452	13408.98568
$B$25	year7	12000	0.532700306	12000	105130.9798	11345.79439
$B$26	year8	10000	0.524288903	10000	144630.1908	10000

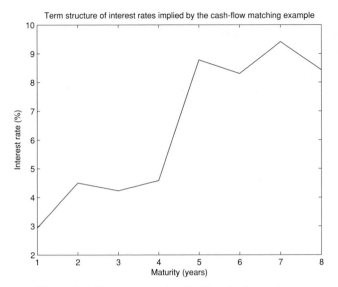

Figure 3.1 Interest rates implied by shadow prices

Exercise 3.16 Analyze the solution tables above and:

(i) interpret the shadow price in year t ($t = 1, \ldots, 8$);
(ii) interpret the reduced cost of bond i ($i = 1, \ldots, 10$);
(iii) interpret the reduced cost of each surplus variable z_t ($t = 0, \ldots, 7$).

Answers:

(i) The shadow price in Year t is the cost of the bond portfolio that can be attributed to a dollar of liability in Year t. For example, each dollar of liability in Year 3 is responsible for $0.883 in the cost of the bond portfolio. Note that, by setting the shadow price in Year t equal to $1/(1 + r_t)^t$, we get a term structure of interest rates. Here $r_3 = 0.0423$. In Figure 3.1 we plot the term structure of interest rates we compute from this solution. If the bonds in the dedication example were risk-free Treasury bonds, this figure should be very similar to the term structure of Treasury rates for the corresponding years.

(ii) The reduced cost of bond i indicates by how much bond i is overpriced for inclusion in the optimal portfolio. For example, bond 2 would have to be $0.83 lower, at $98.17, for inclusion in the optimal portfolio.

Exercise 3.17 Note that the optimal solution has no holdings in Bond 7 which matures in Year 5, despite the $16 000 liability in Year 5. This is likely due to a mispricing of this bond at $100. What would be a more realistic price for this bond?

Answer: Row 7 of the Adjustable Cells table indicates that variable x_7, corresponding to Bond 7 holdings, will become positive only if the price is reduced by

8.786 or more. So, a more realistic price for this bond would be just above $91. By checking the reduced costs, one may sometimes spot errors in the data!

(iii) The reduced cost of the surplus variable z_t indicates what the interest rate on cash reinvested in Year t would have to be in order to keep excess cash in Year t.

Exercise 3.18   Generate the sensitivity report for Exercise 3.12 .

(i) Suppose that the liability in Year 3 is 29 (instead of 28). What would be the increase in cost of the dedicated portfolio?
(ii) Draw a graph of the term structure of interest rates implied by the shadow prices.
(iii) Bond 4 is not included in the optimal portfolio. By how much would the price of Bond 4 have to decrease for Bond 4 to become part of the optimal portfolio?
(iv) The fund manager would like to have 10 000 units of Bond 3 in the portfolio. By how much would this increase the cost of the portfolio?
(v) Is there any bond that looks badly mispriced?
(vi) What interest rate on cash would make it optimal to include cash as part of the optimal portfolio?

3.4 Case study: constructing a dedicated portfolio

Set i to be the year when you are reading this sentence (For example, at the printing of this book $i = 2007$). A municipality sends you today the following liability stream (in million dollars) in years $i + 1$ to $i + 8$:

$6/15/i + 1$	$12/15/i + 1$	$6/15/i + 2$	$12/15/i + 2$	$6/15/i + 3$	$12/15/i + 3$
6	6	9	9	10	10

$6/15/i + 4$	$12/15/i + 4$	$6/15/i + 5$	$12/15/i + 5$	$6/15/i + 6$	$12/15/i + 6$
10	10	8	8	8	8

$6/15/i + 7$	$12/15/i + 7$	$6/15/i + 8$	$12/15/i + 8$
6	6	5	5

Your job:

• Compute the present value of the liability using the Treasury yield curve. You can find current data on numerous websites, such as www.treasury.gov or www.bondsonline.com.

- Identify between 30 and 50 assets that are suitable for a dedicated portfolio (non-callable bonds, treasury bills, or notes). Explain why they are suitable. Current data can be found on the Web or in newspapers such as the *Wall Street Journal*.
- Set up a linear program to identify a lowest-cost dedicated portfolio of assets (so no short selling) and solve with Excel's solver (or any other linear programming software that you prefer). What is the cost of your portfolio? Discuss the composition of your portfolio. Discuss the assets and the liabilities in light of the sensitivity report. What is the term structure of interest rates implied by the shadow prices? Compare with the term structure of Treasury rates. (Hint: Refer to Section 3.3.2.)
- Set up a linear program to identify a lowest-cost portfolio of assets (no short selling) that matches present value, duration, and convexity (or a related measure) between the liability stream and the bond portfolio. Solve the linear program with your favorite software. Discuss the solution. How much would you save by using this immunization strategy instead of dedication? Can you immunize the portfolio against nonparallel shifts of the yield curve? Explain.
- Set up a linear program to identify a lowest-cost portfolio of assets (no short selling) that combines a cash matching strategy for the liabilities in the first three years and an immunization strategy based on present value, duration, and convexity for the liabilities in the last five years. Compare the cost of this portfolio with the cost of the two previous portfolios.
- The municipality would like you to make a second bid: what is your lowest-cost dedicated portfolio of riskfree bonds if short sales are allowed? Discuss the feasibility of your solution. Did you find arbitrage opportunities? Did you model the bid/ask spread? Did you set limits on the transaction amounts?

4

LP models: asset pricing and arbitrage

4.1 Derivative securities and the fundamental theorem of asset pricing

One of the most widely studied problems in financial mathematics is the pricing of *derivative securities*, also known as *contingent claims*. These are securities whose price depends on the value of another *underlying security*. Financial *options* are the most common examples of derivative securities. For example, a European call option gives the holder the right to purchase an underlying security for a prescribed amount (called the *strike price*) at a prescribed time in the future, known as the *expiration* or *exercise date*. The exercise date is also known as the *maturity date* of the derivative security. Recall the similar definitions of European put options as well as American call and put options from Section 1.3.2.

Options are used mainly for two purposes: speculation and hedging. By speculating on the direction of the future price movements of the underlying security, investors can take (bare) positions in options on this security. Since options are often much cheaper than their underlying security, this bet results in much larger earnings in relative terms if the price movements happen in the expected direction compared to what one might earn by taking a similar position in the underlying. Of course, if one guesses the direction of the price movements incorrectly, the losses are also much more severe.

The more common and sensible use of options is for hedging. Hedging refers to the reduction of risk in an investor's overall position by forming a suitable portfolio of assets that are expected to have opposing risks. For example, if an investor holds a share of XYZ and is concerned that the price of this security may fall significantly, she can purchase a put option on XYZ and protect herself against price levels below a certain threshold – the strike price of the put option.

Recall the option example in the simple one-period binomial model of Section 1.3.2. Below, we summarize some of the information from that example.

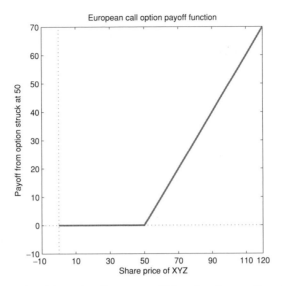

Figure 4.1 Piecewise linear payoff function for a call option

We consider the share price of XYZ stock, which is currently valued at $40. A month from today, we expect the share price of XYZ to either double or halve, with equal probabilities. We also consider a European call option on XYZ with a strike price of $50 which will expire a month from today. The payoff function for the call is shown in Figure 4.1.

We assume that interest rates for cash borrowing or lending are zero and that any amount of XYZ shares can be bought or sold with no commission:

$$S_0 = \$40 \begin{cases} 80 = S_1(u) \\ 20 = S_1(d) \end{cases} \quad \text{and } C_0 = ? \begin{cases} (80 - 50)^+ = 30 \\ (20 - 50)^+ = 0 \end{cases}$$

In Section 1.3.2 we obtained a fair price of $10 for the option using a replication strategy and the no-arbitrage principle. Two portfolios of securities that have identical future payoffs under all possible realizations of the random states must have the same value today. In the example, the first portfolio is the option while the second one is the portfolio of a half share of XYZ and −$10 in cash. Since we know the current value of the second portfolio, we can deduce the fair price of the option. To formalize this approach, we first give a definition of arbitrage:

Definition 4.1 *An* arbitrage *is a trading strategy that:*

- *has a positive initial cash flow and has no risk of a loss later (type A), or*
- *requires no initial cash input, has no risk of a loss, and has a positive probability of making profits in the future (type B).*

In the example, any price other than $10 for the call option would lead to a type A arbitrage – guaranteed profits at the initial time point and no future obligations. We do not need to have a guarantee of profits for type B arbitrage – all we need is a guarantee of no loss, and a positive probability of a gain. Prices adjust quickly so that arbitrage opportunities cannot persist in the markets. Therefore, in pricing arguments it is often assumed that arbitrage opportunities do not exist.

4.1.1 Replication

In the above example, we formulated and solved the following question to determine the fair price of an option: can we form a portfolio of the underlying security (long or short) and cash (borrowed or lent) today, such that the payoff of the portfolio at the expiration date of the option will match the payoff of the option? In other words, can we replicate the option using a portfolio of the underlying security and cash?

Let us work in a slightly more general setting. Let S_0 be the current price of the underlying security and assume that there are two possible outcomes at the end of the period: $S_1^u = S_0 \cdot u$ and $S_1^d = S_0 \cdot d$. Assume $u > d$. We also assume that there is a fixed interest rate of r on cash positions for the given period. Let $R = 1 + r$.

Now we consider a derivative security which has payoffs of C_1^u and C_1^d in the up and down states respectively:

$$S_0 \begin{cases} S_1^u = S_0 \cdot u \\ S_1^d = S_0 \cdot d \end{cases} \qquad C_0 = ? \begin{cases} C_1^u \\ C_1^d \end{cases}$$

To price the derivative security, we will replicate its payoff. For replication consider a portfolio of Δ shares of the underlying and $\$B$ cash. For what values of Δ and B does this portfolio have the same payoffs at the expiration date as the derivative security?

In the "up" state, the replicating portfolio will have value $\Delta S_0 \cdot u + BR$ and in the "down" state it will be worth $\Delta S_0 \cdot d + BR$. Therefore, for perfect replication, we need to solve the following simple system of equations:

$$\Delta S_0 \cdot u + BR = C_1^u$$
$$\Delta S_0 \cdot d + BR = C_1^d.$$

We obtain:

$$\Delta = \frac{C_1^u - C_1^d}{S_0(u - d)}$$
$$B = \frac{uC_1^d - dC_1^u}{R(u - d)}.$$

Since this portfolio is worth $S_0\Delta + B$ today, that should be the price of the derivative security as well:

$$C_0 = \frac{C_1^u - C_1^d}{u - d} + \frac{uC_1^d - dC_1^u}{R(u - d)}$$
$$= \frac{1}{R}\left[\frac{R - d}{u - d}C_1^u + \frac{u - R}{u - d}C_1^d\right].$$

4.1.2 Risk-neutral probabilities

Let

$$p_u = \frac{R - d}{u - d} \quad \text{and} \quad p_d = \frac{u - R}{u - d}.$$

Note that we must have $d < R < u$ to avoid arbitrage opportunities as indicated in the following simple exercise.

Exercise 4.1 Let S_0 be the current price of a security and assume that there are two possible prices for this security at the end of the current period: $S_1^u = S_0 \cdot u$ and $S_1^d = S_0 \cdot d$. (Assume $u > d$.) Also assume that there is a fixed interest rate of r on cash positions for the given period. Let $R = 1 + r$. Show that there is an arbitrage opportunity if $u > R > d$ is not satisfied.

An immediate consequence of this observation is that both $p_u > 0$ and $p_d > 0$. Noting also that $p_u + p_d = 1$ one can interpret p_u and p_d as probabilities. In fact, these are the so-called *risk-neutral probabilities* (RNPs) of up and down states, respectively. Note that they are completely independent from the physical probabilities of these states.

The price of any derivative security can now be calculated as the present value of the expected value of its future payoffs where the expected value is taken using the risk-neutral probabilities.

In our example above $u = 2$, $d = 1/2$ and $r = 0$ so that $R = 1$. Therefore:

$$p_u = \frac{1 - 1/2}{2 - 1/2} = \frac{1}{3} \quad \text{and} \quad p_d = \frac{2 - 1}{2 - 1/2} = \frac{2}{3}.$$

As a result, we have

$$S_0 = 40 = \frac{1}{R}\left(p_u S_1^u + p_d S_1^d\right) = \frac{1}{3}80 + \frac{2}{3}20,$$

$$C_0 = 10 = \frac{1}{R}\left(p_u C_1^u + p_d C_1^d\right) = \frac{1}{3}30 + \frac{2}{3}0,$$

as expected. Using risk neutral probabilities we can also price other derivative securities on the XYZ stock. For example, consider a European put option on the

XYZ stock struck at $60 and with the same expiration date as the call of the example:

$$P_1^u = \max\{0, 60 - 80\} = 0,$$

$$P_0 = ?$$

$$P_1^d = \max\{0, 60 - 20\} = 40.$$

We can easily compute:

$$P_0 = \frac{1}{R}\left(p_u P_1^u + p_d P_1^d\right) = \frac{1}{3}0 + \frac{2}{3}40 = \frac{80}{3},$$

without needing to replicate the option again.

Exercise 4.2   Compute the price of a binary (digital) call option on the XYZ stock that pays $1 if the XYZ price is above the strike price of $50.

Exercise 4.3   Assume that the XYZ stock is currently priced at $40. At the end of the next period, the XYZ price is expected to be in one of the following two states: $S_0 \cdot u$ or $S_0 \cdot d$. We know that $d < 1 < u$ but do not know d or u. The interest rate is zero. If a European call option with strike price $50 is priced at $10 while a European call option with strike price $40 is priced at $13, and we assume that these prices do not contain any arbitrage opportunities, what is the fair price of a European put option with a strike price of $40?

Hint: First note that $u > \frac{5}{4}$ – otherwise the first call would be worthless. Then we must have $10 = p_u(S_0 \cdot u - 50)$ and $13 = p_u(S_0 \cdot u - 40)$. From these equations determine p_u and then u and d.

Exercise 4.4   Assume that the XYZ stock is currently priced at $40. At the end of the next period, the XYZ price is expected to be in one of the following two states: $S_0 \cdot u$ or $S_0 \cdot d$. We know that $d < 1 < u$ but do not know d or u. The interest rate is zero. European call options on XYZ with strike prices of $30, $40, $50, and $60 are priced at $10, $7, $10/3, and $0. Which one of these options is mispriced? Why?

Remark 4.1   *Exercises 4.3 and 4.4 are much simplified and idealized examples of the pricing problems encountered by practitioners. Instead of a set of possible future states for prices that may be difficult to predict, they must work with a set of market prices for related securities. Then, they must extrapolate prices for an unpriced security using no-arbitrage arguments.*

Next we move from our binomial setting to a more general setting and let

$$\Omega = \{\omega_1, \omega_2, \ldots, \omega_m\} \tag{4.1}$$

be the (finite) set of possible future "states." For example, these could be prices for a security at a future date.

For securities S^i, $i = 0, \ldots, n$, let $S_1^i(\omega_j)$ denote the price of this security in state ω_j at time 1. Also let S_0^i denote the current (time 0) price of security S^i. We use $i = 0$ for the "riskless" security that pays the interest rate $r \geq 0$ between time 0 and time 1. It is convenient to assume that $S_0^0 = 1$ and that $S_1^0(\omega_j) = R = 1 + r, \forall j$.

Definition 4.2 *A risk-neutral probability measure on the set $\Omega = \{\omega_1, \omega_2, \ldots, \omega_m\}$ is a vector of positive numbers $(p_1, p_2, \ldots, p_m)$ such that*

$$\sum_{j=1}^{m} p_j = 1$$

and for every security S^i, $i = 0, \ldots, n$,

$$S_0^i = \frac{1}{R}\left(\sum_{j=1}^{m} p_j S_1^i(\omega_j)\right) = \frac{1}{R}\hat{E}[S_1^i].$$

Above, $\hat{E}[S]$ denotes the expected value of the random variable S under the probability distribution $(p_1, p_2, \ldots, p_m)$.

4.1.3 The fundamental theorem of asset pricing

In this section we state the first fundamental theorem of asset pricing and prove it for finite Ω. This proof is a simple exercise in linear programming duality that also utilizes the following well-known result of Goldman and Tucker on the existence of strictly complementary optimal solutions of LPs:

Theorem 4.1 (Goldman and Tucker [31]) *When both the primal and dual linear programming problems*

$$\begin{aligned}
\min_x \ & c^T x \\
& Ax = b \\
& x \geq 0
\end{aligned} \tag{4.2}$$

and

$$\begin{aligned}
\max_y \ & b^T y \\
& A^T y \leq c,
\end{aligned} \tag{4.3}$$

have feasible solutions, they have optimal solutions satisfying strict complementarity, i.e., there exist x^* and y^* optimal for the respective problems such that

$$x^* + (c - A^T y^*) > 0.$$

Now, we are ready to prove the following theorem:

Theorem 4.2 (The first fundamental theorem of asset pricing) *A risk-neutral probability measure exists if and only if there is no arbitrage.*

Proof: We provide the proof for the case when the state space Ω is finite and is given by (4.1). We assume without loss of generality that every state has a positive probability of occuring (since states that have no probability of occuring can be removed from Ω.) Given the current prices S_0^i and the future prices $S_1^i(\omega_j)$ in each state ω_j, for securities 0 to n, consider the following linear program with variables x_i, for $i = 0, \ldots, n$:

$$\min_x \quad \sum_{i=0}^n S_0^i x_i$$
$$\sum_{i=0}^n S_1^i(\omega_j)x_i \geq 0, \quad j = 1, \ldots, m. \tag{4.4}$$

Note that type-A arbitrage corresponds to a feasible solution to this LP with a negative objective value. Since $x = (x_1, \ldots, x_n)$ with $x_i = 0, \forall i$ is a feasible solution, the optimal objective value is always non-positive. Furthermore, since all the constraints are homogeneous, if there exists a feasible solution such that $\sum S_0^i x_i < 0$ (this corresponds to type-A arbitrage), the problem is unbounded. In other words, there is no type-A arbitrage if and only if the optimal objective value of this LP is 0.

Suppose that there is no type-A arbitrage. Then, there is no type-B arbitrage if and only if all constraints are tight for all optimal solutions of (4.4) since every state has a positive probability of occuring. Note that these solutions must have objective value 0.

Consider the dual of (4.4):

$$\max_p \quad \sum_{j=1}^m 0 p_j$$
$$\sum_{j=1}^m S_1^i(\omega_j)p_j = S_0^i, \quad i = 0, \ldots, n, \tag{4.5}$$
$$p_j \geq 0, \quad j = 1, \ldots, m.$$

Since the dual objective function is constant at zero for all dual feasible solutions, any dual feasible solution is also dual optimal.

When there is no type-A arbitrage, (4.4) has an optimal solution. Now, Theorem 2.2 – the strong duality theorem – indicates that the dual must have a feasible solution. If there is no type-B arbitrage either, Goldman and Tucker's theorem indicates that there exists a feasible and therefore optimal dual solution p^* such that $p^* > 0$. This follows from strict complementarity with primal constraints $\sum_{i=1}^n S_1^i(\omega_j)x_i \geq 0$, which are tight. From the dual constraint corresponding to $i = 0$, we have that $\sum_{j=1}^m p_j^* = 1/R$. Multiplying p^* by R one obtains a risk-neutral probability distribution. Therefore, the "no arbitrage" assumption implies the existence of RNPs.

The converse direction is proved in an identical manner. The existence of a RNP measure implies that (4.5) is feasible, and therefore its dual, (4.4) must be bounded, which implies that there is no type-A arbitrage. Furthermore, since we have a strictly feasible (and optimal) dual solution, any optimal solution of the primal must have tight constraints, indicating that there is no type-B arbitrage. ☐

4.2 Arbitrage detection using linear programming

The linear programming (LP) problems (4.4) and (4.5) formulated in the proof of Theorem 4.2 can naturally be used for detection of arbitrage opportunities. As we discussed above, however, this argument works only for finite state spaces. In this section, we discuss how LP formulations can be used to detect arbitrage opportunities without limiting consideration to finite state spaces. The price we pay for this flexibility is the restriction on the selection of the securities: we only consider the prices of a set of derivative securities written on the same underlying with same maturity. This discussion is based on Herzel [40].

Consider an underlying security with a current, time 0, price of S_0 and a random price S_1 at time 1. Consider n derivative securities written on this security that mature at time 1, and have piecewise linear payoff functions $\Psi_i(S_1)$, each with a single breakpoint K_i, for $i = 1, \ldots, n$. The obvious motivation is the collection of calls and puts with different strike prices written on this security. If, for example, the i-th derivative security were a European call with strike price K_i, we would have $\Psi_i(S_1) = (S_1 - K_i)^+$. We assume that the K_i's are in increasing order, without loss of generality. Finally, let S_0^i denote the current price of the i-th derivative security.

Consider a portfolio $x = (x_1, \ldots, x_n)$ of the derivative securities 1 to n and let $\Psi^x(S_1)$ denote the payoff function of the portfolio:

$$\Psi^x(S_1) := \sum_{i=1}^{n} \Psi_i(S_1)x_i. \tag{4.6}$$

The cost of forming the portfolio x at time 0 is given by

$$\sum_{i=1}^{n} S_0^i x_i. \tag{4.7}$$

To determine whether a static arbitrage opportunity exists in the current prices S_0^i, we consider the following problem: what is the cheapest portfolio of the derivative securities 1 to n whose payoff function $\Psi^x(S_1)$ is nonnegative for all $S_1 \in [0, \infty)$? Nonnegativity of $\Psi^x(S_1)$ corresponds to "no future obligations" part of the arbitrage definition. If the minimum initial cost of such a portfolio is negative, then we have a type-A arbitrage.

Since all $\Psi_i(S_1)$'s are piecewise linear, so is $\Psi^x(S_1)$. It will have up to n breakpoints at points K_1 through K_n. Observe that a piecewise linear function is nonnegative over $[0, \infty)$ if and only if it is nonnegative at 0 and at all the break-points, and if the slope of the function is nonnegative to the right of the largest breakpoint. From our notation, $\Psi^x(S_1)$ is nonnegative for all non-negative values of S_1 if and only if:

1. $\Psi^x(0) \geq 0$;
2. $\Psi^x(K_j) \geq 0, \ \forall j$; and
3. $[(\Psi^x)'_+(K_n)] \geq 0$.

Now consider the following linear programming problem:

$$
\begin{aligned}
\min_x \quad & \sum_{i=1}^n S_0^i x_i \\
& \sum_{i=1}^n \Psi_i(0)x_i \geq 0 \\
& \sum_{i=1}^n \Psi_i(K_j)x_i \geq 0, \quad j = 1, \ldots, n, \\
& \sum_{i=1}^n (\Psi_i(K_n + 1) - \Psi_i(K_n))x_i \geq 0.
\end{aligned}
\tag{4.8}
$$

Since all $\Psi_i(S_1)$'s are piecewise linear, the quantity $\Psi_i(K_n + 1) - \Psi_i(K_n)$ gives the right-derivative of $\Psi_i(S_1)$ at K_n. Thus, the expression in the last constraint is the right derivative of $\Psi^x(S_1)$ at K_n. The following observation follows from our arguments above:

Proposition 4.1 *There is no type-A arbitrage in prices S_0^i if and only if the optimal objective value of (4.8) is zero.*

Similar to the previous section, we have the following result:

Proposition 4.2 *Suppose that there are no type-A arbitrage opportunities in prices S_0^i. Then, there are no type-B arbitrage opportunities if and only if the dual of the problem (4.8) has a strictly feasible solution.*

Exercise 4.5 Prove Proposition 4.2 .

Next, we focus on the case where the derivative securities under consideration are European call options with strikes at K_i for $i = 1, \ldots, n$, so that $\Psi_i(S_1) = (S_1 - K_i)^+$. Thus

$$
\Psi_i(K_j) = (K_j - K_i)^+.
$$

In this case, (4.8) reduces to the following problem:

$$
\begin{aligned}
\min_x \ & c^T x \\
& Ax \geq 0,
\end{aligned}
\tag{4.9}
$$

where $c^T = [S_0^1, \ldots, S_0^n]$ and

$$
A = \begin{bmatrix}
K_2 - K_1 & 0 & 0 & \cdots & 0 \\
K_3 - K_1 & K_3 - K_2 & 0 & \cdots & 0 \\
\vdots & \vdots & \vdots & \ddots & \vdots \\
K_n - K_1 & K_n - K_2 & K_n - K_3 & \cdots & 0 \\
1 & 1 & 1 & \cdots & 1
\end{bmatrix}.
\tag{4.10}
$$

This formulation is obtained by removing the first two constraints of (4.8) which are redundant in this particular case.

Using this formulation and our earlier results, one can prove a theorem giving necessary and sufficient conditions for a set of call option prices to contain arbitrage opportunities:

Theorem 4.3 *Let $K_1 < K_2 < \cdots < K_n$ denote the strike prices of European call options written on the same underlying security with the same maturity. There are no arbitrage opportunities if and only if the prices S_0^i satisfy the following conditions:*

1. $S_0^i > 0$, $i = 1, \ldots, n$.
2. $S_0^i > S_0^{i+1}$, $i = 1, \ldots, n-1$.
3. *The function $C(K_i) := S_0^i$ defined on the set $\{K_1, K_2, \ldots, K_n\}$ is a strictly convex function.*

Exercise 4.6 Use Proposition 4.2 to show that there are no arbitrage opportunities for the option prices in Theorem 4.3 if and only if there exists strictly positive scalars $y_1, \ldots, y_n$ satisfying $y_n = S_0^n$, $y_{n-1} = (S_0^{n-1} - S_0^n)/(K^n - K^{n-1})$, and

$$
y_i = \frac{S_0^i - S_0^{i+1}}{K^{i+1} - K^i} - \frac{S_0^{i+1} - S_0^{i+2}}{K^{i+2} - K^{i+1}}, \quad i = 1, \ldots, n-2.
$$

Use this observation to prove Theorem 4.3 .

As an illustration of Theorem 4.3 , consider the scenario in Exercise 4.4 : XYZ stock is currently priced at $40. European call options on XYZ with strike prices of $30, $40, $50, and $60 are priced at $10, $7, $10/3, and $0. Do these prices exhibit an arbitrage opportunity? As we see in Figure 4.2, the option prices violate the third condition of the theorem and therefore must carry an arbitrage opportunity.

Exercise 4.7 Construct a portfolio of the options in the example above that provides a type-A arbitrage opportunity.

4.3 Additional exercises

Exercise 4.8 Consider the linear programming problem (4.9) that we developed to detect arbitrage opportunities in the prices of European call options with a

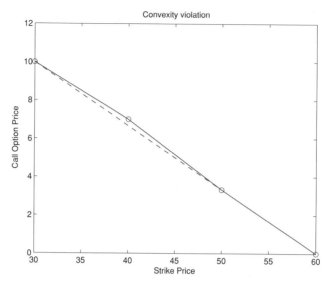

Figure 4.2 Nonconvexity in the call price function indicates arbitrage

common underlying security and common maturity (but different strike prices). This formulation implicitly assumes that the i-th call can be bought or sold at the same current price of S_0^i. In real markets, there is always a gap between the price a buyer pays for a security and the amount the seller collects called the *bid–ask spread*.

Assume that the ask price of the i-th call is given by S_a^i while its bid price is denoted by S_b^i with $S_a^i > S_b^i$. Develop an analogue of the LP (4.9) in the case where we can purchase the calls at their ask prices or sell them at their bid prices. Consider using two variables for each call option in your new LP.

Exercise 4.9   Consider all the call options on the S&P 500 index that expire on the same day, about three months from the current date. Their current prices can be downloaded from the website of the Chicago Board of Options Exchange at www.cboe.com or from several other market quote websites. Formulate the linear programming problem (4.9) (or, rather, the version you developed for Exercise 4.8, since market quotes will include bid and ask prices) to determine whether these prices contain any arbitrage opportunities. Solve this linear programming problem using LP software.

Sometimes, illiquid securities can have misleading prices since the reported price corresponds to the last transaction in that security, which may have happened several days ago, and, if there were to be a new transaction, this value could change dramatically. As a result, it is quite possible that you will discover false "arbitrage opportunities" because of these misleading prices. Repeat the LP formulation and solve it again, this time only using prices of the call options that have had a trading volume of at least 100 on the day you downloaded the prices.

Exercise 4.10

(i) You have $20 000 to invest. Stock XYZ sells at $20 per share today. A European call option to buy 100 shares of stock XYZ at $15 exactly six months from today sells for $1000. You can also raise additional funds which can be immediately invested, if desired, by selling call options with the above characteristics. In addition, a six-month riskless zero-coupon bond with $100 face value sells for $90. You have decided to limit the number of call options that you buy or sell to at most 50.

You consider three scenarios for the price of stock XYZ six months from today: the price will be the same as today, the price will go up to $40, or drop to $12. Your best estimate is that each of these scenarios is equally likely. Formulate and solve a linear program to determine the portfolio of stocks, bonds, and options that maximizes the expected profit.

Answer: First, we define the decision variables:

B = number of bonds purchased,
S = number of shares of stock XYZ purchased,
C = number of call options purchased (if > 0) or sold (if < 0).

The expected profits (per unit of investment) are computed as follows:

Bonds: 10,
Stock XYZ: $\frac{1}{3}(20 + 0 - 8) = 4$,
Call option: $\frac{1}{3}(1500 - 500 - 1000) = 0$.

Therefore, we get the following linear programming formulation:

$$\max 10B + 4S$$
$$90B + 20S + 1000C \leq 20\,000 \quad \text{(budget constraint)}$$
$$C \leq 50 \quad \text{(limit on number of call options purchased)}$$
$$C \geq -50 \quad \text{(limit on number of call options sold)}$$
$$B \geq 0, \ S \geq 0 \quad \text{(nonnegativity)}.$$

Solving (using SOLVER, say), we get the optimal solution $B = 0$, $S = 3500$, $C = -50$ with an expected profit of $14\,000$.

Note that, with this portfolio, the profit is not positive under all scenarios. In particular, if the price of stock XYZ goes to $40, a loss of $5000 will be incurred.

(ii) Suppose that the investor wants a profit of at least $2000 in any of the three scenarios. Write a linear program that will maximize the investor's expected profit under this additional constraint.

Answer: This can be done by introducing three additional variables:

P_i = profit in scenario i.

The formulation is now the following:

$$
\begin{aligned}
\max \quad & \tfrac{1}{3}P_1 + \tfrac{1}{3}P_2 + \tfrac{1}{3}P_3 \\
& 90B + 20S + 1000C \quad \le 20\,000 \\
& 10B + 20S + 1500C \quad = P_1 \\
& 10B \quad\;\; - \quad 500C \quad = P_2 \\
& 10B - \;\; 8S - 1000C \quad = P_3 \\
& \qquad\qquad\qquad P_1 \ge 2000 \\
& \qquad\qquad\qquad P_2 \ge 2000 \\
& \qquad\qquad\qquad P_3 \ge 2000 \\
& \qquad\qquad\qquad C \;\; \le 50 \\
& \qquad\qquad\qquad C \;\; \ge -50 \\
& B \ge 0, \, S \ge 0.
\end{aligned}
$$

(iii) Solve this linear program with SOLVER to find out the expected profit. How does it compare with the earlier figure of $14\,000$?

Answer: The optimum solution is to buy 2800 shares of XYZ and sell 36 call options. The resulting expected worth in six months will be $31\,200$. Therefore, the expected profit is $11\,200$ (= $31\,200 - 20\,000$).

(iv) *Riskless profit* is defined as the largest possible profit that a portfolio is guaranteed to earn, no matter which scenario occurs. What is the portfolio that maximizes riskless profit for the above three scenarios?

Answer: To solve this question, we can use a slight modification of the previous model, by introducing one more variable:

Z = riskless profit.

Here is the formulation:

$$
\begin{aligned}
\max \quad & Z \\
& 90B + 20S + 1000C \quad \le 20\,000 \\
& 10B + 20S + 1500C \quad = P_1 \\
& 10B \quad\;\; - \quad 500C \quad = P_2 \\
& 10B - \;\; 8S - 1000C \quad = P_3 \\
& \qquad\qquad\qquad P_1 \ge Z \\
& \qquad\qquad\qquad P_2 \ge Z \\
& \qquad\qquad\qquad P_3 \ge Z \\
& \qquad\qquad\qquad C \;\; \le 50 \\
& \qquad\qquad\qquad C \;\; \ge -50 \\
& B \ge 0, \, S \ge 0.
\end{aligned}
$$

The result is (obtained using SOLVER) a riskless profit of $7272. This is obtained by buying 2273 shares of XYZ and selling 25.45 call options. The resulting expected profit is $9091 in this case.

Exercise 4.11 *Arbitrage in the currency market:* Consider the global currency market. Given two currencies, say the yen and the US dollar, there is an exchange rate between them (about 118 yen to the dollar in February 2006). It is axiomatic of arbitrage-free markets that there is no method of converting, say, one dollar to yen, then to euros, then pounds, and back to dollars again so that you end up with more than a dollar. How would you recognize when there is an arbitrage opportunity?

The following are actual trades made on February 14, 2002:

from		Dollar	Euro	Pound	Yen
into	Dollar		0.8706	1.4279	0.00750
	Euro	1.1486		1.6401	0.00861
	Pound	0.7003	0.6097		0.00525
	Yen	133.38	116.12	190.45	

For example, one dollar converted into euros yielded 1.1486 euros. It is not obvious from the chart above, but in the absence of any conversion costs, the dollar–pound–yen–dollar conversion actually makes $0.0003 per dollar converted while changing the order to dollar–yen–euro–dollar loses about $0.0002 per dollar converted. How can one formulate a linear program to identify such arbitrage possibilities?

Answer:

```
VARIABLES
DE = quantity of dollars changed into euros
DP = quantity of dollars changed into pounds
DY = quantity of dollars changed into yen
ED = quantity of euros changed into dollars
EP = quantity of euros changed into pounds
EY = quantity of euros changed into yen
PD = quantity of pounds changed into dollars
PE = quantity of pounds changed into euros
PY = quantity of pounds changed into yen
YD = quantity of yen changed into dollars
YE = quantity of yen changed into euros
YP = quantity of yen changed into pounds
D = quantity of dollars generated through arbitrage
OBJECTIVE
Max D
CONSTRAINTS
```

```
Dollar: D+DE+DP+DY-0.8706*ED-1.4279*PD-0.00750*YD = 1
Euro:  ED+EP+EY-1.1486*DE-1.6401*PE-.00861*YE = 0
Pound: PD+PE+PY-0.7003*DP-0.6097*EP-0.00525*YP = 0
Yen:   YD+YE+YP-133.38*DY-116.12*EY-190.45*PY = 0
BOUNDS
D < 10000
END
```

Solving this linear program, we find that, in order to gain \$10 000 in arbitrage, we have to change about \$34 million dollars into euros, then convert these euros into yen and finally change the yen into dollars. There are other solutions as well. The arbitrage opportunity is so tiny (\$0.0003 to the dollar) that, depending on the numerical precision used, some LP solvers do not find it, thus concluding that there is no arbitrage here. An interesting example illustrating the role of numerical precision in optimization solvers!

4.4 Case study: tax clientele effects in bond portfolio management

The goal is to construct an optimal tax-specific bond portfolio, for a given tax bracket, by exploiting the price differential of an after-tax stream of cash flows. This objective is accomplished by purchasing at the ask price "underpriced" bonds (for the specific tax bracket), while simultaneously selling at the bid price "overpriced" bonds. The following model was proposed by Ronn [69]. See also Schaefer [72].

Let

$J = \{1, \ldots, j, \ldots, N\}$ = set of riskless bonds
P_j^a = ask price of bond j
P_j^b = bid price of bond j
X_j^a = amount of bond j bought
X_j^b = amount of bond j sold short.

We make the natural assumption that $P_j^a > P_j^b$. The objective function of the program is

$$Z = \max \sum_{j=1}^{N} P_j^b X_j^b - \sum_{j=1}^{N} P_j^a X_j^a \tag{4.11}$$

since the long side of an arbitrage position must be established at ask prices while the short side of the position must be established at bid prices. Now consider the future cash flows of the portfolio:

$$C_1 = \sum_{j=1}^{N} a_j^1 X_j^a - \sum_{j=1}^{N} a_j^1 X_j^b. \tag{4.12}$$

$$\text{For } t = 2, \ldots, T, \quad C_t = (1 + \rho)C_{t-1} + \sum_{j=1}^{N} a_j^t X_j^a - \sum_{j=1}^{N} a_j^t X_j^b, \tag{4.13}$$

where ρ is the exogenous riskless reinvestment rate, and a_j^t is the coupon and/or principal payment on bond j at time t.

For the portfolio to be riskless, we require

$$C_t \geq 0 \quad t = 1, \ldots, T. \tag{4.14}$$

Since the bid–ask spread has been explicitly modeled, it is clear that $X_j^a \geq 0$ and $X_j^b \geq 0$ are required. Now the resulting linear program admits two possible solutions: either all bonds are priced to within the bid–ask spread, i.e., $Z = 0$; or infinite arbitrage profits may be attained, i.e., $Z = \infty$. Clearly any attempt to exploit price differentials by taking extremely large positions in these bonds would cause price movements: the bonds being bought would appreciate in price; the bonds being sold short would decline in value. In order to provide a finite solution, the constraints $X_j^a \leq 1$ and $X_j^b \leq 1$ are imposed. Thus, with

$$0 \leq X_j^a, \ X_j^b \leq 1 \quad j = 1, \ldots, N, \tag{4.15}$$

the complete problem is now specified as (4.11)–(4.15).

Taxes

The proposed model explicitly accounts for the taxation of income and capital gains for specific investor classes. This means that the cash flows need to be adjusted for the presence of taxes.

For a discount bond (i.e. when $P_j^a < 100$), the after-tax cash flow of bond j in period t is given by

$$a_j^t = c_j^t (1 - \tau),$$

where c_j^t is the coupon payment, and τ is the ordinary income tax rate.

At maturity, the j-th bond yields

$$a_j^t = \left(100 - P_j^a\right)(1 - g) + P_j^a,$$

where g is the capital gains tax rate.

For premium bond (i.e., when $P_j^a > 100$), the premium is amortized against ordinary income over the life of the bond, giving rise to an after-tax coupon payment of

$$a_j^t = \left[c_j^t - \frac{P_j^a - 100}{n_j} \right] (1 - \tau) + \frac{P_j^a - 100}{n_j},$$

where n_j is the number of coupon payments remaining to maturity.

A premium bond also makes a nontaxable repayment of

$$a_j^t = 100$$

at maturity.

Data

The model requires that the data contain bonds with perfectly forecastable cash flows. All callable bonds are excluded from the sample. Thus, all noncallable bonds and notes are deemed appropriate for inclusion in the sample.

Major categories of taxable investors are domestic banks, insurance companies, individuals, nonfinancial corporations, foreigners. In each case, one needs to distinguish the tax rates on capital gains versus ordinary income.

The fundamental question to arise from this study is: does the data reflect tax clientele effects or arbitrage opportunities?

Consider first the class of tax-exempt investors. Using current data, form the optimal "purchased" and "sold" bond portfolios. Do you observe the same tax clientele effect as documented by Schaefer for British government securities; namely, the "purchased" portfolio contains high coupon bonds and the "sold" portfolio is dominated by low coupon bonds? This can be explained as follows: the preferential taxation of capital gains for (most) taxable investors causes them to gravitate towards low coupon bonds. Consequently, for tax-exempt investors, low coupon bonds are "overpriced" and not desirable as investment vehicles.

Repeat the same analysis with the different types of taxable investors.

1. Is there a clientele effect in the pricing of US Government investments, with tax-exempt investors, or those without preferential treatment of capital gains, gravitating towards high coupon bonds?
2. Do you observe that not all high coupon bonds are desirable to investors without preferential treatment of capital gains? Nor are all low coupon bonds attractive to those with preferential treatment of capital gains. Can you find reasons why this may be the case?

The dual price, say u_t, associated with constraint (4.13) represents the present value of an additional dollar at time t. Explain why. It follows that u_t may be used to compute the term structure of spot interest rates R_t, given by the relation

$$R_t = \left(\frac{1}{u_t}\right)^{1/t} - 1.$$

Compute this week's term structure of spot interest rates for tax-exempt investors.

5

Nonlinear programming: theory and algorithms

5.1 Introduction

So far, we have focused on optimization problems with linear constraints and a linear objective function. Linear functions are "nice" – they are smooth and predictable. Consequently, we were able to use specialized and highly efficient techniques for their solution. Many realistic formulations of optimization problems, however, do not fit into this nice structure and require more general methods. In this chapter we study general optimization problems of the form

$$\min_x \ f(x)$$
$$g_i(x) = 0, \quad i \in \mathcal{E}, \tag{5.1}$$
$$g_i(x) \geq 0, \quad i \in \mathcal{I},$$

where f and g_i are functions of $IR^n \to IR$, and $\mathcal{E}$ and $\mathcal{I}$ are index sets for the equality and inequality constraints respectively. Such optimization problems are often called *nonlinear programming problems*, or *nonlinear programs*.

There are many problems where the general framework of nonlinear programming is needed. Here are some illustrations:

1. **Economies of scale:** In many applications costs or profits do not grow linearly with the corresponding activities. In portfolio construction, an individual investor may benefit from economies of scale as fixed costs of trading become negligible for larger trades. Conversely, an institutional investor may suffer from diseconomies of scale if a large trade has an unfavorable market impact on the security traded. Realistic models of such trades must involve nonlinear objective or constraint functions.
2. **Probabilistic elements:** Nonlinearities frequently arise when some of the coefficients in the model are random variables. For example, consider a linear program where the right-hand sides are random. To illustrate, suppose the LP has two constraints:

$$\text{maximize} \quad c_1 x_1 + \cdots + c_n x_n$$
$$a_{11} x_1 + \cdots + a_{1n} x_n \leq b_1$$
$$a_{21} x_1 + \cdots + a_{2n} x_n \leq b_2,$$

where the coefficients b_1 and b_2 are independently distributed and $G_i(y)$ represents the probability that the random variable b_i is at least as large as y. Suppose you want to select the variables $x_1, \ldots, x_n$ so that the joint probability of both the constraints being satisfied is at least β:

$$P[a_{11}x_1 + \cdots + a_{1n}x_n \leq b_1] \times P[a_{21}x_1 + \cdots + a_{2n}x_n \leq b_2] \geq \beta.$$

Then this condition can be written as the following set of constraints:

$$\begin{aligned}
-y_1 \quad + a_{11}x_1 + \cdots + a_{1n}x_n &= 0 \\
-y_2 + a_{21}x_1 + \cdots + a_{2n}x_n &= 0 \\
G_1(y_1) \times G_2(y_2) &\geq \beta,
\end{aligned}$$

where this product leads to nonlinear restrictions on y_1 and y_2.

3. **Value-at-Risk:** The *Value-at-Risk* (VaR) is a risk measure that focuses on rare events. For example, for a random variable X that represents the daily loss from an investment portfolio, VaR would be the largest loss that occurs with a specified frequency, such as once per year. Given a probability level α, say $\alpha = 0.99$, the Value-at-Risk $\text{VaR}_\alpha(X)$ of a random variable X with a continuous distribution function is the value γ such that $P(X \leq \gamma) = \alpha$. As such, VaR focuses on the tail of the distribution of the random variable X. Depending on the distributional assumptions for portfolio returns, the problem of finding a portfolio that minimizes VaR can be a highly nonlinear optimization problem.

4. **Mean-variance optimization:** Markowitz's MVO model introduced in Section 1.3.1 is a quadratic program: the objective function is quadratic and the constraints are linear. In Chapter 7 we will present an interior-point algorithm for this class of nonlinear optimization problems.

5. **Constructing an index fund:** In integer programming applications, such as the model discussed in Section 12.3 for constructing an index fund, the "relaxation" can be written as a multivariate function that is convex but nondifferentiable. Subgradient techniques can be used to solve this class of nonlinear optimization problems.

In contrast to linear programming, where the simplex method can handle most instances and reliable implementations are widely available, there is not a single preferred algorithm for solving general nonlinear programs. Without difficulty, one can find ten or fifteen methods in the literature and the underlying theory of nonlinear programming is still evolving. A systematic comparison between different methods and packages is complicated by the fact that a nonlinear method can be very effective for one type of problem and yet fail miserably for another. In this chapter, we sample a few ideas:

1. the method of steepest descent for unconstrained optimization;
2. Newton's method;
3. the generalized reduced-gradient algorithm;
4. sequential quadratic programming;
5. subgradient optimization for nondifferentiable functions.

We address the solution of two special classes of nonlinear optimization problems, namely quadratic and conic optimization problems in Chapters 7 and 9. For these problem classes, interior-point methods (IPMs) are very effective. While IPMs are heavily used for general nonlinear programs also, we delay their discussion until Chapter 7.

5.2 Software

Some software packages for solving nonlinear programs are:

1. CONOPT, GRG2, Excel's SOLVER (all three are based on the generalized reduced-gradient algorithm);
2. MATLAB optimization toolbox, SNOPT, NLPQL (sequential quadratic programming);
3. MINOS, LANCELOT (Lagrangian approach);
4. LOQO, MOSEK, IPOPT (Interior-point algorithms for the KKT conditions, see Section 5.5).

The *Network Enabled Optimization Server (NEOS)* website we already mentioned in Chapter 2, available at http:neos.mcs.anl.gov/neos/solvers, provides access to many academic and commercial nonlinear optimization solvers. In addition, the *Optimization Software Guide* based on the book by Moré and Wright [58], which is available from www-fp.mcs.anl.gov/otc/Guide/SoftwareGuide/, lists information on more than 30 nonlinear programming packages.

Of course, as is the case for linear programming, one needs a modeling language to work efficiently with large nonlinear models. Two of the most popular are GAMS and AMPL. Most of the optimizers described above accept models written in either of these mathematical programming languages.

5.3 Univariate optimization

Before discussing optimization methods for multivariate and/or constrained problems, we start with a description of methods for solving univariate equations and optimizing univariate functions. These methods, often called *line search* methods are important components to many nonlinear programming algorithms.

5.3.1 Binary search

Binary search is a very simple idea for numerically solving the nonlinear equation $f(x) = 0$, where f is a function of a single variable.

For example, suppose we want to find the maximum of $g(x) = 2x^3 - e^x$. For this purpose we need to identify the *critical points* of the function, namely, those points that satisfy the equation $g'(x) = 6x^2 - e^x = 0$. But there is no closed form solution to this equation. So we solve the equation numerically, through binary search. Letting $f(x) := g'(x) = 6x^2 - e^x$, we first look for two points, say a, b, such that the signs of $f(a)$ and $f(b)$ are opposite. Here $a = 0$ and $b = 1$ would do since $f(0) = -1$ and $f(1) \approx 3.3$. Since f is continuous, we know that there exists an x with $0 < x < 1$ such that $f(x) = 0$. We say that our confidence interval is $[0,1]$. Now let us try the middle point $x = 0.5$. Since $f(0.5) \approx -0.15 < 0$ we know that there is a solution between 0.5 and 1 and we get the new confidence interval $[0.5, 1.0]$. We continue with $x = 0.75$ and since $f(0.75) > 0$ we get the confidence interval $[0.5, 0.75]$. Repeating this, we converge very quickly to a value of x where $f(x) = 0$. Here, after ten iterations, we are within 0.001 of the real value.

In general, if we have a confidence interval of $[a, b]$, we evaluate $f(\frac{a+b}{2})$ to cut the confidence interval in half.

Binary search is fast. It reduces the confidence interval by a factor of 2 for every iteration, so after k iterations the original interval is reduced to $(b - a) \times 2^{-k}$. A drawback is that binary search only finds one solution. So, if g had local extrema in the above example, binary search could converge to any of them. In fact, most algorithms for nonlinear programming are subject to failure for this reason.

Example 5.1   *Binary search can be used to compute the* internal rate of return *(IRR) r of an investment. Mathematically, r is the interest rate that satisfies the equation*

$$\frac{F_1}{1+r} + \frac{F_2}{(1+r)^2} + \frac{F_3}{(1+r)^3} + \cdots + \frac{F_N}{(1+r)^N} - C = 0,$$

where

$$F_t = cash\ flow\ in\ year\ t,$$
$$N = number\ of\ years,$$
$$C = cost\ of\ the\ investment.$$

For most investments, the above equation has a unique solution and therefore the IRR is uniquely defined, but one should keep in mind that this is not always the case. The IRR of a bond is called its yield. *As an example, consider a 4-year non-callable bond with a 10% coupon rate paid annually and a par value of $1000. Such a bond has the following cash flows:*

In Yr. t	F_t
1	$100
2	100
3	100
4	1100

Suppose this bond is now selling for $900. Compute the yield of this bond.

Table 5.1 *Binary search to find the IRR of a non-callable bond*

Iter.	a	c	b	$f(a)$	$f(c)$	$f(b)$
1	0	0.5	1	500	−541.975	−743.75
2	0	0.25	0.5	500	−254.24	−541.975
3	0	0.125	0.25	500	24.85902	−254.24
4	0.125	0.1875	0.25	24.85902	−131.989	−254.24
5	0.125	0.15625	0.1875	24.85902	−58.5833	−131.989
6	0.125	0.140625	0.15625	24.85902	−18.2181	−58.5833
7	0.125	0.132813	0.140625	24.85902	2.967767	−18.2181
8	0.132813	0.136719	0.140625	2.967767	−7.71156	−18.2181
9	0.132813	0.134766	0.136719	2.967767	−2.39372	−7.71156
10	0.132813	0.133789	0.134766	2.967767	0.281543	−2.39372
11	0.133789	0.134277	0.134766	0.281543	−1.05745	−2.39372
12	0.133789	0.134033	0.134277	0.281543	−0.3883	−1.05745

The yield r of the bond is given by the equation

$$\frac{100}{1+r} + \frac{100}{(1+r)^2} + \frac{100}{(1+r)^3} + \frac{1100}{(1+r)^4} - 900 = 0.$$

Let us denote by $f(r)$ the left-hand side of this equation. We find r such that $f(r) = 0$ using binary search.

We start by finding values (a, b) such that $f(a) > 0$ and $f(b) < 0$. In this case, we expect r to be between 0 and 1. Since $f(0) = 500$ and $f(1) = -743.75$, we have our starting values.

Next, we let $c = 0.5$ (the midpoint) and calculate $f(c)$. Since $f(0.5) = -541.975$, we replace our range with $a = 0$ and $b = 0.5$ and repeat. When we continue, we get the table of values shown in Table 5.1.

According to this computation the yield of the bond is approximately $r = 13.4\%$. Of course, this routine sort of calculation can be easily implemented on a computer.

Exercise 5.1   Find a root of the polynomial $f(x) = 5x^4 - 20x + 2$ in the interval $[0,1]$ using binary search.

Exercise 5.2   Compute the yield on a six-year non-callable bond that makes 5% coupon payments in years 1, 3, and 5, coupon payments of 10% in years 2 and 4, and pays the par value in year 6.

Exercise 5.3   The well-known Black–Scholes–Merton option pricing formula has the following form for European call option prices:

$$C(K, T) = S_0 \Phi(d_1) - K e^{-rT} \Phi(d_2),$$

where

$$d_1 = \frac{\log(S_0/K) + (r + \sigma^2/2)T}{\sigma\sqrt{T}},$$

$$d_2 = d_1 - \sigma\sqrt{T},$$

and $\Phi(\cdot)$ is the cumulative distribution function for the standard normal distribution. r in the formula represents the continuously compounded risk-free and constant interest rate and σ is the volatility of the underlying security that is assumed to be constant. S_0 denotes the initial price of the security, K and T are the strike price and maturity of the option. Given the market price of a particular option and an estimate for the interest rate r, the unique value of the volatility parameter σ that satisfies the pricing equation above is called the *implied volatility* of the underlying security. Calculate the implied volatility of a stock currently valued at \$20 if a European call option on this stock with a strike price of \$18 and a maturity of three months is worth \$2.20. Assume a zero interest rate and use binary search.

Golden section search

Golden section search is similar in spirit to binary search. It can be used to solve a univariate equation as above, or to compute the maximum of a function $f(x)$ defined on an interval $[a, b]$. The discussion here is for the optimization version. The main difference between the golden section search and the binary search is in the way the new confidence interval is generated from the old one.

We assume that:

1. f is continuous;
2. f has a unique local maximum in the interval $[a, b]$.

The golden search method consists in computing $f(c)$ and f(d) for $a < d < c < b$.

- If $f(c) > f(d)$, the procedure is repeated with the interval (a, b) replaced by (d, b).
- If $f(c) < f(d)$, the procedure is repeated with the interval (a, b) replaced by (a, c).

Remark 5.1 *The name "golden section" comes from a certain choice of c and d that yields fast convergence, namely $c = a + r(b - a)$ and $d = b + r(a - b)$, where $r = (\sqrt{5} - 1)/2 = 0.618034\ldots$. This is the golden ratio, already known to the ancient Greeks.*

Example 5.2 *Find the maximum of the function $x^5 - 10x^2 + 2x$ in the interval $[0, 1]$.*

In this case, we begin with $a = 0$ and $b = 1$. Using golden section search, that gives $d = 0.382$ and $c = 0.618$. The function values are $f(a) = 0$, $f(d) = -0.687$, $f(c) = -2.493$, and $f(b) = -7$. Since $f(c) < f(d)$, our new range is $a = 0$, $b =$

Table 5.2 *Golden section search in Example 5.2*

Iter.	a	d	c	b	$f(a)$	$f(d)$	$f(c)$	$f(b)$
1	0	0.382	0.618	1	0	−0.6869	−2.4934	−7
2	0	0.2361	0.382	0.618	0	−0.0844	−0.6869	−2.4934
3	0	0.1459	0.2361	0.382	0	0.079	−0.0844	−0.6869
4	0	0.0902	0.1459	0.2361	0	0.099	0.079	−0.0844
5	0	0.0557	0.0902	0.1459	0	0.0804	0.099	0.079
6	0.0557	0.0902	0.1115	0.1459	0.0804	0.099	0.0987	0.079
7	0.0557	0.077	0.0902	0.1115	0.0804	0.0947	0.099	0.0987
8	0.077	0.0902	0.0983	0.1115	0.0947	0.099	0.1	0.0987
9	0.0902	0.0983	0.1033	0.1115	0.099	0.1	0.0999	0.0987
10	0.0902	0.0952	0.0983	0.1033	0.099	0.0998	0.1	0.0999
11	0.0952	0.0983	0.1002	0.1033	0.0998	0.1	0.1	0.0999
12	0.0983	0.1002	0.1014	0.1033	0.1	0.1	0.1	0.0999
13	0.0983	0.0995	0.1002	0.1014	0.1	0.1	0.1	0.1
14	0.0995	0.1002	0.1007	0.1014	0.1	0.1	0.1	0.1
15	0.0995	0.0999	0.1002	0.1007	0.1	0.1	0.1	0.1
16	0.0995	0.0998	0.0999	0.1002	0.1	0.1	0.1	0.1
17	0.0998	0.0999	0.1	0.1002	0.1	0.1	0.1	0.1
18	0.0999	0.1	0.1001	0.1002	0.1	0.1	0.1	0.1
19	0.0999	0.1	0.1	0.1001	0.1	0.1	0.1	0.1
20	0.0999	0.1	0.1	0.1	0.1	0.1	0.1	0.1
21	0.1	0.1	0.1	0.1	0.1	0.1	0.1	0.1

0.618. Recalculating from the new range gives $d = 0.236, c = 0.382$ (note that our current c was our previous d: it is this reuse of calculated values that gives golden section search its speed). We repeat this process to get Table 5.2.

Exercise 5.4   One of the most fundamental techniques of statistical analysis is the method of maximum likelihood estimation. Given a sample set of independently drawn observations from a parametric distribution, the estimation problem is to determine the values of the distribution parameters that maximize the probability that the observed sample set comes from this distribution. See Nocedal and Wright [61], page 255, for example.

Consider, for example, the observations $x_1 = -0.24$, $x_2 = 0.31$, $x_3 = 2.3$, and $x_4 = -1.1$ sampled from a normal distribution. If the mean of the distribution is known to be 0, what is the maximum likelihood estimate of the standard deviation, σ? Construct the log-likelihood function and maximize it using golden section search.

5.3.2 Newton's method

The main workhorse of many optimization algorithms is a centuries-old technique for the solution of nonlinear equations developed by Sir Isaac Newton. We will

discuss the multivariate version of Newton's method later. We focus on the univariate case first. For a given nonlinear function f we want to find an x such that

$$f(x) = 0.$$

Assume that f is continuously differentiable and that we currently have an estimate
x^k of the solution (we will use superscripts for iteration indices in the following
discussion). The first-order (i.e., linear) Taylor series approximation to the function
f around x^k can be written as follows:

$$f(x^k + \delta) \approx \hat{f}(\delta) := f(x^k) + \delta f'(x^k).$$

This is equivalent to saying that we can approximate the function f by the line $\hat{f}(\delta)$
that is tangent to it at x^k. If the first-order approximation $\hat{f}(\delta)$ were perfectly good,
and if $f'(x^k) \neq 0$, the value of δ that satisfies

$$\hat{f}(\delta) = f(x^k) + \delta f'(x^k) = 0$$

would give us the update on the current iterate x^k necessary to get to the solution.
This value of δ is computed easily:

$$\delta = -\frac{f(x^k)}{f'(x^k)}.$$

The expression above is called the Newton update and Newton's method determines
its next estimate of the solution as

$$x^{k+1} = x^k + \delta = x^k - \frac{f(x^k)}{f'(x^k)}.$$

Since $\hat{f}(\delta)$ is only an approximation to $f(x^k + \delta)$, we do not have a guarantee that
$f(x^{k+1})$ is zero, or even small. However, as we discuss below, when x^k is close
enough to a solution of the equation $f(x) = 0$, x^{k+1} is even closer. We can then
repeat this procedure until we find an x^k such that $f(x^k) = 0$, or in most cases, until
$f(x^k)$ becomes reasonably small, say, less than some pre-specified $\varepsilon > 0$.

There is an intuitive geometric explanation of the procedure we just described:
we first find the line that is tangent to the function at the current iterate, then we
calculate the point where this line intersects the x-axis and set the next iterate to
this value and repeat the process. See Figure 5.1 for an illustration.

Example 5.3 *Let us recall Example 5.1 where we computed the IRR of an investment. Here we solve the problem using Newton's method. Recall that the yield r
must satisfy the equation*

$$f(r) = \frac{100}{1+r} + \frac{100}{(1+r)^2} + \frac{100}{(1+r)^3} + \frac{1100}{(1+r)^4} - 900 = 0.$$

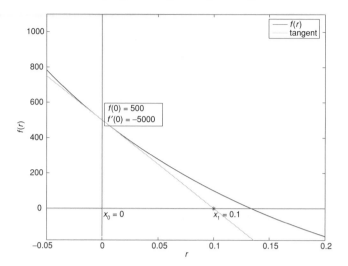

Figure 5.1 First step of Newton's method in Example 5.3

The derivative of $f(r)$ is easily computed:

$$f'(r) = -\frac{100}{(1+r)^2} - \frac{200}{(1+r)^3} - \frac{300}{(1+r)^4} - \frac{4400}{(1+r)^5}.$$

*We need to start Newton's method with an initial guess, let us choose $x^0 = 0$.
Then*

$$x^1 = x^0 - \frac{f(0)}{f'(0)}$$

$$= 0 - \frac{500}{-5000} = 0.1.$$

*We mentioned above that the next iterate of Newton's method is found by calculating
the point where the line tangent to f at the current iterate intersects the axis. This
observation is illustrated in Figure 5.1.*

*Since $f(x^1) = f(0.1) = 100$ is far from zero we continue by substituting x^1
into the Newton update formula to obtain $x^2 = 0.131547080371$ and so on. The
complete iteration sequence is given in Table 5.3.*

A few comments on the speed and reliability of Newton's method are in order.
Under favorable conditions, Newton's method converges very fast to a solution
of a nonlinear equation. Indeed, if x^k is sufficiently close to a solution x^* and if
$f'(x^*) \neq 0$, then the following relation holds:

$$x^{k+1} - x^* \approx C(x^k - x^*)^2 \text{ with } C = \frac{f''(x^*)}{2f'(x^*)}. \tag{5.2}$$

Table 5.3 *Newton's method for Example 5.3*

k	x^k	$f(x^k)$
0	0.000000000000	500.000000000000
1	0.100000000000	100.000000000000
2	0.131547080371	6.464948211497
3	0.133880156946	0.031529863053
4	0.133891647326	0.000000758643
5	0.133891647602	0.000000000000

Equation (5.2) indicates that the error in our approximation $(x^k - x^*)$ is approximately squared in each iteration. This behavior is called the *quadratic convergence* of Newton's method. Note that the number of correct digits is doubled in each iteration of the example above and the method required much fewer iterations than the binary search approach.

However, when the "favorable conditions" we mentioned above are not satisfied, Newton's method may fail to converge to a solution. For example, consider $f(x) = x^3 - 2x + 2$. Starting at 0, one would obtain iterates cycling between 1 and 0. Starting at a point close to 1 or 0, one similarly gets iterates alternating in close neighborhoods of 1 and 0, without ever reaching the root around -1.76. Therefore, it often has to be modified before being applied to general problems. Common modifications of Newton's method include the *line-search* and *trust-region* approaches. We briefly discuss line search approaches in Section 5.3.3. More information on these methods can be found in standard numerical optimization texts such as [61].

Next, we derive a variant of Newton's method that can be applied to univariate optimization problems. If the function to be minimized/maximized has a unique minimizer/maximizer and is twice differentiable, we can do the following. Differentiability and the uniqueness of the optimizer indicate that x^* maximizes (or minimizes) $g(x)$ if and only if $g'(x^*) = 0$. Defining $f(x) = g'(x)$ and applying Newton's method to this function we obtain iterates of the following form:

$$x^{k+1} = x^k - \frac{f(x^k)}{f'(x^k)} = x^k - \frac{g'(x^k)}{g''(x^k)}.$$

Example 5.4 *Let us apply the optimization version of Newton's method to Example 5.2 . Recalling that $f(x) = x^5 - 10x^2 + 2x$, we have $f'(x) = 5x^4 - 20x + 2$ and $f''(x) = 20(x^3 - 1)$. Thus, the Newton update formula is given as*

$$x^{k+1} = x^k - \frac{5(x^k)^4 - 20x^k + 2}{20[(x^k)^3 - 1]}.$$

Starting from 0 and iterating we obtain the sequence given in Table 5.4.

Table 5.4 *Iterates of Newton's method in Example 5.4*

k	x^k	$f(x^k)$	$f'(x^k)$
0	0.000000000000	0.000000000000	2.000000000000
1	0.100000000000	0.100010000000	0.000500000000
2	0.100025025025	0.100010006256	0.000000000188
3	0.100025025034	0.100010006256	0.000000000000

Once again, observe that Newton's method converged very rapidly to the solution and generated several more digits of accuracy than the golden section search. Note, however, that the method would have failed if we had chosen $x^0 = 1$ as our starting point.

Exercise 5.5   Repeat Exercises 5.2, 5.3, and 5.4 using Newton's method.

Exercise 5.6   We derived Newton's method by approximating a given function f using the first two terms of its Taylor series at the current point x_k. When we use Taylor series approximation to a function, there is no a priori reason that tells us to stop at two terms. We can consider, for example, using the first three terms of the Taylor series expansion of the function to get a quadratic approximation. Derive a variant of Newton's method that uses this approximation to determine the roots of the function f. Can you determine the rate of convergence for this new method, assuming that the method converges?

5.3.3 Approximate line search

When we are optimizing a univariate function, sometimes it is not necessary to find the minimizer/maximizer of the function very accurately. This is especially true when the univariate optimization is only one of the steps in an iterative procedure for optimizing a more complicated function. This happens, for example, when the function under consideration corresponds to the values of a multivariate function along a fixed direction. In such cases, one is often satisfied with a new point that provides a sufficient amount of improvement over the previous point. Typically, a point with sufficient improvement can be determined much quicker than the exact minimizer of the function that results in a shorter computation time for the overall algorithm.

The notion of "sufficient improvement" must be formalized to ensure that such an approach will generate convergent iterates. Say we wish to minimize the nonlinear, differentiable function $f(x)$ and we have a current estimate x^k of its minimizer. Assume that $f'(x^k) < 0$, which indicates that the function will decrease by increasing

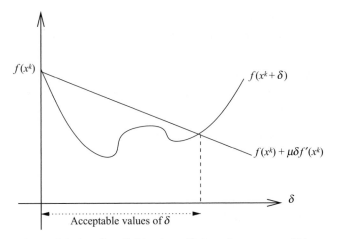

Figure 5.2 Armijo–Goldstein sufficient decrease condition

x^k. Recall the linear Taylor series approximation to the function:

$$f(x^k + \delta) \approx \hat{f}(\delta) := f(x^k) + \delta f'(x^k).$$

The derivative of the function $f'(x^k)$ gives a prediction of the decrease in the function value as we move forward from x^k. If f has a minimizer, we can not expect that it will decrease forever as we increase x^k like its linear approximation above. We can require, however, that we find a new point such that the improvement in the function value is at least a fraction of the improvement predicted by the linear approximation. Mathematically, we can require that

$$f(x^k + \delta) \leq f(x^k) + \mu \delta f'(x^k), \tag{5.3}$$

where $\mu \in (0, 1)$ is the desired fraction. This sufficient decrease requirement is often called the *Armijo–Goldstein* condition. See Figure 5.2 for an illustration.

Among all step sizes satisfying the sufficient decrease condition, one would typically prefer as large a step size as possible. However, trying to find the maximum such step size accurately will often be too time consuming and will beat the purpose of this approximation approach. A typical strategy used in line search is *backtracking*. We start with a reasonably large initial estimate. We check whether this step size satisfies condition (5.3). If it does, we accept this step size, modify our estimate and continue. If not, we backtrack by using a step size that is a fraction of the previous step size we tried. We continue to backtrack until we obtain a step size satisfying the sufficient decrease condition. For example, if the initial step size is 5 and we use the fraction 0.8, first backtracking iteration will use a step size of 4, then 3.2, and so on.

Exercise 5.7 Consider the function $f(x) = (1/4)x^4 - x^2 + 2x - 1$. We want to minimize this function using Newton's method. Verify that starting at a point close to 0 or 1 and using Newton's method, one would obtain iterates alternating between close neighborhoods of 0 and 1 and never converge. Apply Newton's method to this problem with the Armijo–Goldstein condition and backtracking starting from the point 0. Use $\mu = 0.5$ and a backtracking ratio of 0.9. Experiment with other values of $\mu \in (0, 1)$ and the backtracking ratio.

Exercise 5.8 Re-solve Exercise 5.4 using the optimization version of Newton's method with line search and backtracking. Use $\mu = 0.1$ and a backtracking ratio of 0.8.

Exercise 5.9 As Figure 5.2 illustrates, the Armijo–Goldstein condition disallows step sizes that are too large and beyond which the predictive power of the gradient of the function is weak. Backtracking strategy balances this by trying to choose as large an acceptable value of the step size as possible, ensuring that the step size is not too small. Another condition, called the Wolfe condition, rules out step sizes that are too small by requiring that

$$\| f'(x^k + \delta) \| \le \eta \| f'(x^k) \|$$

for some $\eta \in [0, 1]$. The motivation for this condition is that, for a differentiable function f, minimizers (or maximizers) will occur at points where the derivative of the function is zero. The Wolfe condition seeks points whose derivatives are closer to zero than the current point. Interpret the Wolfe condition geometrically on Figure 5.2. For function $f(x) = (1/4)x^4 - x^2 + 2x - 1$ with current iterate $x^k = 0.1$ determine the Newton update and calculate which values of the step size satisfy the Wolfe condition for $\eta = 1/4$ and also for $\eta = 3/4$.

5.4 Unconstrained optimization

We now move on to nonlinear optimization problems with multiple variables. First, we will focus on problems that have no constraints. Typical examples of unconstrained nonlinear optimization problems arise in model fitting and regression. The study of unconstrained problems is also important for constrained optimization as one often solves a sequence of unconstrained problems as subproblems in various algorithms for the solution of constrained problems.

We use the following generic format for unconstrained nonlinear programs we consider in this section:

$$\min f(x), \text{ where } x = (x_1, \dots, x_n).$$

For simplicity, we will restrict our discussion to minimization problems. These ideas can be trivially adapted for maximization problems.

5.4.1 Steepest descent

The simplest numerical method for finding a minimizing solution is based on the idea of going downhill on the graph of the function f. When the function f is differentiable, its gradient always points in the direction of fastest initial increase and the negative gradient is the direction of fastest decrease. This suggests that, if our current estimate of the minimizing point is x^*, moving in the direction of $-\nabla f(x^*)$ is desirable. Once we choose a direction, deciding how far we should move along this direction is determined using line search. The line search problem is a univariate problem that can be solved, perhaps in an approximate fashion, using the methods of the previous section. This will provide a new estimate of the minimizing point and the procedure can be repeated.

We illustrate this approach with the following example:

$$\min \ f(x) = (x_1 - 2)^4 + \exp(x_1 - 2) + (x_1 - 2x_2)^2.$$

The first step is to compute the gradient of the function, namely the vector of the partial derivatives of the function with respect to each variable:

$$\nabla f(x) = \begin{bmatrix} 4(x_1 - 2)^3 + \exp(x_1 - 2) + 2(x_1 - 2x_2) \\ -4(x_1 - 2x_2) \end{bmatrix}. \tag{5.4}$$

Next, we need to choose a starting point. We arbitrarily select the point $x^0 = (0, 3)$. Now we are ready to compute the steepest descent direction at point x^0. It is the direction opposite to the gradient vector computed at x^0, namely

$$d^0 = -\nabla f(x^0) = \begin{bmatrix} 44 + e^{-2} \\ -24 \end{bmatrix}.$$

If we move from x^0 in the direction d^0, using a step size α, we get a new point $x^0 + \alpha d^0$ ($\alpha = 0$ corresponds to staying at x^0). Since our goal is to minimize f, we will try to move to a point $x^1 = x^0 + \alpha d^0$, where α is chosen to approximately minimize the function along this direction. For this purpose, we evaluate the value of the function f along the steepest descent direction as a function of the step size α:

$$\phi(\alpha) := f(x^0 + \alpha d^0) = \{[0 + (44 + e^{-2})\alpha] - 2\}^4 + \exp\{[0 + (44 + e^{-2})\alpha] - 2\}$$
$$+ \{[0 + (44 + e^{-2})\alpha] - 2[3 - 24\alpha]\}^2.$$

Now, the optimal value of α can be found by solving the one-dimensional minimization problem $\min \phi(\alpha)$.

This minimization can be performed through one of the numerical line search procedures of the previous section. Here we use the approximate line search approach with sufficient decrease condition we discussed in Section 5.3.3. We want to choose a step size alpha satisfying

$$\phi(\alpha) \le \phi(0) + \mu\alpha\phi'(0),$$

where $\mu \in (0, 1)$ is the desired fraction for the sufficient decrease condition. We observe that the derivative of the function ϕ at 0 can be expressed as

$$\phi'(0) = \nabla f(x^0)^T d^0.$$

This is the *directional derivative* of the function f at point x^0 and direction d^0. Using this identity the sufficient decrease condition on function ϕ can be written in terms of the original function f as follows:

$$f(x^0 + \alpha d^0) \le f(x^0) + \mu\alpha\nabla f(x^0)^T d^0. \tag{5.5}$$

The condition (5.5) is the multivariate version of the Armijo–Goldstein condition (5.3).

As discussed in Section 5.3.3, the sufficient decrease condition (5.5) can be combined with a backtracking strategy. For this example, we use $\mu = 0.3$ for the sufficient decrease condition and apply backtracking with an initial trial step size of 1 and a backtracking factor of $\beta = 0.8$. Namely, we try step sizes 1, 0.8, 0.64, 0.512, and so on, until we find a step size of the form 0.8^k that satisfied the Armijo–Goldstein condition. The first five iterates of this approach as well as the 20th iterate are given in Table 5.5. For completeness, one also has to specify a termination criterion for the approach. Since the gradient of the function must be the zero vector at an unconstrained minimizer, most implementations will use a termination criterion of the form $\|\nabla f(x)\| \le \varepsilon$, where $\varepsilon > 0$ is an appropriately chosen tolerance parameter. Alternatively, one might stop when successive iterations are getting very close to each other, that is when $\|x^{k+1} - x^k\| \le \varepsilon$ for some $\varepsilon > 0$. This last condition indicates that progress has stalled. While this may be due to the fact that iterates approached the optimizer and can not progress any more, there are instances where the stalling is due to the high degree of nonlinearity in f.

A quick examination of Table 5.5 reveals that the signs of the second coordinate of the steepest descent directions change from one iteration to the next in most cases. What we are observing is the *zigzagging* phenomenon, a typical feature of steepest descent approaches that explain their slow convergence behavior for most problems. When we pursue the steepest descent algorithm for more iterations, the zigzagging phenomenon becomes even more pronounced and the method is slow to converge to the optimal solution $x^* \approx (1.472, 0.736)$. Figure 5.3 shows the steepest descent iterates for our example superimposed on the contour lines of the objective function. Steepest descent directions are perpendicular to the contour lines and zigzag between the two sides of the contour lines, especially when these lines create long and narrow corridors. It takes more than 30 steepest descent iterations in this

Table 5.5 *Steepest descent iterations*

k	(x_1^k, x_2^k)	(d_1^k, d_2^k)	α^k	$\|\nabla f(x^{k+1})\|$
0	(0.000, 3.000)	(43.864, −24.000)	0.055	3.800
1	(2.412, 1.681)	(0.112, −3.799)	0.168	2.891
2	(2.430, 1.043)	(−2.544, 1.375)	0.134	1.511
3	(2.089, 1.228)	(−0.362, −1.467)	0.210	1.523
4	(2.013, 0.920)	(−1.358, 0.690)	0.168	1.163
5	(1.785, 1.036)	(−0.193, −1.148)	0.210	1.188
⋮	⋮	⋮	⋮	⋮
20	(1.472, 0.736)	(−0.001, 0.000)	0.134	0.001

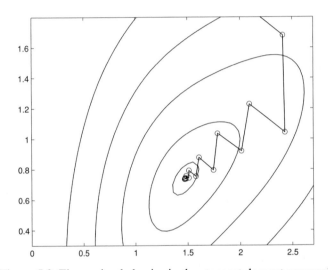

Figure 5.3 Zigzagging behavior in the steepest descent approach

small example to achieve $\|\nabla f(x)\| \leq 10^{-5}$. In summary, while the steepest descent approach is easy to implement and intuitive, and has relatively cheap iterations, it can also be quite slow to converge to solutions.

Exercise 5.10   Consider a differentiable multivariate function $f(x)$ that we wish to minimize. Let x_k be a given estimate of the solution, and consider the first-order Taylor series expansion of the function around x_k:

$$\hat{f}(\delta) = f(x_k) + \nabla f(x)^\top \delta.$$

The quickest decrease in $\hat{f}$ starting from x_k is obtained in the direction that solves

$$\min \hat{f}(\delta)$$
$$\|\delta\| \leq 1.$$

Show that the solution is $\delta^* = \alpha \nabla f(x)$ with some $\alpha < 0$, i.e., the opposite direction to the gradient is the direction of steepest descent.

Exercise 5.11   Recall the maximum likelihood estimation problem we considered in Exercise 5.4 . While we maintain the assumption that the observed samples come from a normal distribution, we will no longer assume that we know the mean of the distribution to be zero. In this case, we have a two-parameter (mean μ and standard deviation σ) maximum likelihood estimation problem. Solve this problem using the steepest descent method.

5.4.2 Newton's method

There are several numerical techniques for modifying the method of steepest descent that reduce the propensity of this approach to zigzag, and thereby speed up convergence. The steepest descent method uses the gradient of the objective function, only a first-order information on the function. Improvements can be expected by employing second-order information on the function, that is by considering its curvature. Methods using curvature information include Newton's method that we have already discussed in the univariate setting. Here, we describe the generalization of this method to multivariate problems.

Once again, we begin with the version of the method for solving equations. We will look at the case where there are several equations involving several variables:

$$
\begin{aligned}
f_1(x_1, x_2, \ldots, x_n) &= 0 \\
f_2(x_1, x_2, \ldots, x_n) &= 0 \\
&\vdots \quad \vdots \\
f_n(x_1, x_2, \ldots, x_n) &= 0.
\end{aligned}
\tag{5.6}
$$

Let us represent this system as

$$
F(x) = 0,
$$

where x is a vector of n variables and $F(x)$ is an $I\!R^n$-valued function with components $f_1(x), \ldots, f_n(x)$. We repeat the procedure in Section 5.3.2: first, we write the first-order Taylor's series approximation to the function F around the current estimate x^k:

$$
F(x^k + \delta) \approx \hat{F}(\delta) := F(x^k) + \nabla F(x^k)\delta.
\tag{5.7}
$$

Above, $\nabla F(x)$ denotes the *Jacobian matrix* of the function F, i.e., $\nabla F(x)$ has rows $(\nabla f_1(x))^\top, \ldots, (\nabla f_n(x))^\top$, the transposed gradients of the functions f_1 through f_n. We denote the components of the n-dimensional vector x using subscripts, i.e.,

$x = (x_1, \ldots, x_n)$. Let us make these statements more precise:

$$\nabla F(x_1, \ldots, x_n) = \begin{bmatrix} \frac{\partial f_1}{\partial x_1} & \cdots & \frac{\partial f_1}{\partial x_n} \\ \vdots & \ddots & \vdots \\ \frac{\partial f_n}{\partial x_1} & \cdots & \frac{\partial f_n}{\partial x_n} \end{bmatrix}.$$

As before, $\hat{F}(\delta)$ is the linear approximation to the function F by the hyperplane that is tangent to it at the current point x^k. The next step is to find the value of δ that would make the approximation equal to zero, i.e., the value that satisfies:

$$F(x^k) + \nabla F(x^k)\delta = 0.$$

Notice that what we have on the right-hand side is a vector of zeros and the equation above represents a system of linear equations. If $\nabla F(x^k)$ is nonsingular, the equality above has a unique solution given by

$$\delta = -\nabla F(x^k)^{-1} F(x^k),$$

and the formula for the Newton update in this case is:

$$x^{k+1} = x^k + \delta = x^k - \nabla F(x^k)^{-1} F(x^k).$$

Example 5.5 *Consider the following problem:*

$$F(x) = F(x_1, x_2) = \begin{pmatrix} f_1(x_1, x_2) \\ f_2(x_1, x_2) \end{pmatrix}$$

$$= \begin{pmatrix} x_1 x_2 - 2x_1 + x_2 - 2 \\ (x_1)^2 + 2x_1 + (x_2)^2 - 7x_2 + 7 \end{pmatrix} = 0.$$

First we calculate the Jacobian:

$$\nabla F(x_1, x_2) = \begin{pmatrix} x_2 - 2 & x_1 + 1 \\ 2x_1 + 2 & 2x_2 - 7 \end{pmatrix}.$$

If our initial estimate of the solution is $x^0 = (0, 0)$, then the next point generated by Newton's method will be:

$$\left(x_1^1, x_2^1\right) = \left(x_1^0, x_2^0\right) - \begin{pmatrix} x_2^0 - 2 & x_1^0 + 1 \\ 2x_1^0 + 2 & 2x_2^0 - 7 \end{pmatrix}^{-1} \begin{pmatrix} x_1^0 x_2^0 - 2x_1^0 + x_2^0 - 2 \\ (x_1^0)^2 + 2x_1^0 + (x_2^0)^2 - 7x_2^0 + 7 \end{pmatrix}$$

$$= (0, 0) - \begin{pmatrix} -2 & 1 \\ 2 & -7 \end{pmatrix}^{-1} \begin{pmatrix} -2 \\ 7 \end{pmatrix}$$

$$= (0, 0) - \left(\frac{7}{12}, -\frac{5}{6}\right) = \left(-\frac{7}{12}, \frac{5}{6}\right).$$

Optimization version

When we use Newton's method for unconstrained optimization of a twice-differentiable function $f(x)$, the nonlinear equality system that we want to solve is the first-order necessary optimality condition $\nabla f(x) = 0$. In this case, the functions $f_i(x)$ in (5.6) are the partial derivatives of the function f. That is,

$$f_i(x) = \frac{\partial f}{\partial x_i}(x_1, x_2, \ldots, x_n).$$

Writing

$$F(x_1, x_2, \ldots, x_n) = \begin{bmatrix} f_1(x_1, x_2, \ldots, x_n) \\ f_2(x_1, x_2, \ldots, x_n) \\ \vdots \\ f_n(x_1, x_2, \ldots, x_n) \end{bmatrix}$$

$$= \begin{bmatrix} \frac{\partial f}{\partial x_1}(x_1, x_2, \ldots, x_n) \\ \frac{\partial f}{\partial x_i}(x_1, x_2, \ldots, x_n) \\ \vdots \\ \frac{\partial f}{\partial x_n}(x_1, x_2, \ldots, x_n) \end{bmatrix} = \nabla f(\mathbf{x}),$$

we observe that the Jacobian matrix $\nabla F(x_1, x_2, \ldots, x_n)$ is nothing but the *Hessian* matrix of function f:

$$\nabla F(x_1, x_2, \ldots, x_n) = \begin{bmatrix} \frac{\partial^2 f}{\partial x_1 \partial x_1} & \cdots & \frac{\partial^2 f}{\partial x_1 \partial x_n} \\ \vdots & \ddots & \vdots \\ \frac{\partial^2 f}{\partial x_n \partial x_1} & \cdots & \frac{\partial^2 f}{\partial x_n \partial x_n} \end{bmatrix} = \nabla^2 f(x).$$

Therefore, the Newton direction at iterate x^k is given by

$$\delta = -\nabla^2 f(x^k)^{-1} \nabla f(x^k) \tag{5.8}$$

and the Newton update formula is

$$x^{k+1} = x^k + \delta = x^k - \nabla f^2(x^k)^{-1} \nabla f(x^k).$$

For illustration and comparison purposes, we apply this technique to the example problem of Section 5.4.1. Recall that the problem was to find

$$\min \; f(x) = (x_1 - 2)^4 + \exp(x_1 - 2) + (x_1 - 2x_2)^2$$

starting from $x^0 = (0, 3)$.

Table 5.6 *Newton iterations*

k	(x_1^k, x_2^k)	(d_1^k, d_2^k)	α^k	$\|\nabla f(x^{k+1})\|$
0	(0.000, 3.000)	(0.662, −2.669)	1.000	9.319
1	(0.662, 0.331)	(0.429, 0.214)	1.000	2.606
2	(1.091, 0.545)	(0.252, 0.126)	1.000	0.617
3	(1.343, 0.671)	(0.108, 0.054)	1.000	0.084
4	(1.451, 0.726)	(0.020, 0.010)	1.000	0.002
5	(1.471, 0.735)	(0.001, 0.000)	1.000	0.000

The gradient of f was given in (5.4) and the Hessian matrix is given below:

$$\nabla^2 f(x) = \begin{bmatrix} 12(x_1 - 2)^2 + \exp(x_1 - 2) + 2 & -4 \\ -4 & 8 \end{bmatrix}. \tag{5.9}$$

Thus, we calculate the Newton direction at $x^0 = (0, 3)$ as follows:

$$\delta = -\nabla^2 f \left(\begin{bmatrix} 0 \\ 3 \end{bmatrix} \right)^{-1} \nabla f \left(\begin{bmatrix} 0 \\ 3 \end{bmatrix} \right)$$

$$= - \begin{bmatrix} 50 + e^{-2} & -4 \\ -4 & 8 \end{bmatrix}^{-1} \begin{bmatrix} -44 + e^{-2} \\ 24 \end{bmatrix} = \begin{bmatrix} 0.662 \\ -2.669 \end{bmatrix}.$$

We list the first five iterates in Table 5.6 and illustrate the rapid progress of the algorithm towards the optimal solution in Figure 5.4. Note that the ideal step size for Newton's method is almost always 1. In our example, this step size always satisfied the sufficient decrease condition and was chosen in each iteration. Newton's method identifies a point with $\|\nabla f(x)\| \le 10^{-5}$ after seven iterations.

Despite its excellent convergence behavior close to a solution, Newton's method is not always the best option, especially for large-scale optimization. Often the Hessian matrix is expensive to compute at each iteration. In such cases, it may be preferable to use an approximation of the Hessian matrix instead. These approximations are usually chosen in such a way that the solution of the linear system in (5.8) is much cheaper that what it would be with the exact Hessian. Such approaches are known as quasi-Newton methods. Most popular variants of quasi-Newton methods are BFGS and DFP methods. These acronyms represent the developers of these algorithms in the late 1960s and early 1970s. Detailed information on quasi-Newton approaches can be found in, for example, [61].

Exercise 5.12 Repeat Exercise 5.11 , this time using the optimization version of Newton's method. Use line search with $\mu = 1/2$ in the Armijo–Goldstein condition and a backtracking ratio of $\beta = 1/2$.

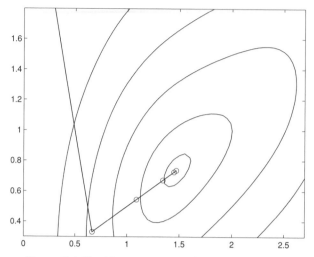

Figure 5.4 Rapid convergence of Newton's method

5.5 Constrained optimization

We now move on to the more general case of nonlinear optimization problems with constraints. Specifically, we consider an optimization problem given by a nonlinear objective function and/or nonlinear constraints. We can represent such problems in the following generic form:

$$
\begin{aligned}
\min_x \; & f(x) \\
& g_i(x) = 0, \quad i \in \mathcal{E}, \\
& g_i(x) \geq 0, \quad i \in \mathcal{I}.
\end{aligned}
\tag{5.10}
$$

In the remainder of this section we assume that f and g_i, $i \in \mathcal{E} \cup \mathcal{I}$, are all continuously differentiable functions.

An important tool in the study of constrained optimization problems is the Lagrangian function. To define this function, one associates a multiplier λ_i – the so-called Lagrange multiplier – with each one of the constraints. For problem (5.10) the Lagrangian is defined as follows:

$$
\mathcal{L}(x, \lambda) := f(x) - \sum_{i \in \mathcal{E} \cup \mathcal{I}} \lambda_i g_i(x).
\tag{5.11}
$$

Essentially, we are considering an objective function that is penalized for violations of the feasibility constraints. For properly chosen values of λ_i, minimizing the unconstrained function $\mathcal{L}(x, \lambda)$ is equivalent to solving the constrained optimization problem (5.10). This equivalence is the primary reason for our interest in the Lagrangian function.

One of the most important theoretical issues related to this problem is the identification of necessary and sufficient conditions for optimality. Collectively, these conditions are called the *optimality conditions* and are the subject of this section.

Before presenting the optimality conditions for (5.10) we first discuss a technical condition called *regularity* that is encountered in the theorems that follow:

Definition 5.1 *Let x be a vector satisfying $g_i(x) = 0$, $i \in \mathcal{E}$ and $g_i(x) \geq 0$, $i \in \mathcal{I}$. Let $\mathcal{J} \subset \mathcal{I}$ be the set of indices for which $g_i(x) \geq 0$ is satisfied with equality. Then, x is a regular point of the constraints of (5.10) if the gradient vectors $\nabla g_i(x)$ for $i \in \mathcal{E} \cup \mathcal{J}$ are linearly independent.*

Constraints corresponding to the set $\mathcal{E} \cup \mathcal{J}$ in the definition above, namely, the constraints for which we have $g_i(x) = 0$, are called the *active* constraints at x.

We discussed two notions of optimality in Chapter 1, *local* and *global*. Recall that a global optimal solution to (5.10) is a vector x^* that is feasible and satisfies $f(x^*) \leq f(x)$ for all feasible x. In contrast, a local optimal solution x^* is feasible and satisfies $f(x^*) \leq f(x)$ for all feasible x in the set $\{x : \|x - x^*\| \leq \varepsilon\}$ for some $\varepsilon > 0$. So, a local solution must be better than all the feasible points in a neighborhood of itself. The optimality conditions we consider below identify local solutions only, which may or may not be global solutions to the problem. Fortunately, there is an important class of problems where local and global solutions coincide, namely convex optimization problems. See Appendix A for a discussion on convexity and convex optimization problems.

Theorem 5.1 (First-order necessary conditions) *Let x^* be a local minimizer of the problem (5.10) and assume that x^* is a regular point for the constraints of this problem. Then, there exists λ_i, $i \in \mathcal{E} \cup \mathcal{I}$ such that*

$$\nabla f(x^*) - \sum_{i \in \mathcal{E} \cup \mathcal{I}} \lambda_i \nabla g_i(x^*) = 0, \tag{5.12}$$

$$\lambda_i \geq 0, \ i \in \mathcal{I}, \tag{5.13}$$

$$\lambda_i g_i(x^*) = 0, \ i \in \mathcal{I}. \tag{5.14}$$

Note that the expression on the left-hand side of (5.12) is the gradient of the Lagrangian function $\mathcal{L}(x, \lambda)$ with respect to the variables x. First-order conditions are satisfied at local minimizers as well as local maximizers and saddle points. When the objective and constraint functions are twice continuously differentiable, one can eliminate maximizers and saddle points using curvature information on the functions. As in Theorem 5.1 , we consider the Lagrangian function $\mathcal{L}(x, \lambda)$ and use the Hessian of this function with respect to the x variables to determine the collective curvature in the objective function as well as the constraint functions at the current point.

Theorem 5.2 (Second-order necessary conditions)   *Assume that f and g_i, $i \in \mathcal{E} \cup \mathcal{I}$ are all twice continuously differentiable functions. Let x^* be a local minimizer of the problem (5.10) and assume that x^* is a regular point for the constraints of this problem. Then, there exists λ_i, $i \in \mathcal{E} \cup \mathcal{I}$ satisfying (5.12)–(5.14) as well as the following condition:*

$$\nabla^2 f(x^*) - \sum_{i \in \mathcal{E} \cup \mathcal{I}} \lambda_i \nabla^2 g_i(x^*) \tag{5.15}$$

is positive semidefinite on the tangent subspace of active constraints at x^*.

The last part of the theorem above can be restated in terms of the Jacobian of the active constraints. Let $A(x^*)$ denote the Jacobian of the active constraints at x^* and let $N(x^*)$ be a null-space basis for $A(x^*)$. Then, the last condition of the theorem above is equivalent to the following condition:

$$N^{\mathrm{T}}(x^*) \left(\nabla^2 f(x^*) - \sum_{i \in \mathcal{E} \cup \mathcal{I}} \lambda_i \nabla^2 g_i(x^*) \right) N(x^*) \tag{5.16}$$

is positive semidefinite.

The satisfaction of the second-order necessary conditions does not always guarantee the local optimality of a given solution vector. The conditions that are sufficient for local optimality are slightly more stringent and a bit more complicated since they need to consider the possibility of *degeneracy*.

Theorem 5.3 (Second-order sufficient conditions)   *Assume that f and g_i, $i \in \mathcal{E} \cup \mathcal{I}$ are all twice continuously differentiable functions. Let x^* be a feasible and regular point for the constraints of the problem (5.10). Let $A(x^*)$ denote the Jacobian of the active constraints at x^* and let $N(x^*)$ be a null-space basis for $A(x^*)$. If there exists λ_i, $i \in \mathcal{E} \cup \mathcal{I}$ satisfying (5.12)–(5.14) as well as*

$$g_i(x^*) = 0, i \in \mathcal{I} \text{ implies } \lambda_i > 0, \tag{5.17}$$

and

$$N^{\mathrm{T}}(x^*) \left(\nabla^2 f(x^*) - \sum_{i \in \mathcal{E} \cup \mathcal{I}} \lambda_i \nabla^2 g_i(x^*) \right) N(x^*) \text{ is positive definite}, \tag{5.18}$$

then x^* is a local minimizer of the problem (5.10).

The conditions listed in Theorems 5.1, 5.2, and 5.3 are often called the Karush–Kuhn–Tucker (KKT) conditions, after their inventors.

Some methods for solving constrained optimization problems formulate a sequence of simpler optimization problems whose solutions are used to generate iterates progressing towards the solution of the original problem. These "simpler" problems can be unconstrained, in which case they can be solved using the techniques we saw in the previous section. We discuss such a strategy in Section 5.5.1. In other cases, the simpler problem solved is a quadratic programming problem and can be solved using the techniques of Chapter 7. The prominent example of this strategy is the sequential quadratic programming method that we discuss in Section 5.5.2.

Exercise 5.13 Recall the definition of the quadratic programming problem given in Chapter 1:

$$\min_x \tfrac{1}{2}x^T Q x + c^T x$$
$$Ax = b \qquad\qquad (5.19)$$
$$x \geq 0,$$

where $A \in I\!R^{m \times n}$, $b \in I\!R^m$, $c \in I\!R^n$, $Q \in I\!R^{n \times n}$ are given, and $x \in I\!R^n$. Assume that Q is symmetric and positive definite. Derive the KKT conditions for this problem. Show that the first-order necessary conditions are also sufficient given our assumptions.

Exercise 5.14 Consider the following optimization problem:

$$\min f(x_1, x_2) = -x_1 - x_2 - x_1 x_2 + \tfrac{1}{2}x_1^2 + x_2^2$$
$$\text{s.t. } x_1 + x_2^2 \leq 3$$
$$\text{and } (x_1, x_2) \geq 0.$$

List the Karush–Kuhn–Tucker optimality conditions for this problem. Verify that $x^* = (2, 1)$ is a local optimal solution to this problem by finding Lagrange multipliers λ_i satisfying the KKT conditions in combination with x^*. Is $x^* = (2, 1)$ a global optimal solution?

5.5.1 The generalized reduced gradient method

In this section, we introduce an approach for solving constrained nonlinear programs. It builds on the method of steepest descent method we discussed in the context of unconstrained optimization. The idea is to reduce the number of variables using the constraints and then to solve this reduced and unconstrained problem using the steepest descent method.

Linear equality constraints
First we consider an example where the constraints are linear equations.

Example 5.6

$$\min f(x) = x_1^2 + x_2 + x_3^2 + x_4$$
$$g_1(x) = x_1 + x_2 + 4x_3 + 4x_4 - 4 = 0$$
$$g_2(x) = -x_1 + x_2 + 2x_3 - 2x_4 + 2 = 0.$$

It is easy to solve the constraint equations for two of the variables in terms of the others. Solving for x_2 and x_3 in terms of x_1 and x_4 gives

$$x_2 = 3x_1 + 8x_4 - 8 \quad and \quad x_3 = -x_1 - 3x_4 + 3.$$

Substituting these expressions into the objective function yields the following reduced problem:

$$\min f(x_1, x_4) = x_1^2 + (3x_1 + 8x_4 - 8) + (-x_1 - 3x_4 + 3)^2 + x_4.$$

This problem is unconstrained and therefore it can be solved using one of the methods presented in Section 5.4.

Nonlinear equality constraints

Now consider the possibility of approximating a problem where the constraints are nonlinear equations by a problem with linear equations. To see how this works, consider the following example, which is similar to the preceding one but has constraints that are nonlinear.

Example 5.7

$$\begin{aligned}
\min f(x) &= x_1^2 + x_2 + x_3^2 + x_4 \\
g_1(x) &= x_1^2 + x_2 + 4x_3 + 4x_4 - 4 = 0 \\
g_2(x) &= -x_1 + x_2 + 2x_3 - 2x_4^2 + 2 = 0.
\end{aligned}$$

We use the Taylor series approximation to the constraint functions at the current point $\bar{x}$:

$$g(x) \approx g(\bar{x}) + \nabla g(\bar{x})(x - \bar{x})^{\mathrm{T}}.$$

This gives

$$g_1(x) \approx \left(\bar{x}_1^2 + \bar{x}_2 + 4\bar{x}_3 + 4\bar{x}_4 - 4 \right) + (2\bar{x}_1, 1, 4, 4) \begin{pmatrix} x_1 - \bar{x}_1 \\ x_2 - \bar{x}_2 \\ x_3 - \bar{x}_3 \\ x_4 - \bar{x}_4 \end{pmatrix}$$

$$\approx 2\bar{x}_1 x_1 + x_2 + 4x_3 + 4x_4 - \left(\bar{x}_1^2 + 4 \right) = 0$$

and

$$g_2(x) \approx -x_1 + x_2 + 2x_3 - 4\bar{x}_4 x_4 + \left(\bar{x}_4^2 + 2 \right) = 0.$$

The idea of the *generalized reduced gradient algorithm (GRG)* is to solve a sequence of subproblems, each of which uses a linear approximation of the constraints. In each iteration of the algorithm, the constraint linearization is recalculated at the point found from the previous iteration. Typically, even though the constraints are only approximated, the subproblems yield points that are progressively closer to the optimal point. A property of the linearization is that, at the optimal point, the linearized problem has the same solution as the original problem.

The first step in applying GRG is to pick a starting point. Suppose that we start with $x^0 = (0, -8, 3, 0)$, which happens to satisfy the original constraints. It is possible to start from an infeasible point as we discuss later on. Using the approximation formulas derived earlier, we form our first approximation problem as follows:

$$\text{min } f(x) = x_1^2 + x_2 + x_3^2 + x_4$$
$$g_1(x) = x_2 + 4x_3 + 4x_4 - 4 = 0$$
$$g_2(x) = -x_1 + x_2 + 2x_3 + 2 = 0.$$

Next we solve the equality constraints of the approximate problem to express two of the variables in terms of the others. Arbitrarily selecting x_2 and x_3, we get

$$x_2 = 2x_1 + 4x_4 - 8 \quad \text{and} \quad x_3 = -\frac{1}{2}x_1 - 2x_4 + 3.$$

Substituting these expressions in the objective function yields the reduced problem

$$\text{min } f(x_1, x_4) = x_1^2 + (2x_1 + 4x_2 - 8) + \left(-\tfrac{1}{2}x_1 - 2x_4 + 3\right)^2 + x_4.$$

Solving this unconstrained minimization problem yields $x_1 = -0.375$, $x_4 = 0.96875$. Substituting in the equations for x_2 and x_3 gives $x_2 = -4.875$ and $x_3 = 1.25$. Thus the first iteration of GRG has produced the new point $x^1 = (-0.375, -4.875, 1.25, 0.968\,75)$.

To continue the solution process, we would re-linearize the constraint functions at the new point, use the resulting system of linear equations to express two of the variables in terms of the others, substitute into the objective to get the new reduced problem, solve the reduced problem for x^2, and so forth. Using the stopping criterion $\|x^{k+1} - x^k\| < T$ where $T = 0.0025$, we get the results summarized in Table 5.7. This is to be compared with the optimum solution, which is

$$x^* = (-0.500, -4.825, 1.534, 0.610)$$

and has an objective value of -1.612. Note that, in Table 5.7, the values of the function $f(x^k)$ are sometimes smaller than the minimum value for $k = 1$, and 2. How is this possible? The reason is that the points x^k computed by GRG are usually not feasible to the constraints. They are only feasible to a linear approximation of these constraints.

Now we discuss the method used by GRG for starting at an infeasible solution: a *phase 1 problem* is solved to construct a feasible one. The objective function for the phase 1 problem is the sum of the absolute values of the violated constraints. The constraints for the phase 1 problem are the nonviolated ones. Suppose we had

Table 5.7 *Summarized results*

k	$(x_1^k, x_2^k, x_3^k, x_4^k)$	$f(x^k)$	$\|x^{k+1} - x^k\|$
0	$(0.000, -8.000, 3.000, 0.000)$	1.000	3.729
1	$(-0.375, -4.875, 1.250, 0.969)$	-2.203	0.572
2	$(-0.423, -5.134, 1.619, 0.620)$	-1.714	0.353
3	$(-0.458, -4.792, 1.537, 0.609)$	-1.610	0.022
4	$(-0.478, -4.802, 1.534, 0.610)$	-1.611	0.015
5	$(-0.488, -4.813, 1.534, 0.610)$	-1.612	0.008
6	$(-0.494, -4.818, 1.534, 0.610)$	-1.612	0.004
7	$(-0.497, -4.821, 1.534, 0.610)$	-1.612	0.002
8	$(-0.498, -4.823, 1.534, 0.610)$	-1.612	

started at the point $x^0 = (1, 1, 0, 1)$ in our example. This point violates the first constraint but satisfies the second, so the phase 1 problem would be

$$\min \ |x_1^2 + x_2 + 4x_3 + 4x_4 - 4|$$
$$-x_1 + x_2 + 2x_3 - 2x_4^2 + 2 = 0.$$

Once a feasible solution has been found by solving the phase 1 problem, the method illustrated above is used to find an optimal solution.

Linear inequality constraints
Finally, we discuss how GRG solves problems having inequality constraints as well as equalities. At each iteration, only the tight inequality constraints enter into the system of linear equations used for eliminating variables (these inequality constraints are said to be *active*). The process is complicated by the fact that active inequality constraints at the current point may need to be released in order to move to a better solution. We illustrate the ideas with the following example:

$$\min \ f(x_1, x_2) = \left(x_1 - \tfrac{1}{2}\right)^2 + \left(x_2 - \tfrac{5}{2}\right)^2$$
$$x_1 - x_2 \geq 0$$
$$x_1 \geq 0$$
$$x_2 \geq 0$$
$$x_2 \leq 2.$$

The feasible set of this problem is shown in Figure 5.5. The arrows in the figure indicate the feasible half-spaces dictated by each constraint. Suppose that we start from $x^0 = (1, 0)$. This point satisfies all the constraints. As can be seen from Figure 5.5, $x_1 - x_2 \geq 0$, $x_1 \geq 0$, and $x_2 \leq 2$, are inactive, whereas the constraint $x_2 \geq 0$ is active. We have to decide whether x_2 should stay at its lower bound or be allowed to leave its bound. We first evaluate the gradient of the objective function

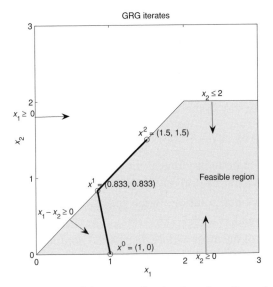

Figure 5.5 Progress of the generalized reduced gradient algorithm

at x^0:

$$\nabla f(x^0) = \left(2x_1^0 - 1, 2x_2^0 - 5\right) = (1, -5).$$

This indicates that we will get the largest decrease in f if we move in the direction $d^0 = -\nabla f(x^0) = (-1, 5)$, i.e., if we decrease x_1 and increase x_2. Since this direction is towards the interior of the feasible region, we decide to release x_2 from its bound. The new point will be $x^1 = x^0 + \alpha^0 d^0$, for some $\alpha^0 > 0$. The constraints of the problem induce an upper bound on α^0, namely $\alpha^0 \leq 0.8333$. Now we perform a line search to determine the best value of α^0 in this range. It turns out to be $\alpha^0 = 0.8333$, so $x^1 = (0.8333, 0.8333)$; see Figure 5.5.

Now, we repeat the process: the constraint $x_1 - x_2 \geq 0$ has become active whereas the others are inactive. Since the active constraint is not a simple upper or lower bound constraint, we introduce a surplus variable, say x_3, and solve for one of the variables in terms of the others. Substituting $x_1 = x_2 + x_3$, we obtain the reduced optimization problem:

$$\begin{aligned} \min \quad & f(x_2, x_3) = \left(x_2 + x_3 - \tfrac{1}{2}\right)^2 + \left(x_2 - \tfrac{5}{2}\right)^2 \\ & 0 \leq x_2 \leq 2 \\ & x_3 \geq 0. \end{aligned}$$

The reduced gradient is

$$\begin{aligned} \nabla f(x_2, x_3) &= (2x_2 + 2x_3 - 1 + 2x_2 - 5, 2x_2 + 2x_3 - 1) \\ &= (-2.667, 0.667) \quad \text{at point } (x_2, x_3)^1 = (0.8333, 0). \end{aligned}$$

Therefore, the largest decrease in f occurs in the direction $(2.667, -0.667)$, that is when we increase x_2 and decrease x_3. But x_3 is already at its lower bound, so we cannot decrease it. Consequently, we keep x_3 at its bound, i.e., we move in the direction $d^1 = (2.667, 0)$ to a new point $(x_2, x_3)^2 = (x_2, x_3)^1 + \alpha^1 d^1$. A line search in this direction yields $\alpha^1 = 0.25$ and $(x_2, x_3)^2 = (1.5, 0)$. The same constraints are still active so we may stay in the space of variables x_2 and x_3. Since

$$\nabla f(x_2, x_3) = (0, 2) \text{ at point } (x_2, x_3)^2 = (1.5, 0)$$

is perpendicular to the boundary line at the current solution x^2 and points towards the exterior of the feasible region, no further decrease in f is possible. Therefore, we have found the optimal solution. In the space of original variables, this optimal solution is $x_1 = 1.5$ and $x_2 = 1.5$.

This is how some of the most widely distributed nonlinear programming solvers, such as Excel's SOLVER, GINO, CONOPT, GRG2, and several others, solve nonlinear programs, with just a few additional details such as the Newton-Raphson direction for line search. Compared with linear programs, the problems that can be solved within a reasonable amount of computational time are typically smaller and the solutions produced may not be very accurate. Furthermore, the potential nonconvexity in the feasible set or in the objective function may generate local optimal solutions that are far from a global solution. Therefore, the interpretation of the output of a nonlinear program requires more care.

Exercise 5.15   Consider the following optimization problem:

$$\min \ f(x_1, x_2) = -x_1 - x_2 - x_1 x_2 + \tfrac{1}{2}x_1^2 + x_2^2$$
$$\text{s.t.} \ \ x_1 + x_2^2 \leq 3$$
$$x_1^2 - x_2 = 3$$
$$(x_1, x_2) \geq 0.$$

Find a solution to this problem using the generalized reduced gradient approach.

5.5.2 Sequential quadratic programming

Consider a general nonlinear optimization problem:

$$\min_x \ f(x)$$
$$g_i(x) = 0, \quad i \in \mathcal{E}, \tag{5.20}$$
$$g_i(x) \geq 0, \quad i \in \mathcal{I}.$$

To solve this problem, one might try to capitalize on the good algorithms available for solving the more structured and easier quadratic programs (see Chapter 7). This is the idea behind *sequential quadratic programming* (SQP). At the current feasible point x^k, the problem (5.20) is approximated by a quadratic program: a quadratic approximation of the Lagrangian function is computed as well as linear approximations of the constraints. The resulting quadratic program is of the form

$$
\begin{aligned}
\min \ \nabla f(x^k)^{\mathrm{T}}(x - x^k) &+ \tfrac{1}{2}(x - x^k)^{\mathrm{T}} B_k(x - x^k) \\
\nabla g_i(x^k)^{\mathrm{T}}(x - x^k) + g_i(x^k) &= 0 \quad \text{for all } i \in \mathcal{E}, \\
\nabla g_i(x^k)^{\mathrm{T}}(x - x^k) + g_i(x^k) &\geq 0 \quad \text{for all } i \in \mathcal{I},
\end{aligned}
\tag{5.21}
$$

where

$$
B^k = \nabla_{xx}^2 \mathcal{L}(x^k, \lambda^k)
$$

is the Hessian of the Lagrangian function (5.11) with respect to the x variables and λ^k is the current estimate of the Lagrange multipliers.

This problem can be solved with one of the specialized algorithms for quadratic programming problems such as the interior-point methods we discuss in Chapter 7. The optimal solution of the quadratic program is used to determine a search direction. Then a line search or trust region procedure is performed to determine the next iterate.

Perhaps the best way to think of sequential quadratic programming is as an extension of the optimization version of Newton's method to constrained problems. Recall that the optimization version of Newton's method uses a quadratic approximation to the objective function and defines the minimizer of this approximation as the next iterate, much like what we described for the SQP method. Indeed, for an unconstrained problem, the SQP is identical to Newton's method. For a constrained problem, the optimality conditions for the quadratic problem we solve in SQP correspond to the Newton direction for the optimality conditions of the original problem at the current iterate.

Sequential quadratic programming iterates until the solution converges. Much like Newton's method, the SQP approaches are very powerful, especially if equipped with line search or trust region methodologies to navigate the nonlinearities and nonconvexities. We refer the reader to the survey of Boggs and Tolle [16] and the text by Nocedal and Wright [61] for further details on the sequential quadratic programming approach.

Exercise 5.16 Consider the following nonlinear optimization problem with equality constraints:

$$
\begin{aligned}
\min \ f(\mathbf{x}) &= \ x_1^2 + x_2 + \ x_3^2 + \ x_4 \\
g_1(\mathbf{x}) &= \ x_1^2 + x_2 + 4x_3 + 4x_4 - 4 = 0 \\
g_2(\mathbf{x}) &= -x_1 + x_2 + 2x_3 - 2x_4^2 + 2 = 0.
\end{aligned}
$$

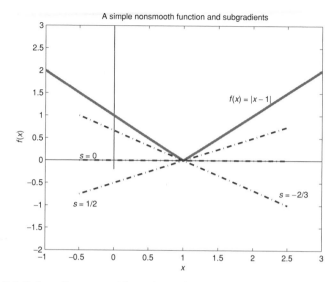

Figure 5.6 Subgradients provide under-estimating approximations to functions

Construct the quadratic programming approximation (5.21) for this problem at point $x^0 = (0, -8, 3, 0)$ and derive the KKT conditions for this quadratic programming problem.

5.6 Nonsmooth optimization: subgradient methods

In this section, we consider unconstrained nonlinear programs of the form

$$\min \quad f(x),$$

where $x = (x_1, \ldots, x_n)$ and f is a nondifferentiable convex function. Optimality conditions based on the gradient are not available since the gradient is not always defined in this case. However, the notion of gradient can be generalized as follows. A *subgradient* of f at point x^* is a vector $s^* = (s_1^*, \ldots, s_n^*)$ such that

$$s^*(x - x^*) \leq f(x) - f(x^*) \text{ for every } x.$$

When the function f is differentiable, the subgradient is identical to the gradient. When f is not differentiable at point x, there are typically many subgradients at x. For example, consider the convex function of one variable

$$f(x) = \max\{1 - x, x - 1\} = |x - 1|.$$

As is evident from Figure 5.6 this function is nondifferentiable at the point $x = 1$ and it is easy to verify that any vector s such that $-1 \leq s \leq 1$ is a subgradient of f at point $x = 1$. Some of these subgradients and the linear approximations defined by them are shown in Figure 5.6. Note that each subgradient of the function at a point defines a linear "tangent" to the function that stays always below the plot of the function – this is the defining property of subgradients.

Consider a nondifferentiable convex function f. The point x^* is a minimum of f if and only if f has a zero subgradient at x^*. In the above example, 0 is a subgradient of f at point $x^* = 1$ and therefore this is where the minimum of f is achieved.

The method of steepest descent can be extended to nondifferentiable convex functions by computing *any* subgradient direction and using the opposite direction to make the next step. Although subgradient directions are not always directions of ascent, one can nevertheless guarantee convergence to the optimum point by choosing the step size appropriately.

A generic *subgradient method* can be stated as follows:

1. **Initialization:** Start from any point x^0. Set $i = 0$.
2. **Iteration i:** Compute a subgradient s^i of f at point x^i. If s^i is 0 or close to 0, stop. Otherwise, let $x^{i+1} = x^i - \alpha_i s^i$, where $\alpha_i > 0$ denotes a step size, and perform the next iteration.

Several choices of the step size α_i have been proposed in the literature. To guarantee convergence to the optimum, the step size α_i needs to be decreased very slowly (for example, $\alpha_i \to 0$ such that $\sum_i \alpha_i = +\infty$ will do). But the slow decrease in α_i results in slow convergence of x_i to the optimum. In practice, in order to get fast convergence, the following choice is popular: start from $\alpha_0 = 2$ and then halve the step size if no improvement in the objective value $f(x^i)$ is observed for k consecutive iterations ($k = 7$ or 8 is often used). This choice is well suited when one wants to get close to the optimum quickly and when finding the exact optimum is not important (this is the case in integer programming applications where subgradient optimization is used to obtain quick bounds in branch-and-bound algorithms). With this in mind, a stopping criterion that is frequently used in practice is a maximum number of iterations (say 200) instead of "s^i is 0 or close to 0."

We will see in Chapter 12 how subgradient optimization is used in a model to construct an index fund.

6

NLP models: volatility estimation

Volatility is a term used to describe how much the security prices, market indices, interest rates, etc., move up and down around their mean. It is measured by the standard deviation of the random variable that represents the financial quantity we are interested in. Most investors prefer low volatility to high volatility and therefore expect to be rewarded with higher long-term returns for holding higher volatility securities.

Many financial computations require volatility estimates. Mean-variance optimization trades off the expected return and volatility of a portfolio of securities. The celebrated option valuation formulas of Black, Scholes, and Merton (BSM) involve the volatility of the underlying security. Risk management revolves around the volatility of the current positions. Therefore, accurate estimation of the volatilities of security returns, interest rates, exchange rates, and other financial quantities is crucial to many quantitative techniques in financial analysis and management.

Most volatility estimation techniques can be classified as either a *historical* or an *implied* method. One either uses historical time series to infer patterns and estimates the volatility using a statistical technique, or considers the known prices of related securities such as options that may reveal the market sentiment on the volatility of the security in question. GARCH models exemplify the first approach while the implied volatilities calculated from the BSM formulas are the best-known examples of the second approach. Both types of techniques can benefit from the use of optimization formulations to obtain more accurate volatility estimates with desirable characteristics such as smoothness. We discuss two examples in this chapter.

6.1 Volatility estimation with GARCH models

Empirical studies analyzing time series data for returns of securities, interest rates, and exchange rates often reveal a clustering behavior for the volatility of the process under consideration. Namely, these time series exhibit high volatility periods alternating with low volatility periods. These observations suggest that future volatility can be estimated with some degree of confidence by relying on historical data.

Currently, describing the evolution of such processes by imposing a stationary model on the conditional distribution of returns is one of the most popular approaches in the econometric modeling of financial time series. This approach expresses the conventional wisdom that models for financial returns should adequately represent the nonlinear dynamics that are demonstrated by the sample autocorrelation and cross-correlation functions of these time series. ARCH (autoregressive conditional heteroscedasticity) and GARCH (generalized ARCH) models of Engle [27] and Bollerslev [17] have been popular and successful tools for future volatility estimation. For the multivariate case, rich classes of stationary models that generalize the univariate GARCH models have also been developed; see, for example, the comprehensive survey by Bollerslev *et al.* [18].

The main mathematical problem to be solved in fitting ARCH and GARCH models to observed data is the determination of the best model parameters that maximize a likelihood function, i.e., an optimization problem. See Nocedal and Wright [61], page 255, for a short discussion of maximum likelihood estimation. Typically, these models are presented as unconstrained optimization problems with recursive terms. In a recent study, Altay-Salih *et al.* [2] argue that because of the recursion equations and the stationarity constraints, these models actually fall into the domain of nonconvex, nonlinearly constrained nonlinear programming. Their study shows that by using a sophisticated nonlinear optimization package (sequential quadratic programming based FILTER method of Fletcher and Leyffer [29] in their case) they are able to significantly improve the log-likelihood functions for multivariate volatility (and correlation) estimation. While their study does not provide a comparison of forecasting effectiveness of the standard approaches to that of the constrained optimization approach, the numerical results suggest that constrained optimization approach provides a better prediction of the extremal behavior of the time series data; see [2]. Here, we briefly review this constrained optimization approach for expository purposes.

We consider a stochastic process Y indexed by natural numbers. Y_t, its value at time t, is an n-dimensional vector of random variables. Autoregressive behavior of these random variables is modeled as

$$Y_t = \sum_{i=1}^{m} \phi_i Y_{t-i} + \varepsilon_t, \qquad (6.1)$$

where m is a positive integer representing the number of periods we look back in our model and ε_t satisfies

$$E[\varepsilon_t | \varepsilon_1, \ldots, \varepsilon_{t-1}] = 0.$$

While these models are of limited value, if at all, in the estimation of the actual time series (Y_t), they have been shown to provide useful information for volatility estimation. For this purpose, GARCH models define

$$h_t := E\left[\varepsilon_t^2 | \varepsilon_1, \ldots, \varepsilon_{t-1}\right]$$

in the univariate case and

$$H_t := E\left[\varepsilon_t \varepsilon_t^T | \varepsilon_1, \ldots, \varepsilon_{t-1}\right]$$

in the multivariate case. Then one models the conditional time dependence of these squared residuals in the univariate case as follows:

$$h_t = c + \sum_{i=1}^{q} \alpha_i \varepsilon_{t-i}^2 + \sum_{j=1}^{p} \beta_j h_{t-j}. \tag{6.2}$$

This model is called GARCH(p, q). Note that ARCH models correspond to choosing $p = 0$.

The generalization of the model (6.2) to the multivariate case can be done in a number of ways. One approach is to use the operator vech to turn the matrices H_t and $\varepsilon_t \varepsilon_t^T$ into vectors. The operator vech takes an $n \times n$ symmetric matrix as an input and produces an $n(n + 1)/2$-dimensional vector as output by stacking the elements of the matrix on and below the diagonal on top of each other. Using this operator, one can write a multivariate generalization of (6.2) as follows:

$$\text{vech}(H_t) = \text{vech}(C) + \sum_{i=1}^{q} A_i \text{vech}\left(\varepsilon_{t-i} \varepsilon_{t-i}^T\right) + \sum_{j=1}^{p} B_j \text{vech}(H_{t-j}). \tag{6.3}$$

In (6.3), the A_i's and B_j's are square matrices of dimension $n(n + 1)/2$ and C is an $n \times n$ symmetric matrix.

After choosing a superstructure for the GARCH model, that is, after choosing p and q, the objective is to determine the optimal parameters ϕ_i, α_i, and β_j. Most often, this is achieved via maximum likelihood estimation. If one assumes a normal distribution for Y_t conditional on the historical observations, the log-likelihood function can be written as follows [2]:

$$-\frac{T}{2} \log 2\pi - \frac{1}{2} \sum_{t=1}^{T} \log h_t - \frac{1}{2} \sum_{t=1}^{T} \frac{\varepsilon_t^2}{h_t}, \tag{6.4}$$

in the univariate case and

$$-\frac{T}{2} \log 2\pi - \frac{1}{2} \sum_{t=1}^{T} \log \det H_t - \frac{1}{2} \sum_{t=1}^{T} \varepsilon_t^T H_t^{-1} \varepsilon_t \tag{6.5}$$

in the multivariate case.

Exercise 6.1 Show that the function in (6.4) is a difference of convex functions by showing that $\log h_t$ is concave and ε_t^2/h_t is convex in ε_t and h_t. Does the same conclusion hold for the function in (6.5)?

Now, the optimization problem to solve in the univariate case is to maximize the log-likelihood function (6.4) subject to the model constraints (6.1) and (6.2) as well as the condition that h_t is nonnegative for all t since $h_t = E[\varepsilon_t^2|\varepsilon_1, \ldots, \varepsilon_{t-1}]$. In the multivariate case we maximize (6.5) subject to the model constraints (6.1) and (6.3) as well as the condition that H_t is a positive semidefinite matrix for all t since H_t defined as $E[\varepsilon_t \varepsilon_t^T|\varepsilon_1, \ldots, \varepsilon_{t-1}]$ must necessarily satisfy this condition. The positive semidefiniteness of the matrices H_t can either be enforced using the techniques discussed in Chapter 9 or using a reparametrization of the variables via Cholesky-type LDL^T decomposition as discussed in [2].

An important issue in GARCH parameter estimation is the stationarity properties of the resulting model. There is a continuing debate about whether it is reasonable to assume that the model parameters for financial time series are stationary over time. It is clear, however, that the estimation and forecasting is easier on stationary models. A sufficient condition for the stationarity of the univariate GARCH model above is that the α_i's and β_j's as well as the scalar c are strictly positive and that

$$\sum_{i=1}^{q} \alpha_i + \sum_{j=1}^{p} \beta_j < 1, \tag{6.6}$$

see, for example, [35]. The sufficient condition for the multivariate case is more involved and we refer the reader to [2] for these details.

Especially in the multivariate case, the problem of maximizing the log-likelihood function with respect to the model constraints is a difficult nonlinear, nonconvex optimization problem. To find a quick solution, more tractable versions of the model (6.3) have been developed where the model is simplified by imposing additional structure on the matrices A_i and B_j such as diagonality. While the resulting problems are easier to solve, the loss of generality from their simplifying assumptions can be costly. As Altay-Salih *et al.* [2] demonstrate, using the full power of state-of-the-art constrained optimization software, one can solve the more general model in reasonable computational time (at least for bivariate and trivariate estimation problems) with much improved log-likelihood values. While the forecasting efficiency of this approach is still to be tested, it is clear that sophisticated nonlinear optimization is emerging as a valuable tool in volatility estimation problems that use historical data.

Exercise 6.2 Consider the model in (6.3) for the bivariate case when $q = 1$ and $p = 0$ (i.e., an ARCH(1) model). Explicitly construct the nonlinear programming problem to be solved in this case. The comparable simplification of the BEKK representation [4] gives

$$H_t = C^\mathsf{T} C + A^\mathsf{T} \varepsilon_{t-1} \varepsilon_{t-1}^t A.$$

Compare these two models and comment on the additional degrees of freedom in the NLP model. Note that the BEKK representation ensures the positive semidefiniteness of H_t by construction at the expense of lost degrees of freedom.

Exercise 6.3   Test the NLP model against the model resulting from the BEKK representation in the previous exercise using daily return data for two market indices, e.g., S&P 500 and FTSE 100, and an NLP solver. Compare the optimal log-likelihood values achieved by both models and comment.

6.2 Estimating a volatility surface

The discussion in this section is largely based on the work of Coleman *et al.* [23, 22].

The Black–Scholes–Merton (BSM) equation for pricing European options is based on a geometric Brownian motion model for the movements of the underlying security. Namely, one assumes that the underlying security price S_t at time t satisfies

$$\frac{\mathrm{d}S_t}{S_t} = \mu\,\mathrm{d}t + \sigma\,\mathrm{d}W_t \qquad (6.7)$$

where μ is the *drift*, σ is the (constant) volatility, and W_t is the standard Brownian motion. Using this equation and some standard assumptions about the absence of frictions and arbitrage opportunities, one can derive the BSM partial differential equation for the value of a European option on this underlying security. Using the boundary conditions resulting from the payoff structure of the particular option, one determines the value function for the option. Recall from Exercise 5.3 that the price of a European call option with strike K and maturity T is given by:

$$C(K, T) = S_0 \Phi(d_1) - K\mathrm{e}^{-r\mathsf{T}} \Phi(d_2), \qquad (6.8)$$

where

$$d_1 = \frac{\log(S_0/K) + (r + \sigma^2/2)T}{\sigma\sqrt{T}},$$

$$d_2 = d_1 - \sigma\sqrt{T},$$

and $\Phi(\cdot)$ is the cumulative distribution function for the standard normal distribution. r in the formula represents the continuously compounded risk-free and constant interest rate and σ is the volatility of the underlying security that is assumed to be constant. Similarly, the European put option price is given by

$$P(K, T) = K\mathrm{e}^{-r\mathsf{T}} \Phi(-d_2) - S_0 \Phi(-d_1). \qquad (6.9)$$

The risk-free interest rate r, or a reasonably close approximation to it, is often available, for example from Treasury bill prices in US markets. Therefore, all one needs to determine the call or put price using these formulas is a reliable estimate of the volatility parameter σ. Conversely, given the market price for a particular European call or put, one can uniquely determine the volatility of the underlying asset implied by this price, called its *implied volatility*, by solving the equations above with the unknown σ. Any one of the univariate equation solving techniques we discussed in Section 5.3 can be used for this purpose.

Empirical evidence against the appropriateness of (6.7) as a model for the movements of most securities is abundant. Most studies refute the assumption of a volatility that does not depend on time or underlying price level. Indeed, studying the prices of options with same maturity but different strikes, researchers observed that the implied volatilities for such options often exhibit a "smile" structure, i.e., higher implied volatilities away from the money in both directions, decreasing to a minimum level as one approaches the at-the-money option from in-the-money or out-of-the-money strikes. This is clearly in contrast with the constant (flat) implied volatilities one would expect had (6.7) been an appropriate model for the underlying price process.

There are many models that try to capture the volatility smile including stochastic volatility models, jump diffusions, etc. Since these models introduce non-traded sources of risk, perfect replication via dynamic hedging as in BSM approach becomes impossible and the pricing problem is more complicated. An alternative that is explored in [23] is the one-factor continuous diffusion model:

$$\frac{\mathrm{d}S_t}{S_t} = \mu(S_t, t)\mathrm{d}t + \sigma(S_t, t)\mathrm{d}W_t, t \in [0, T], \tag{6.10}$$

where the constant parameters μ and σ of (6.7) are replaced by continuous and differentiable functions $\mu(S_t, t)$ and $\sigma(S_t, t)$ of the underlying price S_t and time t. T denotes the end of the fixed time horizon. If the instantaneous risk-free interest rate r is assumed constant and the dividend rate is constant, given a function $\sigma(S, t)$, a European call option with maturity T and strike K has a unique price. Let us denote this price with $C(\sigma(S, t), K, T)$.

While an explicit solution for the price function $C(\sigma(S, t), K, T)$ as in (6.8) is no longer possible, the resulting pricing problem can be solved efficiently via numerical techniques. Since $\mu(S, t)$ does not appear in the generalized BSM partial differential equation, all one needs is the specification of the function $\sigma(S, t)$ and a good numerical scheme to determine the option prices in this generalized framework.

So, how does one specify the function $\sigma(S, t)$? First of all, this function should be consistent with the observed prices of currently or recently traded options on the same underlying security. If we assume that we are given market prices of m call options with strikes K_j and maturities T_j in the form of bid–ask pairs (β_j, α_j) for $j = 1, \ldots, n$, it would be reasonable to require that the volatility function $\sigma(S, t)$ is chosen so that

$$\beta_j \leq C(\sigma(S, t), K_j, T_j) \leq \alpha_j, \ j = 1, \ldots, n. \tag{6.11}$$

To ensure that (6.11) is satisfied as closely as possible, one strategy is to minimize the violations of the inequalities in (6.11):

$$\min_{\sigma(S,t)\in\mathcal{H}} \sum_{j=1}^{n} [\beta_j - C(\sigma(S, t), K_j, T_j)]^+ + [C(\sigma(S, t), K_j, T_j) - \alpha_j]^+. \tag{6.12}$$

Above, $\mathcal{H}$ denotes the space of measurable functions $\sigma(S, t)$ with domain $I\!R^+ \times [0, T]$ and $u^+ = \max\{0, u\}$. Alternatively, using the closing prices C_j for the options under consideration, or choosing the mid-market prices $C_j = (\beta_j + \alpha_j)/2$, we can solve the following nonlinear least-squares problem:

$$\min_{\sigma(S,t)\in\mathcal{H}} \sum_{j=1}^{n} (C(\sigma(S, t), K_j, T_j) - C_j)^2. \tag{6.13}$$

This is a nonlinear least-squares problem since the function $C(\sigma(S, t), K_j, T_j)$ depends nonlinearly on the variables, namely the local volatility function $\sigma(S, t)$.

While the calibration of the local volatility function to the observed prices using the objective functions in (6.12) and (6.13) is important and desirable, there are additional properties that are desirable in the local volatility function. Arguably, the most common feature sought in existing models is smoothness. For example, in [49] the authors try to achieve a smooth volatility function by appending the objective function in (6.13) as follows:

$$\min_{\sigma(S,t)\in\mathcal{H}} \sum_{j=1}^{n} (C(\sigma(S, t), K_j, T_j) - C_j)^2 + \lambda \|\nabla\sigma(S, t)\|_2. \tag{6.14}$$

Here, λ is a positive trade-off parameter and $\| \cdot \|_2$ represents the L^2-norm. Large deviations in the volatility function would result in a high value for the norm of the gradient function and by penalizing such occurrences, the formulation above encourages a smoother solution to the problem. The most appropriate value for the trade-off parameter λ must be determined experimentally. To solve the resulting problem numerically, one must discretize the volatility function on the underlying price and time grid. Even for a relatively coarse discretization of the S_t and t spaces, one can easily end up with an optimization problem with many variables.

An alternative strategy is to build the smoothness into the volatility function by modeling it with spline functions. To define a spline function, the domain of the function is partitioned into smaller subregions and then, the spline function is chosen to be a polynomial function in each subregion. Since polynomials are smooth functions, spline functions are smooth within each subregion by construction and the only possible sources of nonsmoothness are the boundary regions between subregions. When the polynomial is of a high enough degree, the continuity and differentiability of the spline function at the boundaries between subregions can be ensured by properly choosing the polynomial function coefficients. This strategy is similar to the model we consider in more detail in Section 8.4, except that here we model the volatility function rather than the risk-neutral density and also we generate a function that varies over time rather than an estimate at a single point in time. We defer a more detailed discussion of spline functions to Section 8.4. The use of the spline functions not only guarantees the smoothness of the resulting volatility function estimates but also reduces the degrees of freedom in the problem. As a consequence, the optimization problem to be solved has much fewer variables and is easier. This strategy is proposed in [23] and we review it below.

We start by assuming that $\sigma(S, t)$ is a bi-cubic spline. While higher-order splines can also be used, cubic splines often offer a good balance between flexibility and complexity. Next we choose a set of spline knots at points $(\bar{S}_j, \bar{t}_j)$ for $j = 1, \ldots, k$. If the value of the volatility function at these points is given by $\bar{\sigma}_j := \sigma(\bar{S}_j, \bar{t}_j)$, the interpolating cubic spline that goes through these knots and satisfies a particular end condition is uniquely determined. For example, in Section 8.4 we use the *natural spline end condition*, which sets the second derivative of the function at the knots at the boundary of the domain to zero to obtain our cubic spline approximations uniquely. Therefore, to completely determine the volatility function as a natural bi-cubic spline and to determine the resulting call option prices we have k degrees of freedom represented with the choices $\bar{\sigma} = (\bar{\sigma}_1, \ldots, \bar{\sigma}_k)$.

Let $\Sigma(S, t, \bar{\sigma})$ the bi-cubic spline local volatility function obtained setting $\sigma(\bar{S}_j, \bar{t}_j)$'s to $\bar{\sigma}_j$. Let $C(\Sigma(S, t, \bar{\sigma}), S, t)$ denote the resulting call price function. The analog of the objective function (6.13) is then

$$\min_{\bar{\sigma} \in \mathbf{R}^k} \sum_{j=1}^{n} (C(\Sigma(S, t, \bar{\sigma}), K_j, T_j) - C_j)^2. \tag{6.15}$$

One can introduce positive weights w_j for each of the terms in the objective function above to address different accuracies or confidence in the call prices C_j. We can also introduce lower and upper bounds l_i and u_i for the volatilities at each knot to incorporate additional information that may be available from historical data, etc. This way, we form the following nonlinear least-squares problem with k variables:

$$\min_{\bar{\sigma} \in \mathbf{R}^k} \quad f(\bar{\sigma}) := \sum_{j=1}^{n} w_j (C(\Sigma(S, t, \bar{\sigma}), K_j, T_j) - C_j)^2 \tag{6.16}$$

$$\text{s.t.} \quad l \leq \bar{\sigma} \leq u.$$

It should be noted that the formulation above will not be appropriate if there are many more knots than prices, that is if k is much larger than n. In this case, the problem will be underdetermined and solutions may exhibit consequences of "over-fitting." It is better to use fewer knots than available option prices.

The problem (6.16) is a standard nonlinear optimization problem except that the term $C(\Sigma(S, t, \bar{\sigma}), K_j, T_j)$ in the objective function depends on the decision variables $\bar{\sigma}$ in a complicated and nonexplicit manner. Evaluating gradient of f and, therefore, executing any optimization algorithm that requires gradients can be difficult. Without an explicit expression for f, its gradient must be either estimated using a finite difference scheme or using automatic differentiation. Coleman *et al.* [22, 23] implement both alternatives and report that local volatility functions can be estimated very accurately using these strategies. They also test the hedging accuracy of different delta-hedging strategies, one using a constant volatility estimation and another using the local volatility function produced by the strategy above. These tests indicate that the hedges obtained from the local volatility function are significantly more accurate.

Exercise 6.4   The partial derivative $\partial f(x)/\partial x_i$ of the function $f(x)$ with respect to the ith coordinate of the x vector can be estimated as

$$\frac{\partial f(x)}{\partial x_i} \approx \frac{f(x + he_i) - f(x)}{h},$$

where e_i denotes the ith unit vector. Assuming that f is continuously differentiable, provide an upper bound on the estimation error from this finite-difference approximation using a Taylor series expansion for the function f around x. Next, compute a similar bound for the alternative finite-difference formula given by

$$\frac{\partial f(x)}{\partial x_i} \approx \frac{f(x + he_i) - f(x - he_i)}{2h}.$$

Comment on the potential advantages and disadvantages of these two approaches.

7

Quadratic programming: theory and algorithms

7.1 The quadratic programming problem

As we discussed in the introductory chapter, *quadratic programming* (QP) refers to the problem of minimizing a quadratic function subject to linear equality and inequality constraints. In its standard form, this problem is represented as follows:

$$
\min_x \tfrac{1}{2} x^{\mathrm{T}} Q x + c^{\mathrm{T}} x \\
Ax = b \\
x \geq 0,
$$
(7.1)

where $A \in I\!R^{m \times n}$, $b \in I\!R^m$, $c \in I\!R^n$, $Q \in I\!R^{n \times n}$ are given, and $x \in I\!R^n$. QPs are special classes of nonlinear optimization problems and contain linear programming problems as special cases.

Quadratic programming structures are encountered frequently in optimization models. For example, ordinary least-squares problems which are used often in data fitting are QPs with no constraints. Mean-variance optimization problems developed by Markowitz for the selection of *efficient* portfolios are QP problems. In addition, QP problems are solved as subproblems in the solution of general nonlinear optimization problems via *sequential quadratic programming* (SQP) approaches; see Section 5.5.2.

Recall that, when Q is a positive semidefinite matrix, i.e., when $y^{\mathrm{T}} Q y \geq 0$ for all y, the objective function of problem (7.1) is a convex function of x. When this is the case, a local minimizer of this objective function is also a global minimizer. In contrast, when Q is not positive semidefinite (either indefinite or negative semidefinite), the objective function is nonconvex and may have local minimizers that are not global minimizers. This behaviour is illustrated in Figure 7.1 where the contours of a quadratic function with a positive semidefinite Q are contrasted with those of an indefinite Q.

121

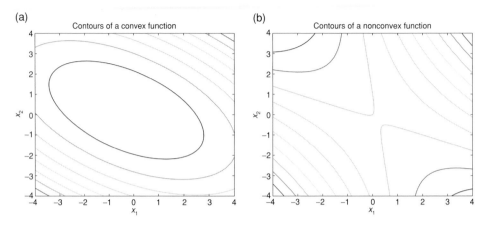

Figure 7.1 Contours of (a) positive semidefinite and (b) indefinite quadratic functions

Exercise 7.1 Consider the quadratic function $f(x) = c^T x + \frac{1}{2} x^T Q x$, where the matrix Q is $n \times n$ and symmetric.

(i) Prove that if $x^T Q x < 0$ for some x, then f is unbounded below.
(ii) Prove that if Q is positive semidefinite (but not positive definite), then either f is unbounded below or it has an infinite number of solutions.
(iii) True or false: f has a unique minimizer if and only if Q is positive definite.

As in linear programming, we can develop a dual of quadratic programming problems. The dual of the problem (7.1) is given below:

$$\max_{x,y,s} \quad b^T y - \frac{1}{2} x^T Q x$$
$$A^T y - Q x + s = c \tag{7.2}$$
$$x \geq 0, \, s \geq 0.$$

Note that, unlike the case of linear programming, the variables of the primal quadratic programming problem also appear in the dual QP.

7.2 Optimality conditions

One of the fundamental tools in the study of optimization problems is the Karush–Kuhn–Tucker theorem, which gives a list of conditions that are necessarily satisfied at any (local) optimal solution of a problem, provided that some mild regularity assumptions are satisfied. These conditions are commonly called KKT conditions and were discussed in the context of general nonlinear optimization problems in Section 5.5.

Applying the KKT theorem to the QP problem (7.1), we obtain the following set of necessary conditions for optimality:

Theorem 7.1 *Suppose that x is a local optimal solution of the QP given in (7.1) so that it satisfies $Ax = b$, $x \geq 0$ and assume that Q is a positive semidefinite matrix. Then, there exist vectors y and s such that the following conditions hold:*

$$A^T y - Qx + s = c \tag{7.3}$$

$$s \geq 0 \tag{7.4}$$

$$x_i s_i = 0, \quad \forall i. \tag{7.5}$$

Furthermore, x is a global optimal solution.

Note that the positive semidefiniteness condition related to the Hessian of the Lagrangian function in the KKT theorem is automatically satisfied for convex quadratic programming problems, and therefore is not included in Theorem 7.1 .

Exercise 7.2 Show that in the case of a positive definite Q, the objective function of (7.1) is strictly convex and, therefore, must have a unique minimizer.

Conversely, if vectors x, y and s satisfy conditions (7.3)–(7.5) as well as primal feasibility conditions

$$Ax = b \tag{7.6}$$

$$x \geq 0, \tag{7.7}$$

then x is a global optimal solution of (7.1). In other words, conditions (7.3)–(7.7) are both necessary and sufficient for x, y, and s to describe a global optimal solution of the QP problem.

In a manner similar to linear programming, optimality conditions (7.3)–(7.7) can be seen as a collection of conditions for:

1. primal feasibility: $Ax = b$, $x \geq 0$;
2. dual feasibility: $A^T y - Qx + s = c$, $s \geq 0$; and
3. complementary slackness: for each $i = 1, \ldots, n$ we have $x_i s_i = 0$.

Using this interpretation, one can develop modifications of the simplex method that can also solve convex quadratic programming problems (Wolfe's method). We do not present this approach here. Instead, we describe an alternative algorithm that is based on Newton's method; see Section 5.4.2.

Exercise 7.3 Consider the following quadratic program:

$$
\begin{aligned}
\min \quad & x_1 x_2 + x_1^2 + \tfrac{3}{2} x_2^2 + 2 x_3^2 \\
& + 2 x_1 + x_2 + 3 x_3 \\
\text{subject to} \quad & x_1 + x_2 + x_3 = 1 \\
& x_1 - x_2 = 0 \\
& x_1 \geq 0, \ x_2 \geq 0, \ x_3 \geq 0.
\end{aligned}
$$

Is the quadratic objective function convex? Show that $x^* = (1/2, 1/2, 0)$ is an optimal solution to this problem by finding vectors y and s that satisfy the optimality conditions jointly with x^*.

7.3 Interior-point methods

7.3.1 Introduction

In 1984, Karmarkar [43] proved that an *interior-point method* (IPM) can solve linear programming problems (LPs) in polynomial time. The two decades that followed the publication of Karmarkar's paper have seen a very intense effort by the optimization research community to study theoretical and practical properties of IPMs. One of the early discoveries was that IPMs can be viewed as modifications of Newton's method that are able to handle inequality constraints. Some of the most important contributions were made by Nesterov and Nemirovski who showed that the IPM machinery can be applied to a much larger class of problems than just LPs [60]. Convex quadratic programming problems, for example, can be solved in polynomial time, as well as many other convex optimization problems using IPMs. For most instances of conic optimization problems we discuss in Chapter 9 and 10, IPMs are by far the best available methods.

Here, we will describe a variant of IPMs for convex quadratic programming. For the QP problem in (7.1) we can write the optimality conditions in matrix form as follows:

$$F(x, y, s) = \begin{bmatrix} A^T y - Qx + s - c \\ Ax - b \\ XSe \end{bmatrix} = \begin{bmatrix} 0 \\ 0 \\ 0 \end{bmatrix}, \quad (x, s) \geq 0. \qquad (7.8)$$

Above, X and S are diagonal matrices with the entries of the x and s vectors, respectively, on the diagonal, i.e., $X_{ii} = x_i$, and $X_{ij} = 0, i \neq j$, and similarly for S. Also, as before, e is an n-dimensional vector of 1s.

The system of equations $F(x, y, s) = 0$ has $n + m + n$ variables and exactly the same number of constraints, i.e., it is a "square" system. Because of the nonlinear equations $x_i s_i = 0$ we cannot solve this system using linear system solution methods such as Gaussian elimination. But, since the system is square we can apply Newton's method. In fact, without the nonnegativity constraints, finding (x, y, s) satisfying these optimality conditions would be a straightforward exercise by applying Newton's method.

The existence of nonnegativity constraints creates a difficulty. The existence and the number of inequality constraints are among the most important factors that contribute to the difficulty of the solution of any optimization problem. Interior-point approaches use the following strategy to handle these inequality constraints:

first identify an initial solution (x^0, y^0, s^0) that satisfies the first two (linear) blocks of equations in $F(x, y, s) = 0$ but not necessarily the third block $XSe = 0$, and also satisfies the nonnegativity constraints *strictly*, i.e., $x^0 > 0$ and $s^0 > 0$. Notice that a point satisfying some inequality constraints strictly lies in the *interior* of the region defined by these inequalities – rather than being on the boundary. This is the reason why the method we are discussing is called an interior-point method.

Once we find such an (x^0, y^0, s^0) we try to generate new points (x^k, y^k, s^k) that also satisfy these same conditions and get progressively closer to satisfying the third block of equations. This is achieved via careful application of a modified Newton's method.

Let us start by defining two sets related to the conditions (7.8):

$$\mathcal{F} := \{(x, y, s) : Ax = b, \ A^{\mathrm{T}}y - Qx + s = c, \ x \geq 0, \ s \geq 0\} \qquad (7.9)$$

is the set of *feasible points*, or simply the *feasible set*. Note that we are using a primal–dual feasibility concept here. More precisely, since x variables come from the primal QP and (y, s) come from the dual QP, we impose both primal and dual feasibility conditions in the definition of $\mathcal{F}$. If $(x, y, s) \in \mathcal{F}$ also satisfy $x > 0$ and $s > 0$ we say that (x, y, s) is a *strictly feasible solution* and define

$$\mathcal{F}^o := \{(x, y, s) : Ax = b, \ A^{\mathrm{T}}y - Qx + s = c, \ x > 0, \ s > 0\} \qquad (7.10)$$

to be the *strictly feasible set*. In mathematical terms, $\mathcal{F}^o$ is the *relative interior* of the set $\mathcal{F}$.

IPMs we discuss here will generate iterates (x^k, y^k, s^k) that all lie in $\mathcal{F}^o$. Since we are generating iterates for both the primal and dual problems, these IPMs are often called *primal–dual interior-point methods*. Using this approach, we will obtain solutions for both the primal and dual problems at the end of the solution procedure. Solving the dual may appear to be a waste of time since we are only interested in the solution of the primal problem. However, years of computational experience demonstrated that primal–dual IPMs lead to the most efficient and robust implementations of the interior-point approach. Intuitively speaking, this happens because having some partial information on the dual problem in the form of the dual iterates (y^k, s^k) helps us make better and faster improvements on the iterates of the primal problem.

Iterative optimization algorithms have two essential components:

- a measure that can be used to evaluate and compare the quality of alternative solutions and search directions;
- a method to generate a better solution, with respect to the measure just mentioned, from a nonoptimal solution.

As we stated before, IPMs rely on Newton's method to generate new estimates of the solutions. Let us discuss this more in depth. Ignore the inequality constraints in

(7.8) for a moment, and focus on the nonlinear system of equations $F(x, y, s) = 0$. Assume that we have a current estimate (x^k, y^k, s^k) of the optimal solution to the problem. The Newton step from this point is determined by solving the following system of linear equations:

$$J(x^k, y^k, s^k) \begin{bmatrix} \Delta x^k \\ \Delta y^k \\ \Delta s^k \end{bmatrix} = -F(x^k, y^k, s^k), \tag{7.11}$$

where $J(x^k, y^k, s^k)$ is the Jacobian of the function F and $[\Delta x^k, \Delta y^k, \Delta s^k]^{\mathrm{T}}$ is the search direction. First, we observe that

$$J(x^k, y^k, s^k) = \begin{bmatrix} -Q & A^{\mathrm{T}} & I \\ A & 0 & 0 \\ S^k & 0 & X^k \end{bmatrix}, \tag{7.12}$$

where X^k and S^k are diagonal matrices with the components of the vectors x^k and s^k along their diagonals. Furthermore, if $(x^k, y^k, s^k) \in \mathcal{F}^o$, then

$$F(x^k, y^k, s^k) = \begin{bmatrix} 0 \\ 0 \\ X^k S^k e \end{bmatrix} \tag{7.13}$$

and the Newton equation reduces to

$$\begin{bmatrix} -Q & A^{\mathrm{T}} & I \\ A & 0 & 0 \\ S^k & 0 & X^k \end{bmatrix} \begin{bmatrix} \Delta x^k \\ \Delta y^k \\ \Delta s^k \end{bmatrix} = \begin{bmatrix} 0 \\ 0 \\ -X^k S^k e \end{bmatrix}. \tag{7.14}$$

Exercise 7.4 Consider the quadratic programming problem given in Exercise 7.3 and the current primal–dual estimate of the solution $x^k = (1/3, 1/3, 1/3)^{\mathrm{T}}$, $y^k = (1, 1/2)^{\mathrm{T}}$, and $s^k = (1/2, 1/2, 2)^{\mathrm{T}}$. Is $(x^k, y^k, s^k) \in \mathcal{F}$? How about $\mathcal{F}^o$? Form and solve the Newton equation for this problem at (x^k, y^k, s^k).

In the standard Newton method, once a Newton step is determined in this manner, it is added to the current iterate to obtain the new iterate. In our case, this action may not be permissible, since the Newton step may take us to a new point that does not necessarily satisfy the nonnegativity constraints $x \geq 0$ and $s \geq 0$. In our modification of Newton's method, we want to avoid such violations and therefore will seek a *step-size parameter* $\alpha_k \in (0, 1]$ such that $x^k + \alpha_k \Delta x^k > 0$ and $s^k + \alpha_k \Delta s^k > 0$. Note that the largest possible value of α_k satisfying these restrictions can be found using a procedure similar to the ratio test in the simplex method. Once we determine the step-size parameter, we choose the next iterate as

$$(x^{k+1}, y^{k+1}, s^{k+1}) = (x^k, y^k, s^k) + \alpha_k (\Delta x^k, \Delta y^k, \Delta s^k).$$

If a value of α_k results in a next iterate $(x^{k+1}, y^{k+1}, s^{k+1})$ that is also in $\mathcal{F}^o$, we say that this value of α_k is *permissible*.

Exercise 7.5 What is the largest permissable step-size α_k for the Newton direction you found in Exercise 7.4 ?

A naive modification of Newton's method as we described above is, unfortunately, not very good in practice since the permissible values of α_k eventually become too small and the progress toward the optimal solution stalls. Therefore, one needs to modify the search direction as well as adjusting the step size along the direction. The usual Newton search direction obtained from (7.14) is called the *pure* Newton direction. We will consider modifications of pure Newton directions called *centered* Newton directions. To describe such directions, we first need to discuss the concept of the *central path*.

7.3.2 The central path

The central path $\mathcal{C}$ is a trajectory in the relative interior of the feasible region $\mathcal{F}^o$ that is very useful for both the theoretical study and also the implementation of IPMs. This trajectory is parameterized by a scalar $\tau > 0$, and the points (x_τ, y_τ, s_τ) on the central path are obtained as solutions of the following system:

$$
F(x_\tau, y_\tau, s_\tau) = \begin{bmatrix} 0 \\ 0 \\ \tau e \end{bmatrix}, \quad (x_\tau, s_\tau) > 0. \tag{7.15}
$$

Then, the central path $\mathcal{C}$ is defined as

$$
\mathcal{C} = \{(x_\tau, y_\tau, s_\tau) : \tau > 0\}. \tag{7.16}
$$

The third block of equations in (7.15) can be rewritten as

$$
(x_\tau)_i (s_\tau)_i = \tau, \quad \forall i.
$$

The similarities between (7.8) and (7.15) are evident. Note that, instead of requiring that x and s are complementary vectors as in the optimality conditions (7.8), we require their component products to be all equal. Note that, as $\tau \to 0$, the conditions (7.15) defining the points on the central path approximate the set of optimality conditions (7.8) more and more closely.

The system (7.15) has a unique solution for every $\tau > 0$, provided that $\mathcal{F}^o$ is nonempty. Furthermore, when $\mathcal{F}^o$ is nonempty, the trajectory (x_τ, y_τ, s_τ) converges to an optimal solution of the problem (7.1). Figure 7.2 depicts a sample feasible set and its central path.

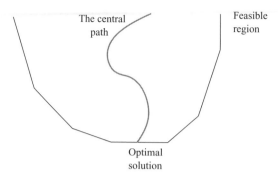

Figure 7.2 The central path

Exercise 7.6 Recall the quadratic programming problem given in Exercise 7.3 and the current primal–dual estimate of the solution $x^k = (1/3, 1/3, 1/3)^T$, $y^k = (1, 1/2)^T$, and $s^k = (1/2, 1/2, 2)^T$. Verify that (x^k, y^k, s^k) is not on the central path. Find a vector $\hat{x}$ such that $(\hat{x}, y^k, s^k)$ is on the central path. What value of τ does this primal–dual solution correspond to?

7.3.3 Path-following algorithms

The observation that points on the central path converge to optimal solutions of the primal–dual pair of quadratic programming problems suggests the following strategy for finding such solutions: in an iterative manner, generate points that approximate central points for decreasing values of the parameter τ. Since the central path converges to an optimal solution of the QP problem, these approximations to central points should also converge to a desired solution. This simple idea is the basis of *path-following* interior-point algorithms for optimization problems.

The strategy we outlined in the previous paragraph may appear confusing in a first reading. For example, one might ask why we do not approximate or find the solutions of the optimality system (7.8) directly rather than generating all these intermediate iterates leading to such a solution. Or, one might wonder why we would want to find approximations to central points, rather than central points themselves. Let us respond to these potential questions. First of all, there is no good and computationally cheap way of solving (7.8) directly since it involves nonlinear equations of the form $x_i s_i = 0$. As we discussed above, if we apply Newton's method to the equations in (7.8), we run into trouble because of the additional nonnegativity constraints. While we also have bilinear equations in the system defining the central points, being somewhat safely away from the boundaries defined by nonnegativity constraints, central points can be computed without most

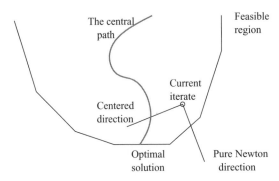

Figure 7.3 Pure and centered Newton directions

of the difficulties encountered in solving (7.8) directly. This is why we use central points for guidance.

Instead of insisting that we obtain a point exactly on the central path, we are often satisfied with an approximation to a central point for reasons of computational efficiency. Central points are also defined by systems of nonlinear equations and additional nonnegativity conditions. Solving these systems exactly (or very accurately) can be as hard as solving the optimality system (7.8) and therefore would not be an acceptable alternative for a practical implementation. It is, however, relatively easy to find a well-defined approximation to central points – see the definition of the neighborhoods of the central path below – especially those that correspond to larger values of τ. Once we identify a point close to a central point on $\mathcal{C}$, we can do a clever and inexpensive search to find another point which is close to another central point on $\mathcal{C}$, corresponding to a smaller value of τ. Furthermore, this idea can be used repeatedly, resulting in approximations to central points with progressively smaller τ values, allowing us to approach an optimal solution of the QP we are trying to solve. This is the essence of the path-following strategies.

7.3.4 Centered Newton directions

We say that a Newton step used in an interior-point method is a *pure Newton step* if it is a step directed toward the optimal point satisfying $F(x, y, s) = [0, 0, 0]^T$. Especially at points close to the boundary of the feasible set, pure Newton directions can be poor search directions as they may point to the exterior of the feasible set and lead to small admissible step sizes. To avoid such behavior, most interior-point methods take a step toward points on the central path $\mathcal{C}$ corresponding to predetermined value of τ. Since such directions are aiming for central points, they are called *centered directions*. Figure 7.3 depicts a pure and centered Newton direction from a sample iterate.

A centered direction is obtained by applying a Newton update to the following system:

$$\hat{F}(x, y, s) = \begin{bmatrix} A^T y - Qx + s - c \\ Ax - b \\ XSe - \tau e \end{bmatrix} = \begin{bmatrix} 0 \\ 0 \\ 0 \end{bmatrix}. \tag{7.17}$$

Since the Jacobian of $\hat{F}$ is identical to the Jacobian of F, proceeding as in equations (7.11)–(7.14), we obtain the following (modified) Newton equation for the centered direction:

$$\begin{bmatrix} -Q & A^T & I \\ A & 0 & 0 \\ S^k & 0 & X^k \end{bmatrix} \begin{bmatrix} \Delta x_c^k \\ \Delta y_c^k \\ \Delta s_c^k \end{bmatrix} = \begin{bmatrix} 0 \\ 0 \\ \tau e - X^k S^k e \end{bmatrix}. \tag{7.18}$$

We used the subscript c with the direction vectors to note that they are centered directions. Notice the similarity between (7.14) and (7.18).

One crucial choice we need to make is the value of τ to be used in determining the centered direction. To illustrate potential strategies for this choice, we first define the following measure, often called the duality gap, or the average complementarity:

$$\mu = \mu(x, s) := \frac{\sum_{i=1}^n x_i s_i}{n} = \frac{x^T s}{n}. \tag{7.19}$$

Note that, when (x, y, s) satisfy the conditions $Ax = b, x \geq 0$ and $A^T y - Qx + s = c, s \geq 0$, then (x, y, s) are optimal if and only if $\mu(x, s) = 0$. If μ is large, then we are far away from the solution. Therefore, μ serves as a measure of optimality for feasible points – the smaller the duality gap, the closer the point to optimality.

For a central point (x_τ, y_τ, s_τ) we have

$$\mu(x_\tau, s_\tau) = \frac{\sum_{i=1}^n (x_\tau)_i (s_\tau)_i}{n} = \frac{\sum_{i=1}^n \tau}{n} = \tau.$$

Because of this identity, we associate the central point (x_τ, y_τ, s_τ) with all feasible points (x, y, s) satisfying $\mu(x, s) = \tau$. All such points can be regarded as being at the same "level" as the central point (x_τ, y_τ, s_τ). When we choose a centered direction from a current iterate (x, y, s), we have the possibility of targeting a central point that is (i) at a lower level than our current point $(\tau < \mu(x, s))$, (ii) at the same level as our current point $(\tau = \mu(x, s))$, or (iii) at a higher level than our current point $(\tau > \mu(x, s))$. In most circumstances, the third option is not a good choice as it targets a central point that is "farther" than the current iterate to the optimal solution. Therefore, we will always choose $\tau \leq \mu(x, s)$ in defining centered directions. Using a simple change in notation, the centered direction can now be described as the solution of the following system:

$$
\begin{bmatrix} -Q & A^{\mathrm{T}} & I \\ A & 0 & 0 \\ S^k & 0 & X^k \end{bmatrix} \begin{bmatrix} \Delta x_c^k \\ \Delta y_c^k \\ \Delta s_c^k \end{bmatrix} = \begin{bmatrix} 0 \\ 0 \\ \sigma^k \mu^k e - X^k S^k e \end{bmatrix}, \tag{7.20}
$$

where $\mu^k := \mu(x^k, s^k) = (x^k)^{\mathrm{T}} s^k / n$ and $\sigma^k \in [0, 1]$ is a user-defined quantity describing the ratio of the duality gap at the target central point and the current point.

When $\sigma^k = 1$ (equivalently, $\tau = \mu^k$ in our earlier notation), we have a *pure centering direction*. This direction does not improve the duality gap and targets the central point whose duality gap is the same as our current iterate. Despite the lack of progress in terms of the duality gap, these steps are often desirable since large step sizes are permissible along such directions and points get well-centered so that the next iteration can make significant progress toward optimality. At the other extreme, we have $\sigma^k = 0$. This, as we discussed before, corresponds to the pure Newton step, also called the *affine-scaling direction*. Practical implementations often choose intermediate values for σ^k.

We are now ready to describe a generic interior-point algorithm that uses centered directions:

Algorithm 7.1 *Generic interior-point algorithm*

0. *Choose* $(x^0, y^0, s^0) \in \mathcal{F}^o$. *For* $k = 0, 1, 2, \ldots$ *repeat the following steps.*
1. *Choose* $\sigma^k \in [0, 1]$, *let* $\mu^k = (x^k)^{\mathrm{T}} s^k / n$. *Solve*

$$
\begin{bmatrix} -Q & A^{\mathrm{T}} & I \\ A & 0 & 0 \\ S^k & 0 & X^k \end{bmatrix} \begin{bmatrix} \Delta x^k \\ \Delta y^k \\ \Delta s^k \end{bmatrix} = \begin{bmatrix} 0 \\ 0 \\ \sigma^k \mu^k e - X^k S^k e \end{bmatrix}.
$$

2. *Choose* α^k *such that*

$$
x^k + \alpha^k \Delta x^k > 0, \quad \text{and } s^k + \alpha^k \Delta s^k > 0.
$$

Set

$$
(x^{k+1}, y^{k+1}, s^{k+1}) = (x^k, y^k, s^k) + \alpha_k (\Delta x^k, \Delta y^k, \Delta s^k),
$$

and $k = k + 1$.

Exercise 7.7 Compute the centered Newton direction for the iterate in Exercise 7.4 for $\sigma^k = 1, 0.5,$ and 0.1. For each σ^k, compute the largest permissible step size along the computed centered direction and compare your findings with that of Exercise 7.5.

7.3.5 Neighborhoods of the central path

Variants of interior-point methods differ in the way they choose the centering parameter σ^k and the step-size parameter α^k in each iteration. Path-following methods aim to generate iterates that are approximations to central points. This is achieved by a careful selection of the centering and step-size parameters. Before we discuss the selection of these parameters we need to make the notion of "approximate central points" more precise.

Recall that central points are those in the set $\mathcal{F}^o$ that satisfy the additional conditions that $x_i s_i = \tau$, $\forall i$, for some positive τ. Consider a central point (x_τ, y_τ, s_τ). If a point (x, y, s) approximates this central point, we would expect the Euclidean distance between these two points to be small, i.e., we would expect

$$\|(x, y, s) - (x_\tau, y_\tau, s_\tau)\|$$

to be small. Then, the set of approximations to (x_τ, y_τ, s_τ) may be defined as:

$$\{(x, y, s) \in \mathcal{F}^o : \|(x, y, s) - (x_\tau, y_\tau, s_\tau)\| \le \varepsilon\}, \tag{7.21}$$

for some $\varepsilon \ge 0$. Note, however, that it is difficult to obtain central points explicitly. Instead, we have their implicit description through the system (7.17). Therefore, a description such as (7.21) is of little practical/algorithmic value when we do not know (x_τ, y_τ, s_τ). Instead, we consider descriptions of sets that *imply* proximity to central points. Such descriptions are often called the *neighborhoods* of the central path. Two of the most commonly used neighborhoods of the central path are:

$$\mathcal{N}_2(\theta) := \left\{(x, y, s) \in \mathcal{F}^o : \|XSe - \mu e\| \le \theta\mu, \ \mu = \frac{x^\mathsf{T} s}{n}\right\}, \tag{7.22}$$

for some $\theta \in (0, 1)$ and

$$\mathcal{N}_{-\infty}(\gamma) := \left\{(x, y, s) \in \mathcal{F}^o : x_i s_i \ge \gamma\mu \ \forall i, \ \mu = \frac{x^\mathsf{T} s}{n}\right\}, \tag{7.23}$$

for some $\gamma \in (0, 1)$. The first neighborhood is called the 2-norm neighborhood while the second one is the one-sided ∞-norm neighborhood (but often called the $-\infty$-norm neighborhood, hence the notation). One can guarantee that the generated iterates are "close" to the central path by making sure that they all lie in one of these neighborhoods. If we choose $\theta = 0$ in (7.22) or $\gamma = 1$ in (7.23), the neighborhoods we defined degenerate to the central path $\mathcal{C}$.

Exercise 7.8   Show that $\mathcal{N}_2(\theta_1) \subset \mathcal{N}_2(\theta_2)$ when $0 < \theta_1 \le \theta_2 < 1$, and that $\mathcal{N}_{-\infty}(\gamma_1) \subset \mathcal{N}_{-\infty}(\gamma_2)$ for $0 < \gamma_2 \le \gamma_1 < 1$.

Exercise 7.9   Show that $\mathcal{N}_2(\theta) \subset \mathcal{N}_{-\infty}(\gamma)$ if $\gamma \le 1 - \theta$.

As hinted in the last exercise, for typical values of θ and γ, the 2-norm neighborhood is often much smaller than the $-\infty$-norm neighborhood. Indeed,

$$\|XSe - \mu e\| \le \theta\mu \Leftrightarrow \left\| \begin{array}{c} \frac{x_1 s_1}{\mu} - 1 \\ \frac{x_2 s_2}{\mu} - 1 \\ \vdots \\ \frac{x_n s_n}{\mu} - 1 \end{array} \right\| \le \theta, \tag{7.24}$$

which, in turn, is equivalent to

$$\sum_{i=1}^{n} \left(\frac{x_i s_i}{\mu} - 1 \right)^2 \le \theta^2.$$

In this last expression, the quantity $(x_i s_i / \mu) - 1 = (x_i s_i - \mu)/\mu$ is the relative deviation of $x_i s_i$'s from their average value μ. Therefore, a point is in the 2-norm neighborhood only if the sum of the squared relative deviations is small. Thus, $\mathcal{N}_2(\theta)$ contains only a small fraction of the feasible points, even when θ is close to 1. On the other hand, for the $-\infty$-norm neighborhood, the only requirement is that each $x_i s_i$ should not be much smaller than their average value μ. For small (but positive) γ, $\mathcal{N}_{-\infty}(\gamma)$ may contain almost the entire set $\mathcal{F}^o$.

In summary, 2-norm neighborhoods are narrow while the $-\infty$-norm neighborhoods are relatively wide. The practical consequence of this observation is that, when we restrict our iterates to be in the 2-norm neighborhood of the central path as opposed to the $-\infty$-norm neighborhood, we have much less room to maneuver and our step-sizes may be cut short. Figure 7.4 illustrates this behavior. For these reasons, algorithms using the narrow 2-norm neighborhoods are often called *short-step path-following methods* while the methods using the wide $-\infty$-norm neighborhoods are called *long-step path-following methods*.

The price we pay for the additional flexibility with wide neighborhoods comes in the theoretical worst-case analysis of convergence. When the iterates are restricted to the 2-norm neighborhood, we have a stronger control of the iterates as they are very close to the central path – a trajectory with many desirable theoretical features. Consequently, we can guarantee that even in the worst case the iterates that lie in the 2-norm neighborhood will converge to an optimal solution relatively fast. In contrast, iterates that are only restricted to a $-\infty$-norm neighborhood can get relatively far away from the central path and may not possess its nice theoretical properties. As a result, iterates may "get stuck" in undesirable corners of the feasible set and the convergence may be slow in these worst-case scenarios. Of course, the worst-case scenarios rarely happen and typically (on average) we see faster convergence with long-step methods than with short-step methods.

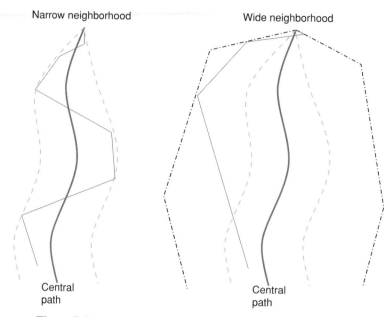

Figure 7.4 Narrow and wide neighborhoods of the central path

Exercise 7.10   Verify that the iterate given in Exercise 7.4 is in $N_{-\infty}(1/2)$. What is the largest γ such that this iterate lies in $N_{-\infty}(\gamma)$?

Exercise 7.11   Recall the centered Newton directions in Exercise 7.7 as well as the pure Newton direction in Exercise 7.5. For each direction, compute the largest α_k such that the updated iterate remains in $N_{-\infty}(1/2)$.

7.3.6 A long-step path-following algorithm

Next, we formally describe a long-step path-following algorithm that specifies some of the parameter choices of the generic algorithm we described above.

Algorithm 7.2   *Long-step path-following algorithm*

0. Given $\gamma \in (0, 1), 0 < \sigma_{\min} < \sigma_{\max} < 1$,   choose $(x^0, y^0, s^0) \in N_{-\infty}(\gamma)$. For $k = 0, 1, 2, \dots$ repeat the following steps.

1. Choose $\sigma^k \in [\sigma_{\min}, \sigma_{\max}]$, let $\mu^k = \frac{(x^k)^{\mathrm{T}} s^k}{n}$. Solve

$$\begin{bmatrix} -Q & A^{\mathrm{T}} & I \\ A & 0 & 0 \\ S^k & 0 & X^k \end{bmatrix} \begin{bmatrix} \Delta x^k \\ \Delta y^k \\ \Delta s^k \end{bmatrix} = \begin{bmatrix} 0 \\ 0 \\ \sigma^k \mu^k e - X^k S^k e \end{bmatrix}.$$

2. Choose α^k such that

$$(x^k, y^k, s^k) + \alpha_k(\Delta x^k, \Delta y^k, \Delta s^k) \in N_{-\infty}(\gamma).$$

Set

$$(x^{k+1}, y^{k+1}, s^{k+1}) = (x^k, y^k, s^k) + \alpha_k(\Delta x^k, \Delta y^k, \Delta s^k),$$

and $k = k + 1$.

7.3.7 Starting from an infeasible point

Both the generic interior-point method and the long-step path-following algorithm we described above require that one starts with a strictly feasible iterate. This requirement is not practical since finding such a starting point is not always a trivial task. Fortunately, however, we can accommodate infeasible starting points in these algorithms with a small modification of the linear system that we solve in each iteration.

For this purpose, we only require that the initial point (x^0, y^0, s^0) satisfy the nonnegativity restrictions strictly: $x^0 > 0$ and $s^0 > 0$. Such points can be generated trivially. We are still interested in solving the following nonlinear system:

$$\hat{F}(x, y, s) = \begin{bmatrix} A^T y - Qx + s - c \\ Ax - b \\ XSe - \tau e \end{bmatrix} = \begin{bmatrix} 0 \\ 0 \\ 0 \end{bmatrix}, \tag{7.25}$$

as well as $x \geq 0$, $s \geq 0$. As in (5.7), the Newton step from an infeasible point (x^k, y^k, s^k) is determined by solving the following system of linear equations:

$$J(x^k, y^k, s^k) \begin{bmatrix} \Delta x^k \\ \Delta y^k \\ \Delta s^k \end{bmatrix} = -\hat{F}(x^k, y^k, s^k), \tag{7.26}$$

which reduces to

$$\begin{bmatrix} -Q & A^T & I \\ A & 0 & 0 \\ S^k & 0 & X^k \end{bmatrix} \begin{bmatrix} \Delta x^k \\ \Delta y^k \\ \Delta s^k \end{bmatrix} = \begin{bmatrix} c + Qx^k - A^T y^k - s^k \\ b - Ax^k \\ \tau e - X^k S^k e \end{bmatrix}. \tag{7.27}$$

We no longer have zeros in the first and second blocks of the right-hand-side vector since we are not assuming that the iterates satisfy $Ax^k = b$ and $A^T y^k - Qx^k + s^k = c$. Replacing the linear system in the two algorithm descriptions above with (7.27) we obtain versions of these algorithms that work with infeasible iterates. In these versions of the algorithms, search for feasibility and optimality are performed simultaneously.

7.4 QP software

As for linear programs, there are several software options for solving practical quadratic programming problems. Many of the commercial software options are very efficient and solve very large QPs within seconds or minutes. A survey of nonlinear programming software, which includes software designed for QPs, can be found at www.lionhrtpub.com/orms/surveys/nlp/nlp.html.

The *Network Enabled Optimization Server (NEOS)* website and the *Optimization Software Guide* website we mentioned when we discussed NLP software are also useful for QP solvers. LOQO is a very efficient and robust interior-point based software for QPs and other nonlinear programming problems. It is available from www.orfe.princeton.edu/~loqo.

OOQP is an object-oriented C++ package, based on a primal-dual interior-point method, for solving convex quadratic programming problems. It contains code that can be used "out of the box" to solve a variety of structured QPs, including general sparse QPs, QPs arising from support vector machines, Huber regression problems, and QPs with bound constraints. It is available for free from www.cs.wisc.edu/~swright/ooqp.

7.5 Additional exercises

Exercise 7.12 In the study of interior-point methods for solving quadratic programming problems we encountered the following matrix:

$$M := \begin{bmatrix} -Q & A^{\mathrm{T}} & I \\ A & 0 & 0 \\ S^k & 0 & X^k \end{bmatrix},$$

where (x^k, y^k, s^k) is the current iterate, X^k and S^k are diagonal matrices with the components of the vectors x^k and s^k along their diagonals. Recall that M is the Jacobian matrix of the function that defines the optimality conditions of the QP problem. This matrix appears in linear systems we need to solve in each interior-point iteration. We can solve these systems only when M is nonsingular. Show that M is necessarily nonsingular when A has full row rank and Q is positive semidefinite. Provide an example with a Q matrix that is not positive semidefinite (but A matrix has full row rank) such that M is singular. (Hint: To prove non-singularity of M when Q is positive semidefinite and A has full row rank, consider a solution of the system

$$\begin{bmatrix} -Q & A^{\mathrm{T}} & I \\ A & 0 & 0 \\ S^k & 0 & X^k \end{bmatrix} \begin{bmatrix} \Delta x \\ \Delta y \\ \Delta s \end{bmatrix} = \begin{bmatrix} 0 \\ 0 \\ 0 \end{bmatrix}.$$

It is sufficient to show that the only solution to this system is $\Delta x = 0$, $\Delta y = 0$, $\Delta s = 0$. To prove this, first eliminate Δs variables from the system, and then eliminate Δx variables.)

Exercise 7.13 Consider the following quadratic programming formulation obtained from a small portfolio selection model:

$$\min_x [x_1 \; x_2 \; x_3 \; x_4] \begin{bmatrix} 0.01 & 0.005 & 0 & 0 \\ 0.005 & 0.01 & 0 & 0 \\ 0 & 0 & 0.04 & 0 \\ 0 & 0 & 0 & 0 \end{bmatrix} \begin{bmatrix} x_1 \\ x_2 \\ x_3 \\ x_4 \end{bmatrix}$$

$$x_1 + x_2 + x_3 = 1$$
$$-x_2 + x_3 + x_4 = 0.1$$
$$x_1 \geq 0, \; x_2 \geq 0, \; x_3 \geq 0, \; x_4 \geq 0.$$

We have the following iterate for this problem:

$$x = \begin{bmatrix} x_1 \\ x_2 \\ x_3 \\ x_4 \end{bmatrix} = \begin{bmatrix} 1/3 \\ 1/3 \\ 1/3 \\ 0.1 \end{bmatrix}, \quad y = \begin{bmatrix} y_1 \\ y_2 \end{bmatrix} = \begin{bmatrix} 0.001 \\ -0.001 \end{bmatrix}, \quad s = \begin{bmatrix} s_1 \\ s_2 \\ s_3 \\ s_4 \end{bmatrix} = \begin{bmatrix} 0.004 \\ 0.003 \\ 0.0133 \\ 0.001 \end{bmatrix}.$$

Verify that $(x, y, s) \in \mathcal{F}^o$. Is this point on the central path? Is it on $\mathcal{N}_{-\infty}(0.1)$? How about $\mathcal{N}_{-\infty}(0.05)$? Compute the pure centering ($\sigma = 1$) and pure Newton ($\sigma = 0$) directions from this point. For each direction, find the largest step-size α that can be taken along that direction without leaving the neighborhood $\mathcal{N}_{-\infty}(0.05)$? Comment on your results.

Exercise 7.14 Implement the long-step path-following algorithm given in Section 7.3.6 using $\sigma_{\min} = 0.2$, $\sigma_{\max} = 0.8$, $\gamma = 0.25$. Solve the quadratic programming problem in Exercise 7.13 starting from the iterate given in that exercise using your implementation. Experiment with alternative choices for $\sigma_{\min}, \sigma_{\max}$, and γ.

8

QP models: portfolio optimization

8.1 Mean-variance optimization

Markowitz' theory of mean-variance optimization (MVO) provides a mechanism for the selection of portfolios of securities (or asset classes) in a manner that trades off expected returns and risk. We explore this model in more detail in this chapter.

Consider assets $S_1, S_2, \ldots, S_n$ ($n \geq 2$) with random returns. Let μ_i and σ_i denote the expected return and the standard deviation of the return of asset S_i. For $i \neq j$, ρ_{ij} denotes the correlation coefficient of the returns of assets S_i and S_j. Let $\mu = [\mu_1, \ldots, \mu_n]^T$, and $\Sigma = (\sigma_{ij})$ be the $n \times n$ symmetric covariance matrix with $\sigma_{ii} = \sigma_i^2$ and $\sigma_{ij} = \rho_{ij}\sigma_i\sigma_j$ for $i \neq j$. Denoting by x_i the proportion of the total funds invested in security i, one can represent the expected return and the variance of the resulting portfolio $x = (x_1, \ldots, x_n)$ as follows:

$$E[x] = \mu_1 x_1 + \cdots + \mu_n x_n = \mu^T x,$$

and

$$\text{Var}[x] = \sum_{i,j} \rho_{ij}\sigma_i\sigma_j x_i x_j = x^T \Sigma x,$$

where $\rho_{ii} \equiv 1$.

Since variance is always nonnegative, it follows that $x^T \Sigma x \geq 0$ for any x, i.e., Σ is positive semidefinite. In this section, we will assume that it is in fact positive definite, which is essentially equivalent to assuming that there are no redundant assets in our collection $S_1, S_2, \ldots, S_n$. We further assume that the set of *admissible* portfolios is a nonempty polyhedral set and represent it as $\mathcal{X} := \{x : Ax = b, Cx \geq d\}$, where A is an $m \times n$ matrix, b is an m-dimensional vector, C is a $p \times n$ matrix, and d is a p-dimensional vector. In particular, one of the constraints in the set $\mathcal{X}$ is

$$\sum_{i=1}^{n} x_i = 1.$$

138

Linear portfolio constraints such as short-sale restrictions or limits on asset/sector allocations are subsumed in our generic notation $\mathcal{X}$ for the polyhedral feasible set.

Recall that a feasible portfolio x is called *efficient* if it has the maximal expected return among all portfolios with the same variance, or, alternatively, if it has the minimum variance among all portfolios that have at least a certain expected return. The collection of efficient portfolios form the *efficient frontier* of the portfolio universe. The efficient frontier is often represented as a curve in a two-dimensional graph where the coordinates of a plotted point corresponds to the standard deviation and the expected return of an efficient portfolio.

When we assume that Σ is positive definite, the variance is a strictly convex function of the portfolio variables and there exists a *unique* portfolio in $\mathcal{X}$ that has the minimum variance; see Exercise 7.2. Let us denote this portfolio with x_{min} and its return $\mu^T x_{min}$ with R_{min}. Note that x_{min} is an efficient portfolio. We let R_{max} denote the maximum return for an admissible portfolio.

Markowitz' *mean-variance optimization* (MVO) problem can be formulated in three different but equivalent ways. We have seen one of these formulations in the first chapter: find the minimum variance portfolio of the securities 1 to n that yields at least a target value of expected return (say b). Mathematically, this formulation produces a quadratic programming problem:

$$\begin{aligned}
\min_x \tfrac{1}{2}x^T\Sigma x \\
\mu^T x \geq R \\
Ax = b \\
Cx \geq d.
\end{aligned} \tag{8.1}$$

The first constraint indicates that the expected return is no less than the target value R. Solving this problem for values of R ranging between R_{min} and R_{max} one obtains all efficient portfolios. As we discussed above, the objective function corresponds to one half the total variance of the portfolio. The constant $1/2$ is added for convenience in the optimality conditions – it obviously does not affect the optimal solution.

This is a convex quadratic programming problem for which the first-order conditions are both necessary and sufficient for optimality. We present these conditions next. x_R is an optimal solution of problem (8.1) if and only if there exists $\lambda_R \in I\!R$, $\gamma_E \in I\!R^m$, and $\gamma_I \in I\!R^p$ satisfying the following KKT conditions:

$$\begin{aligned}
\Sigma x_R - \lambda_R \mu - A^T \gamma_E - C^T \gamma_I = 0, \\
\mu^T x_R \geq R, \quad Ax_R = b, \quad Cx_R \geq d, \\
\lambda_R \geq 0, \quad \lambda_R(\mu^T x_R - R) = 0, \\
\gamma_I \geq 0, \quad \gamma_I^T(Cx_R - d) = 0.
\end{aligned} \tag{8.2}$$

The two other variations of the MVO problem are the following:

$$
\begin{aligned}
\max_x \quad & \mu^T x \\
& x^T \Sigma x \le \sigma^2 \\
& Ax = b \\
& Cx \ge d,
\end{aligned}
\tag{8.3}
$$

and

$$
\begin{aligned}
\max_x \quad & \mu^T x - \frac{\delta}{2} x^T \Sigma x \\
& Ax = b \\
& Cx \ge d.
\end{aligned}
\tag{8.4}
$$

In (8.3), σ^2 is a given upper limit on the variance of the portfolio. In (8.4), the objective function is a *risk-adjusted return* function where the constant δ serves as a risk-aversion constant. While (8.4) is another quadratic programming problem, (8.3) has a convex quadratic constraint and therefore is not a QP. This problem can be solved using the general nonlinear programming solution techniques discussed in Chapter 5. We will also discuss a reformulation of (8.3) as a second-order cone program in Chapter 10. This opens the possibility of using specialized and efficient second-order cone programming methods for its solution.

Exercise 8.1   What are the Karush–Kuhn–Tucker optimality conditions for problems (8.3) and (8.4)?

Exercise 8.2   Consider the following variant of (8.4):

$$
\begin{aligned}
\max_x \quad & \mu^T x - \eta \sqrt{x^T \Sigma x} \\
& Ax = b \\
& Cx \ge d.
\end{aligned}
\tag{8.5}
$$

For each $\eta > 0$, let $x^*(\eta)$ denote the optimal solution of (8.5). Show that there exists a $\delta > 0$ such that $x^*(\eta)$ solves (8.4) for that δ.

8.1.1 Example

We apply Markowitz's MVO model to the problem of constructing a long-only portfolio of US stocks, bonds, and cash. We will use historical return data for these three asset classes to estimate their future expected returns. Note that most models for MVO combine historical data with other indicators such as earnings estimates, analyst ratings, valuation, and growth metrics, etc. Here we restrict our attention to price-based estimates for expositional simplicity. We use the S&P 500 Index for the returns on stocks, the 10-year Treasury Bond Index for the returns on bonds, and we assume that the cash is invested in a money market account whose return is the 1-day federal fund rate. The annual times series for the "total return" are given in Table 8.1 for each asset between 1960 and 2003.

Table 8.1 *Total returns for stocks, bonds, and money market*

Year	Stocks	Bonds	MM	Year	Stocks	Bonds	MM
1960	20.2553	262.935	100.00	1982	115.308	777.332	440.68
1961	25.6860	268.730	102.33	1983	141.316	787.357	482.42
1962	23.4297	284.090	105.33	1984	150.181	907.712	522.84
1963	28.7463	289.162	108.89	1985	197.829	1200.63	566.08
1964	33.4484	299.894	113.08	1986	234.755	1469.45	605.20
1965	37.5813	302.695	117.97	1987	247.080	1424.91	646.17
1966	33.7839	318.197	124.34	1988	288.116	1522.40	702.77
1967	41.8725	309.103	129.94	1989	379.409	1804.63	762.16
1968	46.4795	316.051	137.77	1990	367.636	1944.25	817.87
1969	42.5448	298.249	150.12	1991	479.633	2320.64	854.10
1970	44.2212	354.671	157.48	1992	516.178	2490.97	879.04
1971	50.5451	394.532	164.00	1993	568.202	2816.40	905.06
1972	60.1461	403.942	172.74	1994	575.705	2610.12	954.39
1973	51.3114	417.252	189.93	1995	792.042	3287.27	1007.84
1974	37.7306	433.927	206.13	1996	973.897	3291.58	1061.15
1975	51.7772	457.885	216.85	1997	1298.82	3687.33	1119.51
1976	64.1659	529.141	226.93	1998	1670.01	4220.24	1171.91
1977	59.5739	531.144	241.82	1999	2021.40	3903.32	1234.02
1978	63.4884	524.435	266.07	2000	1837.36	4575.33	1313.00
1979	75.3032	531.040	302.74	2001	1618.98	4827.26	1336.89
1980	99.7795	517.860	359.96	2002	1261.18	5558.40	1353.47
1981	94.8671	538.769	404.48	2003	1622.94	5588.19	1366.73

Let I_{it} denote the above "total return" for asset $i = 1, 2, 3$ and $t = 0, \ldots T$, where $t = 0$ corresponds to 1960 and $t = T$ to 2003. For each asset i, we can convert the raw data $I_{it}, t = 0, \ldots, T$, given in Table 8.1 into rates of returns $r_{it}, t = 1, \ldots, T$, using the formula

$$r_{it} = \frac{I_{i,t} - I_{i,t-1}}{I_{i,t-1}}.$$

These rates of returns are shown in Table 8.2. Let R_i denote the random rate of return of asset i. From the above historical data, we can compute the arithmetic mean rate of return for each asset:

$$\bar{r}_i = \frac{1}{T} \sum_{t=1}^{T} r_{it},$$

which gives:

	Stocks	Bonds	MM
Arithmetic mean $\bar{r}_i$	12.06%	7.85%	6.32%

Table 8.2 *Rates of return for stocks, bonds and money market*

Year	Stocks	Bonds	MM	Year	Stocks	Bonds	MM
1961	26.81	2.20	2.33	1983	22.56	1.29	9.47
1962	−8.78	5.72	2.93	1984	6.27	15.29	8.38
1963	22.69	1.79	3.38	1985	31.17	32.27	8.27
1964	16.36	3.71	3.85	1986	18.67	22.39	6.91
1965	12.36	0.93	4.32	1987	5.25	−3.03	6.77
1966	−10.10	5.12	5.40	1988	16.61	6.84	8.76
1967	23.94	−2.86	4.51	1989	31.69	18.54	8.45
1968	11.00	2.25	6.02	1990	−3.10	7.74	7.31
1969	−8.47	−5.63	8.97	1991	30.46	19.36	4.43
1970	3.94	18.92	4.90	1992	7.62	7.34	2.92
1971	14.30	11.24	4.14	1993	10.08	13.06	2.96
1972	18.99	2.39	5.33	1994	1.32	−7.32	5.45
1973	−14.69	3.29	9.95	1995	37.58	25.94	5.60
1974	−26.47	4.00	8.53	1996	22.96	0.13	5.29
1975	37.23	5.52	5.20	1997	33.36	12.02	5.50
1976	23.93	15.56	4.65	1998	28.58	14.45	4.68
1977	−7.16	0.38	6.56	1999	21.04	−7.51	5.30
1978	6.57	−1.26	10.03	2000	−9.10	17.22	6.40
1979	18.61	−1.26	13.78	2001	−11.89	5.51	1.82
1980	32.50	−2.48	18.90	2002	−22.10	15.15	1.24
1981	−4.92	4.04	12.37	2003	28.68	0.54	0.98
1982	21.55	44.28	8.95				

Since the rates of return are multiplicative over time, we prefer to use the geometric mean instead of the arithmetic mean. The *geometric mean* is the constant yearly rate of return that needs to be applied in years $t = 0$ through $t = T - 1$ in order to get the compounded total return I_{iT}, starting from I_{i0}. The formula for the geometric mean is:

$$\mu_i = \left(\prod_{t=1}^{T} (1 + r_{it}) \right)^{1/T} - 1.$$

We get the following results:

	Stocks	Bonds	MM
Geometric mean μ_i	10.73%	7.37%	6.27%

We also compute the covariance matrix:

$$\text{cov}(R_i, R_j) = \frac{1}{T} \sum_{t=1}^{T} (r_{it} - \bar{r}_i)(r_{jt} - \bar{r}_j).$$

Covariance	Stocks	Bonds	MM
Stocks	0.02778	0.00387	0.00021
Bonds	0.00387	0.01112	−0.00020
MM	0.00021	−0.00020	0.00115

It is interesting to compute the volatility of the rate of return on each asset $\sigma_i = \sqrt{\text{cov}(R_i, R_i)}$:

	Stocks	Bonds	MM
Volatility	16.67%	10.55%	3.40%

and the correlation matrix $\rho_{ij} = \frac{\text{cov}(R_i, R_j)}{\sigma_i \sigma_j}$:

Correlation	Stocks	Bonds	MM
Stocks	1	0.2199	0.0366
Bonds	0.2199	1	−0.0545
MM	0.0366	−0.0545	1

Setting up the QP for portfolio optimization:

$$\min 0.02778x_S^2 + 2 \cdot 0.00387x_S x_B + 2 \cdot 0.00021x_S x_M$$
$$+0.01112x_B^2 - 2 \cdot 0.00020x_B x_M + 0.00115x_M^2$$
$$0.1073x_S + 0.0737x_B + 0.0627x_M \geq R \qquad (8.6)$$
$$x_S + x_B + x_M = 1$$
$$x_S \geq 0, x_B \geq 0, x_M \geq 0,$$

and solving it for $R = 6.5\%$ to $R = 10.5\%$ with increments of 0.5%, we get the optimal portfolios shown in Table 8.3 and the corresponding variance. The optimal allocations on the efficient frontier are also depicted in Figure 8.1(b).

Based on the first two columns of Table 8.3, Figure 8.1(a) plots the maximum expected rate of return R of a portfolio as a function of its volatility (standard deviation). This curve is the *efficient frontier* we discussed earlier. Every possible portfolio consisting of long positions in stocks, bonds, and money market investments is represented by a point lying on or below the efficient frontier in the standard deviation/expected return plane.

Exercise 8.3 Solve Markowitz' MVO model for constructing a portfolio of US stocks, bonds, and cash using arithmetic means, instead of geometric means as above. Vary R from 6.5% to 12% with increments of 0.5%. Compare with the results obtained above.

Table 8.3 *Efficient portfolios*

Rate of return R	Variance	Stocks	Bonds	MM
0.065	0.0010	0.03	0.10	0.87
0.070	0.0014	0.13	0.12	0.75
0.075	0.0026	0.24	0.14	0.62
0.080	0.0044	0.35	0.16	0.49
0.085	0.0070	0.45	0.18	0.37
0.090	0.0102	0.56	0.20	0.24
0.095	0.0142	0.67	0.22	0.11
0.100	0.0189	0.78	0.22	0
0.105	0.0246	0.93	0.07	0

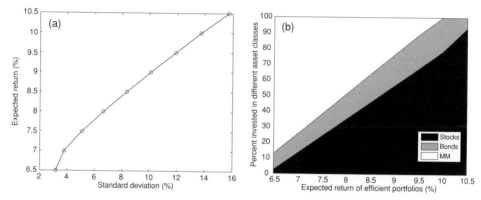

Figure 8.1 Efficient frontier and the composition of efficient portfolios

Exercise 8.4   In addition to the three securities given earlier (S&P 500 Index, 10-year Treasury Bond Index, and Money Market), consider a fourth security (the NASDAQ Composite Index) with the "total return" shown in Table 8.4.

Construct a portfolio consisting of the S&P 500 Index, the NASDAQ Index, the 10-year Treasury Bond Index and cash, using Markowitz's MVO model. Solve the model for different values of R.

Exercise 8.5   Repeat the previous exercise, this time assuming that one can leverage the portfolio up to 50% by borrowing at the money market rate. How do the risk/return profiles of optimal portfolios change with this relaxation? How do your answers change if the borrowing rate for cash is expected to be 1% higher than the lending rate?

8.1.2 Large-scale portfolio optimization

In this section, we consider practical issues that arise when the mean-variance model is used to construct a portfolio from a large underlying family of assets.

Table 8.4 *Total returns for the NASDAQ Composite Index*

Year	NASDAQ	Year	NASDAQ	Year	NASDAQ
1960	34.461	1975	77.620	1990	373.84
1961	45.373	1976	97.880	1991	586.34
1962	38.556	1977	105.05	1992	676.95
1963	46.439	1978	117.98	1993	776.80
1964	57.175	1979	151.14	1994	751.96
1965	66.982	1980	202.34	1995	1052.1
1966	63.934	1981	195.84	1996	1291.0
1967	80.935	1982	232.41	1997	1570.3
1968	101.79	1983	278.60	1998	2192.7
1969	99.389	1984	247.35	1999	4069.3
1970	89.607	1985	324.39	2000	2470.5
1971	114.12	1986	348.81	2001	1950.4
1972	133.73	1987	330.47	2002	1335.5
1973	92.190	1988	381.38	2003	2003.4
1974	59.820	1989	454.82		

For concreteness, let us consider a portfolio of stocks constructed from a set of n stocks with known expected returns and covariance matrix, where n may be in the hundreds or thousands.

Diversification

In general, there is no reason to expect that solutions to the Markowitz model will be well diversified portfolios. In fact, this model tends to produce portfolios with unreasonably large weights in certain asset classes and, when short positions are allowed, unintuitively large short positions. This issue is well documented in the literature, including the paper by Green and Hollifield [36] and is often attributed to estimation errors. Estimates that may be slightly "off" may lead the optimizer to chase phantom low-risk high-return opportunities by taking large positions. Hence, portfolios chosen by this quadratic program may be subject to idiosyncratic risk. Practitioners often use additional constraints on the x_i's to insure themselves against estimation and model errors and to ensure that the chosen portfolio is well diversified. For example, a limit m may be imposed on the size of each x_i, say

$$x_i \leq m \quad \text{for } i = 1, \ldots, n.$$

One can also reduce sector risk by grouping together investments in securities of a sector and setting a limit on the exposure to this sector. For example, if m_k is the maximum that can be invested in sector k, we add the constraint

$$\sum_{i \text{ in sector } k} x_i \leq m_k.$$

Note, however, that the more constraints one adds to a model, the more the objective value deteriorates. So the above approach to producing diversification, at least *ex ante*, can be quite costly.

Transaction costs

We can add a portfolio turnover constraint to ensure that the change between the current holdings x^0 and the desired portfolio x is bounded by h. This constraint is essential when solving large mean-variance models since the covariance matrix is almost singular in most practical applications and hence the optimal decision can change significantly with small changes in the problem data. To avoid big changes when reoptimizing the portfolio, turnover constraints are imposed. Let y_i be the amount of asset i bought and z_i the amount sold. We write

$$x_i - x_i^0 \le y_i, \quad y_i \ge 0,$$
$$x_i^0 - x_i \le z_i, \quad z_i \ge 0,$$

$$\sum_{i=1}^{n} (y_i + z_i) \le h.$$

Instead of a turnover constraint, we can introduce transaction costs directly into the model. Suppose that there is a transaction cost t_i proportional to the amount of asset i bought, and a transaction cost t_i' proportional to the amount of asset i sold. Suppose that the portfolio is reoptimized once per period. As above, let x^0 denote the current portfolio. Then a reoptimized portfolio is obtained by solving

$$\min \sum_{i=1}^{n} \sum_{j=1}^{n} \sigma_{ij} x_i x_j$$

subject to

$$\sum_{i=1}^{n} (\mu_i x_i - t_i y_i - t_i' z_i) \ge R$$

$$\sum_{i=1}^{n} x_i = 1$$

$$x_i - x_i^0 \le y_i \quad \text{for } i = 1, \dots, n,$$
$$x_i^0 - x_i \le z_i \quad \text{for } i = 1, \dots, n,$$
$$y_i \ge 0 \quad\quad\quad \text{for } i = 1, \dots, n,$$
$$z_i \ge 0 \quad\quad\quad \text{for } i = 1, \dots, n,$$
$$x_i \text{ unrestricted for } i = 1, \dots, n.$$

Parameter estimation

The Markowitz model gives us an *optimal* portfolio *assuming* that we have perfect information on the μ_i's and σ_{ij}'s for the assets that we are considering. Therefore, an important practical issue is the estimation of the μ_i's and σ_{ij}'s.

A reasonable approach for estimating these data is to use time series of past returns (r_{it} = return of asset i from time $t-1$ to time t, where $i = 1, \ldots, n$, $t = 1, \ldots, T$). Unfortunately, it has been observed that small changes in the time series r_{it} lead to changes in the μ_i's and σ_{ij}'s that often lead to significant changes in the "optimal" portfolio.

Markowitz recommends using the β's of the securities to calculate the μ_i's and σ_{ij}'s as follows. Let

$$r_{it} = \text{ return of asset } i \text{ in period } t, i = 1, \ldots, n, \text{ and } t = 1, \ldots, T,$$

$$r_{mt} = \text{ market return in period } t,$$

$$r_{ft} = \text{ return of risk-free asset in period } t.$$

We estimate β_i by a linear regression based on the capital asset pricing model

$$r_{it} - r_{ft} = \beta_i(r_{mt} - r_{ft}) + \varepsilon_{it}$$

where the vector ε_i represents the idiosyncratic risk of asset i. We assume that $\text{cov}(\varepsilon_i, \varepsilon_j) = 0$. The β's can also be purchased from financial research groups and risk model providers.

Knowing β_i, we compute μ_i by the relation

$$\mu_i - E(r_f) = \beta_i(E(r_m) - E(r_f)),$$

and σ_{ij} by the relation

$$\sigma_{ij} = \beta_i \beta_j \sigma_m^2 \quad \text{for } i \neq j,$$
$$\sigma_{ii} = \beta_i^2 \sigma_m^2 + \sigma_{\varepsilon_i}^2,$$

where σ_m^2 denotes the variance of the market return and $\sigma_{\varepsilon_i}^2$ the variance of the idiosyncratic return.

But the fundamental weakness of the Markowitz model remains, no matter how cleverly the μ_i's and σ_{ij}'s are computed: the solution is extremely sensitive to small changes in the data. Only one small change in one μ_i may produce a totally different portfolio x. What can be done in practice to overcome this problem, or at least reduce it? Michaud [57] recommends to resample returns from historical data to generate alternative μ and σ estimates, to solve the MVO problem repeatedly with inputs generated this way, and then to combine the optimal portfolios obtained in this manner. Robust optimization approaches provide an alternative strategy to mitigate the input sensitivity in MVO models; we discuss some examples in Chapters 19 and 20. Another interesting approach is considered in the next section.

Exercise 8.6   Express the following restrictions as linear constraints:

(i) The β of the portfolio should be between 0.9 and 1.1.
(ii) Assume that the stocks are partitioned by capitalization: large, medium, and small. We want the portfolio to be divided evenly between large and medium cap stocks, and the investment in small cap stocks to be between two and three times the investment in large cap stocks.

Exercise 8.7   Using historical returns of the stocks in the DJIA, estimate their mean μ_i and covariance matrix. Let R be the median of the μ_i's.

(i) Solve Markowitz' MVO model to construct a portfolio of stocks from the DJIA that has expected return at least R.
(ii) Generate a random value uniformly in the interval $[0.95\mu_i, 1.05\mu_i]$, for each stock i. Resolve Markowitz' MVO model with these mean returns, instead of μ_i's as in (i). Compare the results obtained in (i) and (ii).
(iii) Repeat three more times and average the five portfolios found in (i), (ii) and (iii). Compare this portfolio with the one found in (i).

8.1.3 The Black–Litterman model

Black and Litterman [14] recommend to combine the investor's view with the market equilibrium, as follows.

The expected return vector μ is assumed to have a probability distribution that is the product of two multivariate normal distributions. The first distribution represents the returns at market equilibrium, with mean π and covariance matrix $\tau\Sigma$, where τ is a small constant and $\Sigma = (\sigma_{ij})$ denotes the covariance matrix of asset returns. (Note that the factor τ should be small since the variance $\tau\sigma_i^2$ of the random variable μ_i is typically much smaller than the variance σ_i^2 of the underlying asset returns.) The second distribution represents the investor's view about the μ_i's. These views are expressed as

$$P\mu = q + \varepsilon,$$

where P is a $k \times n$ matrix and q is a k-dimensional vector that are provided by the investor and ε is a normally distributed random vector with mean 0 and diagonal covariance matrix Ω (the stronger the investor's view, the smaller the corresponding $\omega_i = \Omega_{ii}$).

The resulting distribution for μ is a multivariate normal distribution with mean

$$\bar{\mu} = [(\tau \Sigma)^{-1} + P^{\mathrm{T}} \Omega^{-1} P]^{-1} [(\tau \Sigma)^{-1} \pi + P^{\mathrm{T}} \Omega^{-1} q]. \tag{8.7}$$

Black and Litterman use $\bar{\mu}$ as the vector of expected returns in the Markowitz model.

Example 8.1 *Let us illustrate the Black–Litterman approach on the example of Section 8.1.1. The expected returns on Stocks, Bonds, and Money Market were computed to be*

	Stocks	Bonds	MM
Market rate of return	10.73%	7.37%	6.27%

This is what we use for the vector π representing market equilibrium. In practice, π is obtained from the vector of shares of global wealth invested in different asset classes via reverse optimization. We need to choose the value of the small constant τ. We take $\tau = 0.1$. We have two views that we would like to incorporate into the model. First, we hold a strong view that the Money Market rate will be 2% next year. Second, we also hold the view that S&P 500 will outperform 10-year Treasury Bonds by 5% but we are not as confident about this view. These two views can be expressed as follows

$$\mu_M = 0.02 \quad \text{strong view: } \omega_1 = 0.00001, \tag{8.8}$$

$$\mu_S - \mu_B = 0.05 \quad \text{weaker view: } \omega_2 = 0.001.$$

Thus $P = \begin{pmatrix} 0 & 0 & 1 \\ 1 & -1 & 0 \end{pmatrix}$, $q = \begin{pmatrix} 0.02 \\ 0.05 \end{pmatrix}$ *and* $\Omega = \begin{pmatrix} 0.00001 & 0 \\ 0 & 0.001 \end{pmatrix}$.

Applying formula (8.7) to compute $\bar{\mu}$, we get

	Stocks	Bonds	MM
Mean rate of return $\bar{\mu}$	11.77%	7.51%	2.34%

We solve the same QP as in (8.6) except for the modified expected return constraint:

$$\min 0.02778 x_S^2 + 2 \cdot 0.00387 x_S x_B + 2 \cdot 0.00021 x_S x_M$$

$$+0.01112 x_B^2 - 2 \cdot 0.00020 x_B x_M + 0.00115 x_M^2$$

$$0.1177 x_S + 0.0751 x_B + 0.0234 x_M \geq R \tag{8.9}$$

$$x_S + x_B + x_M = 1$$

$$x_S \geq 0, x_B \geq 0, x_M \geq 0$$

Solving for $R = 4.0\%$ to $R = 11.5\%$ with increments of 0.5% we now get the optimal portfolios and the efficient frontier depicted in Table 8.5 and Figure 8.2.

Table 8.5 *Black–Litterman efficient portfolios*

Rate of return R	Variance	Stocks	Bonds	MM
0.040	0.0012	0.08	0.17	0.75
0.045	0.0015	0.11	0.21	0.68
0.050	0.0020	0.15	0.24	0.61
0.055	0.0025	0.18	0.28	0.54
0.060	0.0032	0.22	0.31	0.47
0.065	0.0039	0.25	0.35	0.40
0.070	0.0048	0.28	0.39	0.33
0.075	0.0059	0.32	0.42	0.26
0.080	0.0070	0.35	0.46	0.19
0.085	0.0083	0.38	0.49	0.13
0.090	0.0096	0.42	0.53	0.05
0.095	0.0111	0.47	0.53	0
0.100	0.0133	0.58	0.42	0
0.105	0.0163	0.70	0.30	0
0.110	0.0202	0.82	0.18	0
0.115	0.0249	0.94	0.06	0

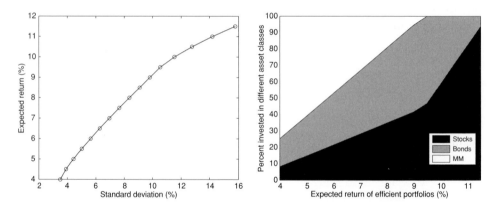

Figure 8.2 Efficient frontier and the composition of efficient portfolios using the Black–Litterman approach

Exercise 8.8 Repeat the example above, with the same investor's views, but adding the fourth security of Exercise 8.4 (the NASDAQ Composite Index).

Black and Litterman give the following intuition for their approach using the following example. Suppose we know the true structure of the asset returns: for each asset, the return is composed of an equilibrium risk premium plus a common factor and an independent shock:

$$R_i = \pi_i + \gamma_i Z + v_i$$

where

R_i = the return on the ith asset,

π_i = the equilibrium risk premium on the ith asset,

Z = a common factor,

γ_i = the impact of Z on the ith asset,

ν_i = an independent shock to the ith asset.

The covariance matrix Σ of asset returns is assumed to be known. The expected returns of the assets are given by:

$$\mu_i = \pi_i + \gamma_i E[Z] + E[\nu_i].$$

While a consideration of the equilibrium motivates the Black–Litterman model, they do not assume that $E[Z]$ and $E[\nu_i]$ are equal to 0, which would indicate that the expected excess returns are equal to the equilibrium risk premiums. Instead, they assume that the expected excess returns μ_i are unobservable random variables whose distribution is determined by the distribution of $E[Z]$ and $E[\nu_i]$. Their additional assumptions imply that the covariance matrix of expected returns is $\tau \Sigma$ for some small positive scalar τ. All this information is assumed to be known to all investors.

Investors differ in the additional, subjective informative they have about future returns. They express this information as their "views" such as "I expect that asset A will outperform asset B by 2%." Coupled with a measure of confidence, such views can be incorporated into the equilibrium returns to generate conditional distribution of the expected returns. For example, if we assume that the equilibrium distribution of μ is given by the normal distribution $N(\pi, \tau \Sigma)$ and views are represented using the constraint $P\mu = q$ (with 100% confidence), the mean $\bar{\mu}$ of the normal distribution conditional on this view is obtained as the optimal solution of the following quadratic optimization problem:

$$\begin{aligned} &\min \ (\mu - \pi)^T (\tau \Sigma)^{-1} (\mu - \pi) \\ &\text{s.t.} \ \ P\mu = q. \end{aligned} \tag{8.10}$$

Using the KKT optimality conditions presented in Section 5.5, the solution to the above minimization problem can be shown to be

$$\bar{\mu} = \pi + (\tau \Sigma) P^T [P(\tau \Sigma) P^T]^{-1} (q - P\pi). \tag{8.11}$$

Exercise 8.9 Prove that $\bar{\mu}$ in (8.11) solves (8.10) using KKT conditions.

Of course, an investor rarely has 100% confidence in his/her views. In the more general case, the views are expressed as $P\mu = q + \varepsilon$ where P and q are given by the investor as above and ε is an unobservable normally distributed random vector with mean 0 and diagonal covariance matrix Ω. A diagonal Ω corresponds to the assumption that the views are independent. When this is the case, $\bar{\mu}$ is given by the Black–Litterman formula

$$\bar{\mu} = [(\tau\Sigma)^{-1} + P^{\mathrm{T}}\Omega^{-1}P]^{-1}[(\tau\Sigma)^{-1}\pi + P^{\mathrm{T}}\Omega^{-1}q],$$

as stated earlier. We refer to the Black and Litterman paper for additional details and an example of an international portfolio [14].

Exercise 8.10   Repeat Exercise 8.4 , this time using the Black–Litterman methodology outlined above. Use the expected returns you computed in Exercise 8.4 as equilibrium returns and incorporate the view that NASDAQ stocks will outperform the S&P 500 stocks by 4% and that the average of NASDAQ and S&P 500 returns will exceed bond returns by 3%. Both views are relatively strong and are expressed with $\omega_1 = \omega_2 = 0.0001$.

8.1.4 Mean-absolute deviation to estimate risk

Konno and Yamazaki [46] propose a linear programming model instead of the classical quadratic model. Their approach is based on the observation that different measures of risk, such as volatility and L_1-risk, are closely related, and that alternate measures of risk are also appropriate for portfolio optimization.

The volatility of the portfolio return is

$$\sigma = \sqrt{E\left[\left(\sum_{i=1}^{n}(R_i - \mu_i)x_i\right)^2\right]},$$

where R_i denotes the random return of asset i and μ_i denotes its mean.

The L_1-risk of the portfolio return is defined as

$$w = E\left[\left|\sum_{i=1}^{n}(R_i - \mu_i)x_i\right|\right].$$

Theorem 8.1 (Konno and Yamazaki [43])   *If $(R_1, \ldots, R_n)$ are multivariate normally distributed random variables, then $w = \sqrt{2/\pi}\sigma$.*

Proof: Let $(\mu_1, \ldots, \mu_n)$ be the mean of $(R_1, \ldots, R_n)$. Also let $\Sigma = (\sigma_{ij}) \in I\!R^{n \times n}$ be the covariance matrix of $(R_1, \ldots, R_n)$. Then $\sum R_i x_i$ is normally distributed [66] with mean $\sum \mu_i x_i$ and standard deviation

$$\sigma(x) = \sqrt{\sum_i \sum_j \sigma_{ij} x_i x_j}.$$

Therefore $w = E[|U|]$ where $U \sim N(0, \sigma)$, and

$$w(x) = \frac{1}{\sqrt{2\pi}\sigma(x)} \int_{-\infty}^{+\infty} |u| e^{-\frac{u^2}{2\sigma^2(x)}} du = \frac{2}{\sqrt{2\pi}\sigma(x)} \int_0^{+\infty} u e^{-\frac{u^2}{2\sigma^2(x)}} du = \sqrt{\frac{2}{\pi}} \sigma(x).$$

$\square$

This theorem implies that minimizing σ is equivalent to minimizing w when $(R_1, \ldots, R_n)$ is multivariate normally distributed. With this assumption, the Markowitz model can be formulated as

$$\min E\left[\left|\sum_{i=1}^n (R_i - \mu_i) x_i\right|\right]$$

subject to

$$\sum_{i=1}^n \mu_i x_i \geq R$$

$$\sum_{i=1}^n x_i = 1$$

$$0 \leq x_i \leq m_i \quad \text{for } i = 1, \ldots, n.$$

Whether $(R_1, \ldots, R_n)$ has a multivariate normal distribution or not, the above mean-absolute deviation (MAD) model constructs efficient portfolios for the L_1-risk measure. Let r_{it} be the realization of random variable R_i during period t for $t = 1, \ldots, T$, which we assume to be available through the historical data or from future projection. Then

$$\mu_i = \frac{1}{T} \sum_{t=1}^T r_{it}.$$

Furthermore

$$E\left[\left|\sum_{i=1}^n (R_i - \mu_i) x_i\right|\right] = \frac{1}{T} \sum_{t=1}^T \left|\sum_{i=1}^n (r_{it} - \mu_i) x_i\right|.$$

Note that the absolute value in this expression makes it nonlinear. But it can be linearized using additional variables. Indeed, one can replace $|x|$ by $y + z$ where $x = y - z$ and $y, z \geq 0$. When the objective is to minimize $y + z$, at most one of y

Table 8.6 *Konno–Yamazaki efficient portfolios*

Rate of return R	Variance	Stocks	Bonds	MM
0.065	0.0011	0.05	0.01	0.94
0.070	0.0015	0.15	0.04	0.81
0.075	0.0026	0.25	0.11	0.64
0.080	0.0046	0.32	0.28	0.40
0.085	0.0072	0.42	0.32	0.26
0.090	0.0106	0.52	0.37	0.11
0.095	0.0144	0.63	0.37	0
0.100	0.0189	0.78	0.22	0
0.105	0.0246	0.93	0.07	0

or z will be positive. Therefore the model can be rewritten as

$$\min \sum_{t=1}^{T} y_t + z_t$$

subject to

$$y_t - z_t = \sum_{i=1}^{n} (r_{it} - \mu_i)x_i \quad \text{for } t = 1, \ldots, T,$$

$$\sum_{i=1}^{n} \mu_i x_i \geq R$$

$$\sum_{i=1}^{n} x_i = 1$$

$$0 \leq x_i \leq m_i \quad \text{for } i = 1, \ldots, n,$$

$$y_t \geq 0, \ z_t \geq 0 \quad \text{for } t = 1, \ldots, T.$$

This is a linear program! Therefore this approach can be used to solve large-scale portfolio optimization problems.

Example 8.2   *We illustrate the approach on our three-asset example, using the historical data on stocks, bonds, and cash given in Section 8.1.1. Solving the linear program for R = 6.5% to R = 10.5% with increments of 0.5% we get the optimal portfolios and the efficient frontier depicted in Table 8.6 and Figure 8.3.*

In Table 8.6, we computed the variance of the MAD portfolio for each level R of the rate of return. These variances can be compared with the results obtained in Section 8.1.1 for the MVO portfolio. As expected, the variance of a MAD portfolio is always at least as large as that of the corresponding MVO portfolio. Note, however, that the difference is small. This indicates that, although the normality assumption

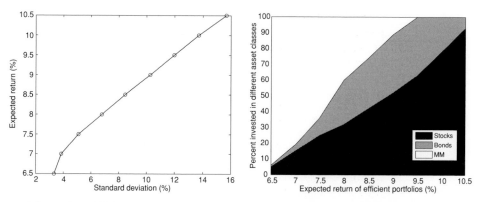

Figure 8.3 Efficient frontier and the composition of efficient portfolios using the Konno–Yamazaki approach

of Theorem 8.1 does not hold, minimizing the L_1-risk (instead of volatility) produces comparable portfolios.

Exercise 8.11   Add the fourth security of Exercise 8.4 (the NASDAQ Composite Index) to the three-asset example. Solve the resulting MAD model for varying values of R. Compare with the portfolios obtained in Exercise 8.4 .

We note that the portfolios generated using the mean-absolute deviation criteria has the additional property that they are never *stochastically dominated* [71]. This is an important property as a portfolio has second-order stochastic dominance over another one if and only if it is preferred to the other by any concave (risk-averse) utility function. By contrast, mean-variance optimization may generate optimal portfolios that are stochastically dominated. This and other criticisms of Markowitz' mean-variance optimization model we mentioned above led to the development of alternative formulations including the Black–Litterman and Konno–Yamazaki models as well as the robust optimization models we consider in Chapter 20. Steinbach provides an excellent review of Markowitz' mean-variance optimization model, its many variations and its extensions to multi-period optimization setting [77].

8.2 Maximizing the Sharpe ratio

Consider the setting in Section 8.1. Recall that we denote with $R_{\min}$ and $R_{\max}$ the minimum and maximum expected returns for efficient portfolios. Let us define the function

$$\sigma(R) : [R_{\min}, R_{\max}] \to I\!R, \quad \sigma(R) := \left(x_R^T \Sigma x_R\right)^{1/2},$$

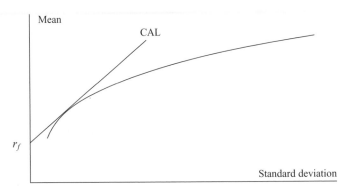

Figure 8.4 Capital allocation line

where x_R denotes the unique solution of problem (8.1). Since we assumed that Σ is positive definite, it is easy to show that the function $\sigma(R)$ is strictly convex in its domain. The efficient frontier is the graph

$$E = \{(R, \sigma(R)) : R \in [R_{\min}, R_{\max}]\}.$$

We now consider a *riskless asset* whose return is $r_f \geq 0$ with probability 1. We will assume that $r_f < R_{\min}$, which is natural since the portfolio $x_{\min}$ has a positive risk associated with it while the riskless asset does not.

Return/risk profiles of different combinations of a risky portfolio with the riskless asset can be represented as a straight line – a *capital allocation line (CAL)* – on the standard deviation vs. mean graph; see Figure 8.4. The optimal CAL is the CAL that lies above all the other CALs for $R > r_f$, since the corresponding portfolios will have the lowest standard deviation for any given value of $R > r_f$. Then, it follows that this optimal CAL goes through a point on the efficient frontier and never goes below a point on the efficient frontier. The point where the optimal CAL touches the efficient frontier corresponds to the optimal risky portfolio.

Alternatively, one can think of the optimal CAL as the CAL with the largest slope. Mathematically, this can be expressed as the portfolio x that maximizes the quantity

$$h(x) = \frac{\mu^{\mathrm{T}}x - r_f}{(x^{\mathrm{T}}\Sigma x)^{1/2}},$$

among all $x \in S$. This quantity is precisely the *reward-to-volatility ratio* introduced by Sharpe to measure the performance of mutual funds [76]. This quantity is now more commonly known as the Sharpe ratio. The portfolio that maximizes the Sharpe ratio is found by solving the following problem:

$$\max_x \frac{\mu^T x - r_f}{(x^T \Sigma x)^{1/2}}$$
$$Ax = b \tag{8.12}$$
$$Cx \geq d.$$

In this form, this problem is not easy to solve. Although it has a nice polyhedral feasible region, its objective function is somewhat complicated, and it is possibly non-concave. Therefore, (8.12) is not a convex optimization problem. The standard strategy to find the portfolio maximizing the Sharpe ratio, often called the *optimal risky portfolio*, is the following: First, one traces out the efficient frontier on a two dimensional return vs. standard deviation graph. Then, the point on this graph corresponding to the optimal risky portfolio is found as the tangency point of the line going through the point representing the riskless asset and is tangent to the efficient frontier. Once this point is identified, one can recover the composition of this portfolio from the information generated and recorded while constructing the efficient frontier.

Here, we describe a direct method to obtain the optimal risky portfolio by constructing a convex quadratic programming problem equivalent to (8.12). We need two assumptions: first, we assume that $\sum_{i=1}^{n} x_i = 1$ for any feasible portfolio x. This is a natural assumption since the x_i's are the proportions of the portfolio in different asset classes. Second, we assume that a feasible portfolio $\hat{x}$ exists with $\mu^T \hat{x} > r_f$ since, if all feasible portfolios have expected return bounded by the risk-free rate, there is no need to optimize: the risk-free investment dominates all others.

Proposition 8.1 *Given a set $\mathcal{X}$ of feasible portfolios with the properties that $e^T x = 1$, $\forall x \in \mathcal{X}$ and $\exists \hat{x} \in \mathcal{X}$, $\mu^T \hat{x} > r_f$, the portfolio x^* with the maximum Sharpe ratio in this set can be found by solving the following problem*

$$\min y^T \Sigma y \text{ s.t. } (y, \kappa) \in \mathcal{X}^+, \ (\mu - r_f e)^T y = 1, \tag{8.13}$$

where

$$\mathcal{X}^+ := \left\{ x \in \mathbb{R}^n, \kappa \in \mathbb{R} : \ \kappa > 0, \frac{x}{\kappa} \in \mathcal{X} \right\} \cup (0, 0). \tag{8.14}$$

If (y, κ) solves (8.13), then $x^* = y/\kappa$. □

Problem (8.13) is a quadratic program that can be solved using the methods discussed in Chapter 7.

Proof: By our second assumption, it suffices to consider only those x for which $(\mu - r_f e)^T x > 0$. Let us make the following change of variables in (8.12):

$$\kappa = \frac{1}{(\mu - r_f e)^T x}$$
$$y = \kappa x.$$

Then, $\sqrt{x^T \Sigma x} = (1/\kappa)\sqrt{y^T \Sigma y}$ and the objective function of (8.12) can be written as $1/\sqrt{y^T \Sigma y}$ in terms of the new variables. Note also that

$$(\mu - r_f e)^T x > 0, \, x \in \mathcal{X} \Leftrightarrow \kappa > 0, \, \frac{y}{\kappa} \in \mathcal{X},$$

and

$$\kappa = \frac{1}{(\mu - r_f e)^T x} \Leftrightarrow (\mu - r_f e)^T y = 1,$$

given $y/\kappa = x$. Thus, (8.12) is equivalent to

$$\min y^T \Sigma y \text{ s.t. } \kappa > 0, \, (y, \kappa) \in \mathcal{X}, \, (\mu - r_f e)^T y = 1.$$

Since $(\mu - r_f e)^T y = 1$ rules out $(0,0)$ as a solution, replacing $\kappa > 0, \, (y, \kappa) \in \mathcal{X}$ with $(y, \kappa) \in \mathcal{X}^+$ does not affect the solutions – it just makes the feasible set a closed set. □

Exercise 8.12   Show that $\mathcal{X}^+$ is a cone. If $\mathcal{X} = \{x: \, Ax \geq b, \, Cx = d\}$, show that $\mathcal{X}^+ = \{(x, \kappa): \, Ax - b\kappa \geq 0, \, Cx - d\kappa = 0, \, \kappa \geq 0\}$. What if $\mathcal{X} = \{x: \, \|x\| \leq 1\}$?

Exercise 8.13   Find the Sharpe ratio maximizing portfolio of the four assets in Exercise 8.4 assuming that the risk-free return rate is 3% by solving the QP (8.13) resulting from its reformulation. Verify that the CAL passing through the point representing the standard deviation and the expected return of this portfolio is tangent to the efficient frontier.

8.3 Returns-based style analysis

In two very influential articles, Sharpe described how constrained optimization techniques can be used to determine the effective asset mix of a fund using only the return time series for the fund and contemporaneous time series for returns of a number of carefully chosen asset classes [74, 75]. Often, passive indices or index funds are used to represent the chosen asset classes and one tries to determine a portfolio of these funds and indices whose returns provide the best match for the returns of the fund being analyzed. The allocations in the portfolio can be interpreted as the fund's style and consequently, this approach has become to known as *returns-based style analysis*, or RBSA.

RBSA provides an inexpensive and timely alternative to *fundamental analysis* of a fund to determine its style/asset mix. Fundamental analysis uses the information on actual holdings of a fund to determine its asset mix. When all the holdings are known, the asset mix of the fund can be inferred easily. However, this information is rarely available, and when it is available, it is often quite expensive and several weeks or months old. Since RBSA relies only on returns data which is immediately available for publicly traded funds, and well-known optimization techniques, it can be employed in circumstances where fundamental analysis cannot be used.

The mathematical model for RBSA is surprisingly simple. It uses the following generic linear factor model: let R_t denote the return of a security – usually a mutual fund, but can be an index, etc. – in period t for $t = 1, \ldots, T$, where T corresponds to the number of periods in the modeling window. Furthermore, let F_{it} denote the return on factor i in period t, for $i = 1, \ldots, n, t = 1, \ldots, T$. Then, R_t can be represented as follows:

$$R_t = w_{1t} F_{1t} + w_{2t} F_{2t} + \cdots + w_{nt} F_{nt} + \varepsilon_t \qquad (8.15)$$
$$= F_t w_t + \varepsilon_t, \ t = 1, \ldots, T.$$

In this equation, w_{it} quantities represent the sensitivities of R_t to each one of the n factors, and ε_t represents the non-factor return. We use the notation $w_t = [w_{1t}, \ldots, w_{nt}]^T$ and $F_t = [F_{1t}, \ldots, F_{nt}]$.

The linear factor model (8.15) has the following convenient interpretation when the factor returns F_{it} correspond to the returns of passive investments, such as those in an index fund for an asset class: one can form a benchmark portfolio of the passive investments (with weights w_{it}), and the difference between the fund return R_t and the return of the benchmark portfolio $F_t w_t$ is the non-factor return contributed by the fund manager using stock selection, market timing, etc. In other words, ε_t represents the additional return resulting from active management of the fund. Of course, this additional return can be negative.

The benchmark portfolio return interpretation for the quantity $F_t w_t$ suggests that one should choose the sensitivities (or weights) w_{it} such that they are all nonnegative and sum to one. With these constraints in mind, Sharpe proposes to choose w_{it} to minimize the variance of the non-factor return ε_t. In his model, Sharpe restricts the weights to be constant over the period in consideration so that w_{it} does not depend on t. In this case, we use $w = [w_1, \ldots, w_n]^T$ to denote the time-invariant factor weights and formulate the following quadratic programming problem:

$$\min_{w \in \mathbf{R}^n} \ \text{var}(\varepsilon_t) = \text{var}(R_t - F_t w)$$
$$\text{s.t.} \qquad \sum_{i=1}^n w_i = 1 \qquad (8.16)$$
$$w_i \geq 0, \forall i.$$

The objective of minimizing the variance of the non-factor return ε_t deserves some comment. Since we are essentially formulating a tracking problem, and since ε_t represents the "tracking error," one may wonder why we do not minimize the magnitude of this quantity rather than its variance. Since the Sharpe model interprets the quantity ε_t as a consistent management effect, the objective is to determine a benchmark portfolio such that the difference between fund returns and the benchmark returns is as close to constant (i.e., variance 0) as possible. So, we want the fund return and benchmark return graphs to show two almost parallel lines with the distance between these lines corresponding to manager's consistent contribution to the fund return. This objective is almost equivalent to choosing weights in order to maximize the R-square of this regression model. The equivalence is not exact since we are using constrained regression and this may lead to correlation between ε_t and asset class returns.

The objective function of this QP can easily be computed:

$$\operatorname{var}(R_t - F_t w) = \frac{1}{T} \sum_{t=1}^{T} (R_t - F_t w)^2 - \left(\frac{\sum_{t=1}^{T} (R_t - F_t w)}{T} \right)^2$$

$$= \frac{1}{T} \| R - F w \|^2 - \left(\frac{e^T (R - F w)}{T} \right)^2$$

$$= \left(\frac{\| R \|^2}{T} - \frac{(e^T R)^2}{T^2} \right) - 2 \left(\frac{R^T F}{T} - \frac{e^T R}{T^2} e^T F \right) w$$

$$+ w^T \left(\frac{1}{T} F^T F - \frac{1}{T^2} F^T e e^T F \right) w.$$

Above, we introduced and used the notation

$$R = \begin{bmatrix} R_1 \\ \vdots \\ R_T \end{bmatrix}, \text{ and } F = \begin{bmatrix} F_1 \\ \cdots \\ F_T \end{bmatrix} = \begin{bmatrix} F_{11} & \cdots & F_{n1} \\ \vdots & \ddots & \vdots \\ F_{1T} & \cdots & F_{nT} \end{bmatrix},$$

and e denotes a vector of 1s of appropriate size. Convexity of this quadratic function of w can be easily verified. Indeed,

$$\frac{1}{T} F^T F - \frac{1}{T^2} F^T e e^T F = \frac{1}{T} F^T \left(I - \frac{e e^T}{T} \right) F, \tag{8.17}$$

and the symmetric matrix $M = I - e e^T / T$ in the middle of the right-hand-side expression above is a positive semidefinite matrix with only two eigenvalues: 0 (multiplicity 1) and 1 (multiplicity $T - 1$). Since M is positive semidefinite, so is $F^T M F$, and therefore the variance of ε_t is a convex quadratic function of w. Therefore, the problem (8.16) is a convex quadratic programming problem and is easily solvable using well-known optimization techniques such as interior-point methods, which we discussed in Chapter 7.

Exercise 8.14 Implement the returns-based style analysis approach to determine the effective asset mix of your favorite mutual fund. Use the following asset classes as your "factors": large-growth stocks, large-value stocks, small-growth stocks, small-value stocks, international stocks, and fixed-income investments. You should obtain time series of returns representing these asset classes from online resources. You should also obtain a corresponding time series of returns for the mutual fund you picked for this exercise. Solve the problem using 30 periods of data (i.e., $T = 30$).

8.4 Recovering risk-neural probabilities from options prices

Recall our discussion on risk-neutral probability measures in Section 4.1.2. There, we considered a one-period economy with n securities. Current prices of these securities are denoted by S_0^i for $i = 1, \ldots, n$. At the end of the current period, the economy will be in one of the states from the state space Ω. If the economy reaches state $\omega \in \Omega$ at the end of the current period, security i will have the payoff $S_1^i(\omega)$. We assume that we know all S_0^i's and $S_1^i(\omega)$'s but do not know the particular terminal state ω, which will be determined randomly.

Let r denote the one-period (riskless) interest rate and let $R = 1 + r$. A risk neutral probability measure (RNPM) is defined as the probability measure under which the present value of the expected value of future payoffs of a security equals its current price. More specifically,

- (**discrete case:**) on the state space $\Omega = \{\omega_1, \omega_2, \ldots, \omega_m\}$, an RNPM is a vector of positive numbers $p_1, p_2, \ldots, p_m$ such that
  1. $\sum_{j=1}^m p_j = 1$,
  2. $S_0^i = \frac{1}{R} \sum_{j=1}^m p_j S_1^i(\omega_j)$, $\forall i$.
- (**continuous case:**) on the state space $\Omega = (a, b)$ an RNPM is a density function $p : \Omega \to \mathbb{R}_+$ such that
  1. $\int_a^b p(\omega) d\omega = 1$,
  2. $S_0^i = \frac{1}{R} \int_a^b p(\omega) S_1^i(\omega) d\omega$, $\forall i$.

Also recall the following result from Section 4.1.2 that is often called the *first fundamental theorem of asset pricing*:

Theorem 8.2 *A risk-neutral probability measure exists if and only if there are no arbitrage opportunities.*

If we can identify a risk-neutral probability measure associated with a given state space and a set of observed prices we can price any security for which we can determine the payoffs for each state in the state space. Therefore, a fundamental problem in asset pricing is the identification of a RNPM consistent with a given set of prices. Of course, if the number of states in the state space is much larger than the number of observed prices, this problem becomes under-determined and we cannot obtain a sensible solution without introducing some additional structure into the RNPM we seek. In this section, we outline a strategy that guarantees the smoothness of the RNPM by constructing it through cubic splines. We first describe spline functions briefly.

Consider a function $f : [a, b] \to \mathbb{R}$ to be estimated using its values $f_i = f(x_i)$ given on a set of points $\{x_i\}$, $i = 1, \ldots, m+1$. It is assumed that $x_1 = a$ and $x_{m+1} = b$.

A *spline function*, or *spline*, is a *piecewise* polynomial approximation $S(x)$ to the function f such that the approximation agrees with f on each node x_i, i.e., $S(x_i) = f(x_i), \forall i$.

The graph of a spline function S contains the data points (x_i, f_i) (called *knots*) and is continuous on $[a, b]$.

A spline on $[a, b]$ is of order n if (i) its first $n - 1$ derivatives exist on each interior knot, (ii) the highest degree for the polynomials defining the spline function is n.

A cubic (third-order) spline uses cubic polynomials of the form $f_i(x) = \alpha_i x^3 + \beta_i x^2 + \gamma_i x + \delta_i$ to estimate the function in each interval $[x_i, x_{i+1}]$ for $i = 1, \ldots, m$. A cubic spline can be constructed in such a way that it has second derivatives at each node. For $m + 1$ knots ($x_1 = a, \ldots x_{m+1} = b$) in $[a, b]$ there are m intervals and, therefore $4m$ unknown constants to evaluate. To determine these $4m$ constants we use the following $4m$ equations:

$$f_i(x_i) = f(x_i), \quad i = 1, \ldots, m, \quad \text{and} \quad f_m(x_{m+1}) = f(x_{m+1}), \qquad (8.18)$$
$$f_{i-1}(x_i) = f_i(x_i), \quad i = 2, \ldots, m, \qquad (8.19)$$
$$f'_{i-1}(x_i) = f'_i(x_i), \quad i = 2, \ldots, m, \qquad (8.20)$$
$$f''_{i-1}(x_i) = f''_i(x_i), \quad i = 2, \ldots, m, \qquad (8.21)$$
$$f''_1(x_1) = 0 \text{ and } f''_m(x_{m+1}) = 0. \qquad (8.22)$$

The last condition leads to a so-called *natural* spline.

We now formulate a quadratic programming problem with the objective of finding a risk-neutral probability density function (described by cubic splines) for future values of an underlying security that best fits the observed option prices on this security.

We choose a security for consideration, say a stock or an index. We then fix an exercise date – later than the date for which we will obtain a probability density function of the price of our security. Finally, we fix a range $[a, b]$ for possible terminal values of the price of the underlying security at the exercise date of the options and an interest rate r for the period between now and the exercise date. The inputs to our optimization problem are current market prices C_K of call options and P_K for put options on the chosen underlying security with strike price K and the chosen expiration date. This data is easily available from newspapers and online sources. Let C and P, respectively, denote the set of strike prices K for which reliable market prices C_K and P_K are available. For example, C may denote the strike prices of call options that were traded on the day the problem is formulated.

Next, we fix a superstructure for the spline approximation to the risk-neutral density, meaning that we choose how many knots to use, where to place the knots and what kind of polynomial (quadratic, cubic, etc.) functions to use. For example, we may decide to use cubic splines and $m + 1$ equally spaced knots. The parameters of the polynomial functions that comprise the spline function will be the variables of the optimization problem we are formulating. For cubic splines with $m + 1$ knots, we will have $4m$ variables $(\alpha_i, \beta_i, \gamma_i, \delta_i)$ for $i = 1, \ldots, m$. Collectively, we will represent these variables with y. For all y chosen so that the corresponding polynomial functions f_i satisfy the equations (8.19)–(8.22) above, we will have a particular choice of a natural spline function defined on the interval $[a, b]$.[1] Let $p_y(\cdot)$ denote this function. Imposing the following additional restrictions we make sure that p_y is a probability density function:

$$p_y(x) \geq 0, \forall x \in [a, b], \tag{8.23}$$

$$\int_a^b p_y(\omega) d\omega = 1. \tag{8.24}$$

The constraint (8.24) is a linear constraint on the variables $(\alpha_i, \beta_i, \gamma_i, \delta_i)$ of the problem and can be enforced as follows:

$$\sum_{s=1}^{n_s} \int_{x_s}^{x_{s+1}} f_s(\omega) d\omega = 1. \tag{8.25}$$

On the other hand, enforcing condition (8.23) is not straightforward as it requires the function to be nonnegative for *all* values of x in $[a, b]$. Here, we relax condition (8.23), and require the cubic spline approximation to be nonnegative only at the knots:

$$p_y(x_i) \geq 0, \ i = 1, \ldots, m. \tag{8.26}$$

While this relaxation simplifies the problem greatly, we cannot guarantee that the spline approximation we generate will be nonnegative in its domain. We will discuss in Chapter 10.3 a more sophisticated technique that rigorously enforces condition (8.23).

---

[1] Note that we do not impose the conditions (8.18), because the values of the probability density function we are approximating are unknown and will be determined as a solution of an optimization problem.

Next, we define the discounted expected value of the terminal value of each option using p_y as the risk-neutral density function:

$$C_K(y) := \frac{1}{1+r} \int_a^b (\omega - K)^+ p_y(\omega) d\omega, \tag{8.27}$$

$$P_K(y) := \frac{1}{1+r} \int_a^b (K - \omega)^+ p_y(\omega) d\omega. \tag{8.28}$$

Then, $C_K(y)$ is the theoretical option price if p_y is the true risk-neutral probability measure and

$$(C_K - C_K(y))^2$$

is the squared difference between the actual option price and this theoretical value. Now consider the aggregated error function for a given y:

$$E(y) := \sum_{K \in \mathcal{C}} (C_K - C_K(y))^2 + \sum_{K \in \mathcal{P}} (P_K - P_K(y))^2.$$

The objective now is to choose y such that conditions (8.19)–(8.22) of the spline function description as well as (8.26) and (8.24) are satisfied and $E(y)$ is minimized. This is essentially a constrained least-squares problem.

We choose the number of knots and their locations so that the knots form a superset of $\mathcal{C} \cup \mathcal{P}$. Let $x_0 = a, x_1, \ldots, x_m = b$ denote the locations of the knots. Now, consider a call option with strike K and assume that K coincides with the location of the jth knot, i.e., $x_j = K$. Recall that y denotes the collection of variables $(\alpha_i, \beta_i, \gamma_i, \delta_i)$ for $i = 1, \ldots, m$. Now, we can derive a formula for $C_K(y)$:

$$(1+r)C_K(y) = \int_a^b S_y(\omega)(\omega - K)^+ d\omega$$

$$= \sum_{i=1}^m \int_{x_{i-1}}^{x_i} S_y(\omega)(\omega - K)^+ d\omega$$

$$= \sum_{i=j+1}^m \int_{x_{i-1}}^{x_i} S_y(\omega)(\omega - K) d\omega$$

$$= \sum_{i=j+1}^m \int_{x_{i-1}}^{x_i} (\alpha_i \omega^3 + \beta_i \omega^2 + \gamma_i \omega + \delta_i)(\omega - K) d\omega.$$

It is easily seen that this expression for $C_K(y)$ is a linear function of the components $(\alpha_i, \beta_i, \gamma_i, \delta_i)$ of y. A similar formula can be derived for $P_K(y)$. The reason for choosing the knots at the strike prices is the third equation in the sequence above – we can immediately ignore some of the terms in the summation and the $(\cdot)^+$ function is linear (and not piecewise linear) in each integral.

Now, it is clear that the problem of minimizing $E(y)$ subject to spline function conditions (8.26) and (8.24) is a quadratic optimization problem and can be solved using the techniques of the previous chapter.

8.5 Additional exercises

Exercise 8.15 Recall the mean-variance optimization problem we considered in Section 8.1:

$$
\begin{aligned}
\min_x \ & x^T \Sigma x \\
& \mu^T x \geq R \\
& Ax = b \\
& Cx \geq d.
\end{aligned} \tag{8.29}
$$

Now, consider the problem of finding the feasible portfolio with smallest overall variance, without imposing any expected return constraint:

$$
\begin{aligned}
\min_x \ & x^T \Sigma x \\
& Ax = b \\
& Cx \geq d.
\end{aligned} \tag{8.30}
$$

(i) Does the optimal solution to (8.30) give an efficient portfolio? Why?
(ii) Let x_R, $\lambda_R \in I\!R$, $\gamma_E \in I\!R^m$, and $\gamma_I \in I\!R^p$ satisfy the optimality conditions of (8.29) (see system (8.2)). If $\lambda_R = 0$, show that x_R is an optimal solution to (8.30). (Hint: What are the optimality conditions for (8.30)? How are they related to (8.2)?)

Exercise 8.16 Classification problems are among the important classes of problems in financial mathematics that can be solved using optimization models and techniques. In a classification problem we have a vector of "features" describing an entity and the objective is to analyze the features to determine which one of the two (or more) "classes" each entity belongs to. For example, the classes might be "growth stocks" and "value stocks," and the entities (stocks) may be described by a feature vector that may contain elements such as stock price, price-earnings ratio, growth rate for the previous periods, growth estimates, etc.

Mathematical approaches to classification often start with a "training" exercise. One is supplied with a list of entities, their feature vectors, and the classes they belong to. From this information, one tries to extract a mathematical structure for the entity classes so that additional entities can be classified using this mathematical structure and their feature vectors. For two-class classification, a hyperplane is probably the simplest mathematical structure that can be used to "separate" the feature vectors of these two different classes. Of course, there may not be any hyperplane that separates two sets of vectors. When such a hyperplane exists, we say that the two sets can be linearly separated.

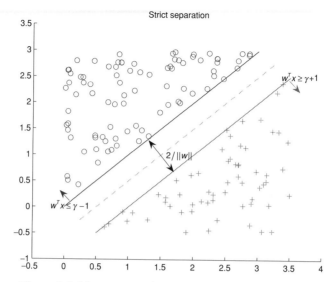

Figure 8.5 Linear separation of two classes of data points

Consider feature vectors $a_i \in \mathbb{R}^n$ for $i = 1, \ldots, k_1$ corresponding to class 1, and vectors $b_i \in \mathbb{R}^n$ for $i = 1, \ldots, k_2$ corresponding to class 2. If these two vector sets can be linearly separated, a hyperplane $w^{\mathrm{T}}x = \gamma$ exists with $w \in \mathbb{R}^n$, $\gamma \in \mathbb{R}$ such that

$$w^{\mathrm{T}}a_i \geq \gamma, \quad \text{for } i = 1, \ldots, k_1,$$
$$w^{\mathrm{T}}b_i \leq \gamma, \quad \text{for } i = 1, \ldots, k_2.$$

To have a "strict" separation, we often prefer to obtain w and γ such that

$$w^{\mathrm{T}}a_i \geq \gamma + 1, \quad \text{for } i = 1, \ldots, k_1,$$
$$w^{\mathrm{T}}b_i \leq \gamma - 1, \quad \text{for } i = 1, \ldots, k_2.$$

In this manner, we find two parallel lines ($w^{\mathrm{T}}x = \gamma + 1$ line and $w^{\mathrm{T}}x = \gamma - 1$) that form the boundary of the class 1 and class 2 portion of the vector space. This type of separation is shown in Figure 8.5.

There may be several such parallel lines that separate the two classes. Which one should we choose? A good criterion is to choose the lines that have the largest margin (distance between the lines).

(i) Consider the following quadratic problem:

$$\min_{w,\gamma} \|w\|_2^2$$
$$a_i^{\mathrm{T}}w \geq \gamma + 1, \quad \text{for } i = 1, \ldots, k_1, \tag{8.31}$$
$$b_i^{\mathrm{T}}w \leq \gamma - 1, \quad \text{for } i = 1, \ldots, k_2.$$

Show that the objective function of this problem is equivalent to maximizing the margin between the lines $w^{\mathrm{T}}x = \gamma + 1$ and $w^{\mathrm{T}}x = \gamma - 1$.

(ii) The linear separation idea we presented above can be used even when the two vector sets $\{a_i\}$ and $\{b_i\}$ are not linearly separable. (Note that linearly inseparable sets will result in an infeasible problem in (8.31).) This is achieved by introducing a nonnegative "violation" variable for each constraint of (8.31). Then, we have two objectives: to minimize the sum of the constraint violations and to maximize the margin. Develop a quadratic programming model that combines these two objectives using an adjustable parameter that can be chosen in a way to put more weight on violations or margin, depending on one's preference.

Exercise 8.17 The classification problems we discussed in the previous exercise can also be formulated as linear programming problems, if one agrees to use 1-norm rather than 2-norm of w in the objective function. Recall that $\|w\|_1 = \sum_i |w_i|$. Show that if we replace $\|w\|_2^2$ with $\|w\|_1$ in the objective function of (8.31), we can write the resulting problem as an LP. Show also that this new objective function is equivalent to maximizing the distance between $w^Tx = \gamma + 1$ and $w^Tx = \gamma - 1$ if one measures the distance using ∞-norm ($\|g\|_\infty = \max_i |g_i|$).

8.6 Case study: constructing an efficient portfolio

Investigate the performance of one of the variations on the classical Markowitz model proposed by Michaud, or Black–Litterman or Konno–Yamazaki; see Sections 8.1.2–8.1.4.

Possible suggestions:

- Choose 30 stocks and retrieve their historical returns over a meaningful horizon.
- Use the historical information to compute expected returns and the variance–covariance matrix for these stock returns.
- Set up the model and solve it with MATLAB or Excel's Solver for different levels R of expected return. Allow for short sales and include no diversification constraints.
- Recompute these portfolios with no short sales and various diversification constraints.
- Compare portfolios constructed in period t (based on historical data up to period t) by observing their performance in period $t + 1$, i.e., compute the actual portfolio return in period $t + 1$. Repeat this experiment several times. Comment.
- Investigate how sensitive the optimal portfolios that you obtained are to small changes in the data. For example, how sensitive are they to a small change in the expected return of the assets?
- You currently own the following portfolio: $x_i^0 = 0.20$ for $i = 1, \ldots, 5$ and $x_i^0 = 0$ for $i = 6, \ldots, 30$. Include turnover constraints to reoptimize the portfolio for a fixed level R of expected return and observe the dependency on h, the total turnover allowed for reoptimization.
- You currently own the following portfolio: $x_i^0 = 0.20$ for $i = 1, \ldots, 5$ and $x_i^0 = 0$ for $i = 6, \ldots, 30$. Reoptimize the portfolio considering transaction costs for buying and selling. Solve for a fixed level R of expected return and observe the dependency on transaction costs.

9

Conic optimization tools

9.1 Introduction

In this chapter and the next, we address conic optimization problems and their applications in finance. Conic optimization refers to the problem of minimizing or maximizing a linear function over a set defined by linear equalities and cone membership constraints. Cones are defined and discussed in Appendix B. While they are not as well known or as widely used as their close relatives linear and quadratic programming, conic optimization problems continue to grow in importance thanks to their wide applicability and the availability of powerful methods for their solution.

We recall the definition of a standard form conic optimization problem that was provided in Chapter 1:

$$
\min_x c^\mathsf{T} x \\
Ax = b \qquad\qquad (9.1) \\
x \in C,
$$

where C denotes a closed convex cone in a finite-dimensional vector space X.

When $X = I\!R^n$ and $C = I\!R^n_+$, this problem is the standard form linear programming problem. Therefore, conic optimization is a generalization of linear optimization. In fact, it is much more general than linear programming since we can use non-polyhedral (i.e., nonlinear) cones C in the description of these problems and formulate certain classes of nonlinear convex objective functions and nonlinear convex constraints. In particular, conic optimization provides a powerful and unifying framework for problems in linear programming (LP), second-order cone programming (SOCP), and semidefinite programming (SDP). We describe these two new and important classes of conic optimization problems in more detail.

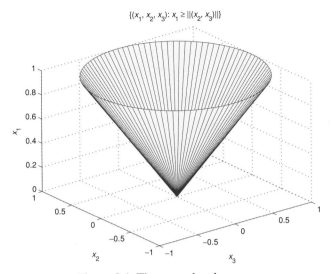

Figure 9.1 The second-order cone

9.2 Second-order cone programming

SOCPs involve the second-order cone, which is defined by the property that for each of its members the first element is at least as large as the Euclidean norm of the remaining elements. This corresponds to the case where C is the second-order cone (also known as the quadratic cone, Lorenz cone, and the ice-cream cone):

$$C_q := \{x = (x_1, x_2, \ldots, x_n) \in I\!R^n : x_1 \geq \|(x_2, \ldots, x_n)\|\}. \tag{9.2}$$

A portion of the second-order cone in three dimensions for $x_1 \in [0, 1]$ is depicted in Figure 9.1. As seen from the figure, the second-order cone in three dimensions resembles an ice-cream cone that stretches to infinity. We observe that by "slicing" the second-order cone, i.e., by intersecting it with a hyperplane at different angles, we can obtain spherical and ellipsoidal sets. Any convex quadratic constraint can be expressed using the second-order cone (or its rotations) and one or more hyperplanes.

Exercise 9.1 Another important cone that appears in conic optimization formulations is the *rotated quadratic cone*, which is defined as follows:

$$C_q^r := \left\{(x_1, x_2, x_3, \ldots, x_n) : 2x_1x_2 \geq \sum_{j=3}^{n} x_j^2, x_1, x_2 \geq 0.\right\}. \tag{9.3}$$

Show that $x = (x_1, x_2, x_3, \ldots, x_n) \in C_q^r$ if and only if $y = (y_1, y_2, y_3, \ldots, y_n) \in C_q$ where $y_1 = (1/\sqrt{2})(x_1 + x_2)$, $y_2 = (1/\sqrt{2})(x_1 - x_2)$, and $y_j = x_j, j =$

$3, \ldots, n$. The vector y given here is obtained by rotating the vector x by 45 degrees in the plane defined by the first two coordinate axes. In other words, each element of the cone C_q^r can be mapped to a corresponding element of C_q through a 45-degree rotation (why?). This is why the cone C_q^r is called the rotated quadratic cone.

Exercise 9.2   Show that the problem

$$\min \qquad x^{3/2}$$
$$\text{s.t.} \quad x \geq 0, \ x \in S$$

is equivalent to the following problem:

$$\min \qquad t$$
$$\text{s.t.} \quad x \geq 0, \ x \in S$$
$$x^2 \leq tu$$
$$u^2 \leq x.$$

Express the second problem as an SOCP using C_q^r.

Exercise 9.3   Consider the following optimization problem:

$$\min \quad c_1 x_1 + c_2 x_2 + d_1 x_1^{3/2} + d_2 x_2^{3/2}$$
$$\text{s.t.} \quad a_{11} x_1 + a_{12} x_2 = b_1,$$
$$x_1, x_2 \geq 0,$$

where $d_1, d_2 > 0$. The nonlinear objective function of this problem is a convex function. Write this problem as a conic optimization problem with a linear objective function and convex cone constraints. [Hint: Use the previous exercise.]

A review of second-order cone programming models and methods is provided in [1]. One of the most common uses of second-order cone programs in financial applications is in the modeling and treatment of parameter uncertainties in optimization problems. After generating an appropriate description of the uncertainties, robust optimization models seek to find solutions to such problems that will perform well under many scenarios. As we will see in Chapter 19, ellipsoidal sets are among the most popular structures used for describing uncertainty in such problems and the close relationship between ellipsoidal sets and second-order cones make them particularly useful. We illustrate this approach in the following subsection.

9.2.1 Ellipsoidal uncertainty for linear constraints

Consider the following single-constraint linear program:

$$\min \qquad c^T x$$
$$\text{s.t.} \quad a^T x + b \geq 0.$$

We consider the setting where the objective function is certain but the constraint coefficients are uncertain. We assume that the constraint coefficients $[a; b]$ belong to an ellipsoidal uncertainty set:

$$\mathcal{U} = \left\{ [a; b] = [a^0; b^0] + \sum_{j=1}^{k} u_j[a^j; b^j], \|u\| \le 1 \right\}.$$

Our objective is to find a solution that minimizes the objective function among the vectors that are feasible for all $[a; b] \in \mathcal{U}$. In other words, we want to solve

$$\begin{aligned} \min \quad & c^T x \\ \text{s.t.} \quad & a^T x + b \ge 0, \forall [a; b] \in \mathcal{U}. \end{aligned}$$

For a fixed x, the "robust" version of the constraint is satisfied by x if and only if

$$0 \le \min_{[a;b] \in \mathcal{U}} a^T x + b \equiv \min_{u:\|u\| \le 1} \alpha + u^T \beta, \tag{9.4}$$

where $\alpha = (a^0)^T x + b^0$ and $\beta = (\beta_1, \ldots, \beta_k)$ with $\beta_j = (a^j)^T x + b^j$.

The second minimization problem in (9.4) is easy. Since α is constant, all we need to do is to minimize $u^T \beta$ subject to the constraint $\|u\| \le 1$. Recall that for the angle θ between vectors u and β the following trigonometric equality holds:

$$\cos \theta = \frac{u^T \beta}{\|u\| \|\beta\|},$$

or $u^T \beta = \|u\| \|\beta\| \cos \theta$. Since $\|\beta\|$ is constant, this expression is minimized when $\|u\| = 1$ and $\cos \theta = -1$. This means that u points in the opposite direction from β, namely $-\beta$. Normalizing to satisfy the bound constraint we obtain $u^* = -\beta/\|\beta\|$ as shown in Figure 9.2. Substituting this value we find

$$\min_{[a;b] \in \mathcal{U}} a^T x + b = \alpha - \|\beta\| = (a^0)^T x + b^0 - \sqrt{\sum_{j=1}^{k} ((a^j)^T x + b^j)^2}, \tag{9.5}$$

and we obtain the robust version of the inequality $a^T x + b \ge 0$ as

$$(a^0)^T x + b^0 - \sqrt{\sum_{j=1}^{k} ((a^j)^T x + b^j)^2} \ge 0. \tag{9.6}$$

Now observe that (9.6) can be written equivalently as:

$$z_j = (a^j)^T x + b^j, \quad j = 0, \ldots k,$$

$$(z_0, z_1, \ldots, z_k) \in C_q,$$

where C_q is the second-order cone (9.2).

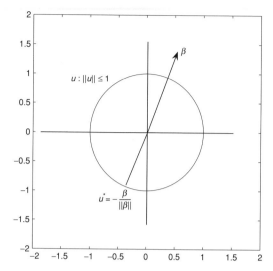

Figure 9.2 Minimization of a linear function over a circle

The approach outlined above generalizes to multiple constraints as long as the uncertainties are *constraint-wise*, that is, the uncertainty sets of parameters in different constraints are unrelated. Thus, robust optimization models for uncertain linear constraints with ellipsoidal uncertainty lead to SOCPs. The strategy outlined above is well-known and is used in, for example, [8].

9.2.2 Conversion of quadratic constraints into second-order cone constraints

The second-order cone membership constraint $(x_0, x_1, \ldots, x_k) \in C_q$ can be written equivalently as the combination of a linear and a quadratic constraint:

$$x_0 \geq 0, \quad x_0^2 - x_1^2 - \cdots - x_k^2 \geq 0.$$

Conversely, any convex quadratic constraint of an optimization problem can be rewritten using second-order cone membership constraints. When we have access to a reliable solver for second-order cone optimization, it may be desirable to convert convex quadratic constraints to second-order cone constraints. Fortunately, a simple recipe is available for these conversions.

Consider the following quadratic constraint:

$$x^T Q x + 2 p^T x + \gamma \leq 0. \tag{9.7}$$

This is a convex constraint if the function on the left-hand side is convex which is true if and only if Q is a positive semidefinite matrix. Let us assume Q is positive definite for simplicity. In that case, there exists an invertible matrix, say R, satisfying

$Q = RR^T$. For example, the Cholesky factor of Q satisfies this property. Then, (9.7) can be written as

$$(R^Tx)^T(R^Tx) + 2p^Tx + \gamma \leq 0. \tag{9.8}$$

Define $y = (y_1, \ldots, y_k)^T = R^Tx + R^{-1}p$. Then, we have

$$y^Ty = (R^Tx)^T(R^Tx) + 2p^Tx + p^TQ^{-1}p.$$

Thus, (9.8) is equivalent to

$$\exists y \text{ s.t. } y = R^Tx + R^{-1}p, \, y^Ty \leq p^TQ^{-1}p - \gamma.$$

From this equivalence, we observe that the constraint (9.7) can be satisfied only if $p^TQ^{-1}p - \gamma \geq 0$. We will assume that this is the case.

Now, it is straightforward to note that (9.7) is equivalent to the following set of linear equations coupled with a second-order cone constraint:

$$\begin{bmatrix} y_1 \\ \vdots \\ y_k \end{bmatrix} = R^Tx + R^{-1}p,$$

$$y_0 = \sqrt{p^TQ^{-1}p - \gamma},$$

$$(y_0, y_1, \ldots, y_k) \in C_q.$$

Exercise 9.4 Rewrite the following convex quadratic constraint in "conic form," i.e., as the intersection of linear equality constraints and a second-order cone constraint:

$$10x_1^2 + 2x_1x_2 + 5x_2^2 + 4x_1 + 6x_2 + 1 \leq 0.$$

Exercise 9.5 Discuss how the approach outlined in this section must be modified to address the case when Q is positive semidefinite but not positive definite. In this case there still exists a matrix R satisfying $Q = RR^T$. But R is no longer invertible and we can no longer define the vector y as above.

9.3 Semidefinite programming

In SDPs, the set of variables are represented by a symmetric matrix that is required to be in the cone of positive semidefinite matrices in addition to satisfying a system of linear equations. We say that a matrix $M \in I\!R^{n \times n}$ is positive semidefinite if $y^TMy \geq 0$ for all $y \in I\!R^n$. When M is symmetric, this is equivalent to M having eigenvalues that are all nonnegative. A stronger condition is positive definiteness. M is positive definite if $y^TMy > 0$ for all $y \in I\!R^n$ with the exception of $y = 0$.

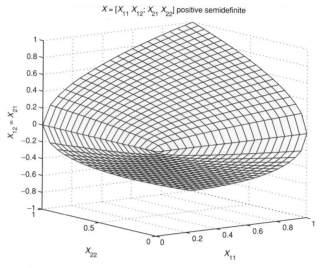

Figure 9.3 The cone of positive semidefinite matrices

For symmetric M, positive definiteness is equivalent to the positivity of all of its eigenvalues.

Since multiplication by a positive number preserves the positive semidefiniteness property, the set of positive semidefinite matrices is a cone. In fact, it is a convex cone. The cone of positive semidefinite matrices of a fixed dimension (say n) is defined as follows:

$$C_s^n := \left\{ X = \begin{bmatrix} x_{11} & \cdots & x_{1n} \\ \vdots & \ddots & \vdots \\ x_{n1} & \cdots & x_{nn} \end{bmatrix} \in I\!R^{n \times n} : X \succeq 0 \right\}. \tag{9.9}$$

Above, the notation $X \succeq 0$ means that X is a symmetric positive semidefinite matrix. We provide a depiction of the cone of positive semidefinite matrices of dimension 2 in Figure 9.3. The diagonal elements X_{11} and X_{22} of a two-dimensional symmetric matrix are shown on the horizontal axes while the off-diagonal element $X_{12} = X_{21}$ is on the vertical axis. Symmetric two-dimensional matrices whose elements lie inside the shaded region are positive semidefinite matrices. As the nonnegative orthant and the second-order cone, the cone of positive semidefinite matrices has a *point* or a *corner* at the origin. Also note the convexity of the cone and the nonlinearity of its boundary.

Semidefinite programming problems arise in a variety of disciplines. The review by Todd provides an excellent introduction to their solution methods and the rich set of applications [79]. One of the common occurrences of semidefiniteness constraints results from the so-called S-procedure, which is a generalization of the well-known S-lemma [65]:

Lemma 9.1   *Let* $F_i(x) = x^{\mathrm{T}} A_i x + 2 + b_i^{\mathrm{T}} x + c_i, \ i = 0, 1, \ldots, p$ *be quadratic functions of* $x \in I\!\!R^n$. *Then,*

$$F_i(x) \geq 0, i = 1, \ldots, p \Rightarrow F_0(x) \geq 0$$

if there exist $\lambda_i \geq 0$ *such that*

$$\begin{bmatrix} A_0 & b_0 \\ b_0^{\mathrm{T}} & c_0 \end{bmatrix} - \sum_{i=1}^{p} \lambda_i \begin{bmatrix} A_i & b_i \\ b_i^{\mathrm{T}} & c_i \end{bmatrix} \succeq 0.$$

If $p = 1$, *the converse also holds as long as* $\exists x_0$ *s.t.* $F_1(x_0) > 0$.

The S-procedure provides a sufficient condition for the implication of a quadratic inequality by other quadratic inequalities. Furthermore, this condition is also a necessary condition in certain special cases. This equivalence can be exploited in robust modeling of quadratic constraints as we illustrate next.

9.3.1 Ellipsoidal uncertainty for quadratic constraints

This time we consider a convex-quadratically constrained problem where the objective function is certain but the constraint coefficients are uncertain:

$$\begin{aligned} \min \quad & c^{\mathrm{T}} x \\ \text{s.t.} \quad & -x^{\mathrm{T}} (A^{\mathrm{T}} A) x + 2 b^{\mathrm{T}} x + \gamma \geq 0, \ \forall [A; b; \gamma] \in \mathcal{U}, \end{aligned}$$

where $A \in I\!\!R^{m \times n}$, $b \in I\!\!R^n$, and γ is a scalar. We again consider the case where the uncertainty set is ellipsoidal:

$$\mathcal{U} = \left\{ [A; b; \gamma] = [A^0; b^0; \gamma^0] + \sum_{j=1}^{k} u_j [A^j; b^j; \gamma^j], \ \|u\| \leq 1 \right\}.$$

To reformulate the robust version of this problem we use the S-procedure described above. The robust version of our convex quadratic inequality can be written as

$$[A; b; \gamma] \in \mathcal{U} \Rightarrow -x^{\mathrm{T}} (A^{\mathrm{T}} A) x + 2 b^{\mathrm{T}} x + \gamma \geq 0. \qquad (9.10)$$

This is equivalent to the following expression:

$$\begin{aligned} \|u\| \leq 1 \Rightarrow -x^{\mathrm{T}} \left(A^0 + \sum_{j=1}^{k} A^j u_j \right) \left(A^0 + \sum_{j=1}^{k} A^j u_j \right)^{\mathrm{T}} x \\ + 2 \left(b^0 + \sum_{j=1}^{k} b^j u_j \right)^{\mathrm{T}} x + \left(\gamma^0 + \sum_{j=1}^{k} \gamma^j u_j \right) \geq 0. \quad (9.11) \end{aligned}$$

Defining $A(x) : I\!R^n \rightarrow I\!R^{m \times k}$ as

$$A(x) = [A^1 x | A^2 x | \cdots | A^k x],$$

$b(x) : I\!R^n \rightarrow I\!R^k$ as

$$b(x) = [x^{\mathrm{T}} b^1 \; x^{\mathrm{T}} b^2 \cdots x^{\mathrm{T}} b^k]^{\mathrm{T}} + \frac{1}{2} [\gamma^1 \; \gamma^2 \cdots \gamma^k]^{\mathrm{T}} - A(x)^{\mathrm{T}} A^0 x,$$

and

$$\gamma(x) = \gamma^0 + 2(b^0)^{\mathrm{T}} x - x^{\mathrm{T}} (A^0)^{\mathrm{T}} A^0 x,$$

and rewriting $\|u\| \leq 1$ as $-u^{\mathrm{T}} I u + 1 \geq 0$, we can simplify (9.11) as follows:

$$-u^{\mathrm{T}} I u + 1 \geq 0 \Rightarrow -u^{\mathrm{T}} A(x)^{\mathrm{T}} A(x) u + 2b(x)^{\mathrm{T}} u + \gamma(x) \geq 0. \tag{9.12}$$

Now we can apply Lemma 9.1 with $p = 1$, $A_1 = I$, $b_1 = 0$, $c_1 = 1$ and $A_0 = A(x)^{\mathrm{T}} A(x)$, $b_0 = b(x)$, and $c_0 = \gamma(x)$. Thus, the robust constraint (9.12) can be written as

$$\exists \lambda \geq 0 \text{ s.t. } \begin{bmatrix} \gamma(x) - \lambda & b(x)^{\mathrm{T}} \\ b(x) & A(x)^{\mathrm{T}} A(x) - \lambda I \end{bmatrix} \succeq 0. \tag{9.13}$$

Thus, we transformed the robust version of the quadratic constraint into a semidefiniteness constraint for a matrix that depends on the variables x and also a new variable λ. However, because of the term $A(x)^{\mathrm{T}} A(x)$, this results in a non-linear semidefinite optimization problem, which is difficult and beyond the immediate territory of most conic optimization algorithms. Fortunately, however, the semidefiniteness condition above is equivalent to the following semidefiniteness condition:

$$\exists \lambda \geq 0 \text{ s.t. } \begin{bmatrix} \gamma'(x) - \lambda & b'(x)^{\mathrm{T}} & (A^0 x)^{\mathrm{T}} \\ b'(x) & \lambda I & A(x)^{\mathrm{T}} \\ A^0 x & A(x) & I \end{bmatrix} \succeq 0, \tag{9.14}$$

where

$$b'(x) = [x^{\mathrm{T}} b^1 \; x^{\mathrm{T}} b^2 \; \cdots \; x^{\mathrm{T}} b^k]^{\mathrm{T}} + \frac{1}{2} [\gamma^1 \; \gamma^2 \; \cdots \; \gamma^k]^{\mathrm{T}}$$

and

$$\gamma'(x) = \gamma^0 + 2(b^0)^{\mathrm{T}} x.$$

Since all of $A(x)$, $b'(x)$, and $\gamma'(x)$ are linear in x, we obtain a linear semidefinite optimization problem from the reformulation of the robust quadratic constraint via the S-procedure. For details of this technique and many other useful results for reformulation of robust constraints, we refer the reader to [8].

Exercise 9.6   Verify that (9.11) is equivalent to (9.10).

Exercise 9.7   Verify that (9.13) and (9.14) are equivalent.

9.4 Algorithms and software

Since most conic optimization problem classes are special cases of nonlinear programming problems, they can be solved using general nonlinear optimization strategies we discussed in Chapter 5. As in linear and quadratic programming problems, the special structure of conic optimization problems allows the use of specialized and more efficient methods that take advantage of this structure. In particular, many conic optimization problems can be solved efficiently using the generalizations of sophisticated interior-point algorithms for linear and quadratic programming problems. These generalizations of interior-point methods are based on the groundbreaking work of Nesterov and Nemirovski [60] as well as the theoretical and computational advances that followed their work.

During the past decade, an intense theoretical and algorithmic study of conic optimization problems produced a number of increasingly sophisticated software products for several problem classes including SeDuMi [78] and SDPT3 [81] which are freely available. Interested readers can obtain additional information on such software by following the software link of the following page dedicated to semidefinite programming and maintained by Christoph Helmberg: www-user.tu-chemnitz.de/~helmberg/semidef.html.

There are also commercial software products that address conic optimization problems. For example, MOSEK (www.mosek.com) provides a powerful engine for second-order and linear cone optimization. AXIOMA's (www.axiomainc.com) portfolio optimization software employs a conic optimization solver that handles convex quadratic constraints as well as ellipsoidal uncertainties among other things.

10

Conic optimization models in finance

Conic optimization problems are encountered in a wide array of fields including truss design, control and system theory, statistics, eigenvalue optimization, and antenna array weight design. Robust optimization formulations of many convex programming problems also lead to conic optimization problems, see, e.g., [8, 9]. Furthermore, conic optimization problems arise as relaxations of hard combinatorial optimization problems such as the max-cut problem. Finally, some of the most interesting applications of conic optimization are encountered in financial mathematics and we will address several examples in this chapter.

10.1 Tracking error and volatility constraints

In most quantitative asset management environments, portfolios are chosen with respect to a carefully selected benchmark. Typically, the benchmark is a market index, reflecting a particular market (e.g., domestic or foreign), or a segment of the market (e.g., large cap growth) the investor wants to invest in. Then, the portfolio manager's problem is to determine an *index-tracking* portfolio with certain desirable characteristics. An index-tracking portfolio intends to track the movements of the underlying index closely with the ultimate goal of adding value by beating the index. Since this goal requires departures from the underlying index, one needs to balance the expected excess return (i.e., expected return in excess of the benchmark return) with the variance of the excess returns.

The tracking error for a given portfolio with a given benchmark refers to the difference between the returns of the portfolio and the benchmark. If the return vector is given by r, the weight vector for the benchmark portfolio is denoted by x_{BM}, and the weight vector for the portfolio is x, then this difference is given as

$$r^{\mathrm{T}}x - r^{\mathrm{T}}x_{BM} = r^{\mathrm{T}}(x - x_{BM}).$$

While some references in the literature define tracking error as this quantity, we will prefer to refer to it as the *excess return*. Using the common conventions, we define tracking error as a measure of variability of excess returns. The *ex-ante*, or predicted, tracking error of the portfolio (with respect to the risk model given by Σ) is defined as follows:

$$\text{TE}(x) := \sqrt{(x - x_{BM})^{\text{T}} \Sigma (x - x_{BM})}. \tag{10.1}$$

In contrast, the *ex-post*, or realized, tracking error is a statistical dispersion measure for the realized excess returns, typically the standard deviation of regularly (e.g., daily) observed excess returns.

In benchmark relative portfolio optimization, we solve mean-variance optimization (MVO) problems where expected absolute return and standard deviation of returns are replaced by expected excess return and the predicted tracking error. For example, the variance constrained MVO problem (8.3) is replaced by the following formulation:

$$\begin{aligned} \max_x \quad & \mu^{\text{T}}(x - x_{BM}) \\ & (x - x_{BM})^{\text{T}} \Sigma (x - x_{BM}) \leq \text{TE}^2 \\ & Ax = b \\ & Cx \geq d, \end{aligned} \tag{10.2}$$

where $x = (x_1, \ldots, x_n)$ is the variable vector whose components x_i denote the proportion of the total funds invested in security i, μ and Σ are the expected return vector and the covariance matrix, and A, b, C, and d are the coefficients of the linear equality and inequality constraints that define feasible portfolios. The objective is to maximize the expected excess return while limiting the portfolio tracking error to a predefined value of TE.

Unlike the formulations (8.1) and (8.4), which have only linear constraints, this formulation is not in standard quadratic programming form and therefore can not be solved directly using efficient and widely available QP algorithms. The reason for this is the existence of a nonlinear constraint, namely the constraint limiting the portfolio tracking error. So, if all MVO formulations are essentially equivalent as we argued before, why would anyone use the "harder" formulations with the risk constraint?

As Jorion [42] observes, *ex-post* returns are "enormously noisy measures of expected returns" and therefore investors may not be able or willing to determine minimum acceptable expected return levels, or risk-aversion constants – inputs required for problems (8.1) and (8.4) – with confidence. Jorion notes that "it is much easier to constrain the risk profile, either before or after the fact – which is no doubt why investors give managers tracking error constraints."

Fortunately, the tracking error constraint is a convex quadratic constraint, which means that we can rewrite this constraint in conic form as we saw in the previous chapter. If the remaining constraints are linear as in (10.2), the resulting problem is a second-order cone optimization problem that can be solved with specialized methods.

Furthermore, in situations where the control of multiple measures of risk is desired the conic reformulations can become very useful. In [42], Jorion observes that MVO with only a tracking error constraint may lead to portfolios with high overall variance. He considers a model where a variance constraint as well as a tracking error constraint is imposed for optimizing the portfolio. When no additional constraints are present, Jorion is able to solve the resulting problem since analytic solutions are available. His approach, however, does not generalize to portfolio selection problems with additional constraints such as no-shorting limitations, or exposure limitations to such factors as size, beta, sectors, or industries. The strength of conic optimization models, and, in this particular case, of second-order cone programming approaches, is that the algorithms developed for them will work for any combination of linear equality, linear inequality, and convex quadratic inequality constraints. Consider, for example, the following generalization of the models in [42]:

$$\max_x \quad \mu^T x$$
$$\sqrt{x^T \Sigma x} \leq \sigma$$
$$\sqrt{(x - x_{BM})^T \Sigma (x - x_{BM})} \leq TE \qquad (10.3)$$
$$Ax = b$$
$$Cx \geq d.$$

This problem can be rewritten as a second-order cone programming problem using the conversions outlined in Section 9.2.2. Since Σ is positive semidefinite, there exists a matrix R such that $\Sigma = RR^T$. Defining

$$y = R^T x$$
$$z = R^T x - R^T x_{BM},$$

we see that the first two constraints of (10.3) are equivalent to $(y_0, y) \in C_q, (z_0, z) \in C_q$ with $y_0 = \sigma$ and $z_0 = TE$. Thus, (10.3) is equivalent to the following second-order cone program:

$$\max_x \quad \mu^T x$$
$$Ax = b$$
$$Cx \geq d$$
$$R^T x - y = 0$$
$$R^T x - z = R^T x_{BM} \qquad (10.4)$$
$$y_0 = \sigma$$
$$z_0 = TE$$
$$(y_0, y) \in C_q, (z_0, z) \in C_q.$$

Exercise 10.1   Second-order cone formulations can also be used for modeling a tracking error constraint under different risk models. For example, if we had k alternative estimates of the covariance matrix denoted by $\Sigma_1, \ldots, \Sigma_k$ and wanted to limit the tracking error with respect to each estimate we would have a sequence of constraints of the form

$$\sqrt{(x - x_{BM})^{\mathrm{T}} \Sigma_i (x - x_{BM})} \le TE_i, \ i = 1, \ldots, k.$$

Show how these constraints can be converted to second-order cone constraints.

Exercise 10.2   Using historical returns of the stocks in the DJIA, estimate their mean μ_i and covariance matrix. Let R be the median of the μ_i's. Find an expected return maximizing long-only portfolio of Dow Jones constituents that has (i) a tracking error of 10% or less, and (ii) a volatility of 20% or less.

10.2 Approximating covariance matrices

The covariance matrix of a vector of random variables is one of the most important and widely used statistical descriptors of the joint behavior of these variables. Co-variance matrices are encountered frequently in financial mathematics, for example, in mean-variance optimization, in forecasting, in time-series modeling, etc.

Often, true values of covariance matrices are not observable and one must rely on estimates. Here, we do not address the problem of estimating covariance matrices and refer the reader, e.g., to Chapter 16 in [52]. Rather, we consider the case where a covariance matrix estimate is already provided and one is interested in determining a modification of this estimate that satisfies some desirable properties. Typically, one is interested finding the smallest distortion of the original estimate that achieves these desired properties.

Symmetry and positive semidefiniteness are structural properties shared by all "proper" covariance matrices. A correlation matrix satisfies the additional property that its diagonal consists of all 1s. Recall that a symmetric and positive semidefinite matrix $M \in I\!R^{n \times n}$ satisfies the property that

$$x^{\mathrm{T}} M x \ge 0, \forall x \in I\!R^n.$$

This property is equivalently characterized by the nonnegativity of the eigenvalues of the matrix M.

In some cases, for example when the estimation of the covariance matrix is performed entry-by-entry, the resulting estimate may not be a positive semidefinite matrix, that is, it may have negative eigenvalues. Using such an estimate would suggest that some linear combinations of the underlying random variables have *negative* variance and possibly result in disastrous results in mean-variance optimization. Therefore, it is important to correct such estimates before they are used in any financial decisions.

Even when the initial estimate is symmetric and positive semidefinite, it may be desirable to modify this estimate without compromising these properties. For example, if some pairwise correlations or covariances appear counter-intuitive to a financial analyst's trained eye, the analyst may want to modify such entries in the matrix. All these variations of the problem of obtaining a desirable modification of an initial covariance matrix estimate can be formulated within the powerful framework of semidefinite optimization and can be solved with standard software available for such problems.

We start the mathematical treatment of the problem by assuming that we have an estimate $\hat{\Sigma} \in \mathcal{S}^n$ of a covariance matrix and that $\hat{\Sigma}$ is not necessarily positive semidefinite. Here, $\mathcal{S}^n$ denotes the space of symmetric $n \times n$ matrices. An important question in this scenario is the following: what is the "closest" positive semidefinite matrix to $\hat{\Sigma}$? For concreteness, we use the Frobenius norm of the distortion matrix as a measure of closeness:

$$d_F(\Sigma, \hat{\Sigma}) := \sqrt{\sum_{i,j}(\Sigma_{ij} - \hat{\Sigma}_{ij})^2}.$$

Now we can state the *nearest covariance matrix problem* as follows: given $\hat{\Sigma} \in \mathcal{S}^n$,

$$\min_{\Sigma} d_F(\Sigma, \hat{\Sigma}) \\ \Sigma \in C_s^n, \tag{10.5}$$

where C_s^n is the cone of $n \times n$ symmetric and positive semidefinite matrices as defined in (9.9). Notice that the decision variable in this problem is represented as a *matrix* rather than a vector as in all previous optimization formulations we considered.

Furthermore, by introducing a dummy variable t, we can rewrite the last problem above as

$$\min \quad t \\ d_F(\Sigma, \hat{\Sigma}) \le t \\ \Sigma \in C_s^n.$$

It is easy to see that the inequality $d_F(\Sigma, \hat{\Sigma}) \le t$ can be written as a second-order cone constraint, and therefore, the formulation above can be transformed into a conic optimization problem.

Variations of this formulation can be obtained by introducing additional linear constraints. As an example, consider a subset E of all (i, j) covariance pairs and

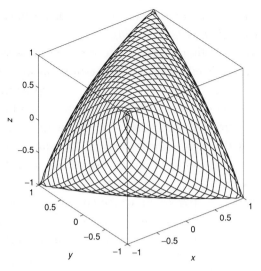

Figure 10.1 The feasible set of the nearest correlation matrix problem in three-dimensions

lower/upper limits $l_{ij}, u_{ij} \forall (i, j) \in E$ that we wish to impose on these entries. Then, we would need to solve the following problem:

$$
\begin{aligned}
\min \quad & d_F(\Sigma, \hat{\Sigma}) \\
l_{ij} \leq \quad & \Sigma_{ij} \quad \leq u_{ij}, \forall (i, j) \in E \\
& \Sigma \quad \in C_s^n.
\end{aligned}
\tag{10.6}
$$

When E consists of all the diagonal (i, i) elements and $l_{ii} = u_{ii} = 1, \forall i$, we get the correlation matrix version of the original problem. For example, three-dimensional correlation matrices have the following form:

$$
\Sigma = \begin{bmatrix} 1 & x & y \\ x & 1 & z \\ y & z & 1 \end{bmatrix}, \quad \Sigma \in C_s^3.
$$

The feasible set for this instance is shown in Figure 10.1.

Example 10.1 *We consider the following estimate of the correlation matrix of four securities:*

$$
\hat{\Sigma} = \begin{bmatrix} 1.0 & 0.8 & 0.5 & 0.2 \\ 0.8 & 1.0 & 0.9 & 0.1 \\ 0.5 & 0.9 & 1.0 & 0.7 \\ 0.2 & 0.1 & 0.7 & 1.0 \end{bmatrix}.
\tag{10.7}
$$

This, in fact, is not a valid correlation matrix; its smallest eigenvalue is negative:
$\lambda_{min} = -0.1337$. *Note, for example, the high correlations between assets 1 and 2
as well as assets 2 and 3. This suggests that 1 and 3 should be highly correlated
as well, but they are not. Which entry should one adjust to find a valid correlation
matrix?*

*We can approach this problem using formulation (10.6) with E consisting of all
the diagonal* (i, i) *elements and* $l_{ii} = u_{ii} = 1, \forall i$. *Solving the resulting problem,
for example, using SDPT3 [81], we obtain (approximately) the following nearest
correction to* $\hat{\Sigma}$:

$$
\Sigma = \begin{bmatrix}
1.00 & 0.76 & 0.53 & 0.18 \\
0.76 & 1.00 & 0.82 & 0.15 \\
0.53 & 0.82 & 1.00 & 0.65 \\
0.18 & 0.15 & 0.65 & 1.00
\end{bmatrix}.
$$

Exercise 10.3   Use a semidefinite optimization software package to verify that Σ
given above is the solution to (10.5) when $\hat{\Sigma}$ is given by (10.7).

Exercise 10.4   Resolve the problem above, this time imposing the constraint that
$\Sigma_{23} = \Sigma_{32} \geq 0.85$.

One can consider several variations on the "plain vanilla" version of the nearest
correlation matrix problem. For example, if we would rather keep some of the
entries of the matrix $\hat{\Sigma}$ constant, we can expand the set E to contain those elements
with matching lower and upper bounds. Another possibility is to weight the changes
in different entries, for example, if estimates of some entries are more trustworthy
than others.

Another important variation of the original problem is obtained by placing lower
limits on the smallest eigenvalue of the correlation matrix. Even when we have a
valid (positive semidefinite) correlation matrix estimate, having small eigenvalues
in the matrix can be undesirable as they lead to unstable portfolios. Indeed, the valid
correlation matrix we obtained above has a positive but very small eigenvalue, which
would in fact be exactly zero in exact arithmetic. Hauser and Zuev consider models
where minimum eigenvalue of the covariance matrix is maximized and use the
matrices in a robust optimization setting [39].

Exercise 10.5   We want to find the nearest symmetric matrix to $\hat{\Sigma}$ in (10.7) whose
smallest eigenvalue is at least 0.25. Express this problem as a semidefinite optimization problem. Solve it using an SDP software package.

All these variations are easily handled using semidefinite programming formulations and solved using semidefinite optimization software. As such, semidefinite
optimization presents a new tool for asset managers that was not previously available

at this level of sophistication and flexibility. While these tools are not yet available as commercial software packages, many academic products are freely available; see the links given in Section 9.4.

10.3 Recovering risk-neutral probabilities from options prices

In this section, we revisit our study of the risk-neutral density estimation problem in Section 8.4. Recall that the objective of this problem is to estimate an implied risk-neutral density function for the future price of an underlying security using the prices of options written on that security. Representing the density function using cubic splines to ensure its smoothness, and using a least-squares type objective function for the fit of the estimate with the observed option prices, we formulated an optimization problem in Section 8.4.

One issue that we left open in Section 8.4 is the rigorous enforcement of the nonnegativity of the risk-neutral density estimate. While we heuristically handled this issue by enforcing the nonnegativity of the cubic splines at the knots, it is clear that a cubic function that is nonnegative at the endpoints of an interval can very well become negative in between and therefore, the heuristic technique of Section 8.4 may be inadequate. Here we discuss an alternative formulation that is based on necessary and sufficient conditions for ensuring the nonnegativity of a single variable polynomial in intervals. This characterization is due to Bertsimas and Popescu [11] and is stated in the next proposition.

Proposition 10.1 (Proposition 1(d), [11])   *The polynomial $g(x) = \sum_{r=0}^{k} y_r x^r$ satisfies $g(x) \geq 0$ for all $x \in [a, b]$ if and only if there exists a positive semidefinite matrix $X = [x_{ij}]_{i,j=0,\ldots,k}$ such that*

$$\sum_{i,j:i+j=2\ell-1} x_{ij} = 0, \quad \ell = 1, \ldots, k, \tag{10.8}$$

$$\sum_{i,j:i+j=2\ell} x_{ij} = \sum_{m=0}^{\ell} \sum_{r=m}^{k+m-\ell} y_r \binom{r}{m} \binom{k-r}{\ell-m} a^{r-m} b^m, \tag{10.9}$$

$$\ell = 0, \ldots, k, \tag{10.10}$$

$$X \succeq 0. \tag{10.11}$$

In the statement of the proposition above, the notation $\binom{r}{m}$ stands for $\frac{r!}{m!(r-m)!}$ and $X \succeq 0$ indicates that the matrix X is symmetric and positive semidefinite. For the cubic polynomials $f_s(x) = \alpha_s x^3 + \beta_s x^2 + \gamma_s x + \delta_s$ that are used in the formulation of Section 8.4, the result can be simplified as follows:

Corollary 10.1   *The   polynomial $f_s(x) = \alpha_s x^3 + \beta_s x^2 + \gamma_s x + \delta_s$ satisfies $f_s(x) \geq 0$ for all $x \in [x_s, x_{s+1}]$ if and only if there exists a 4×4 matrix*

$X^s = [x_{ij}^s]_{i,j=0,\ldots,3}$ *such that*

$$x_{ij}^s = 0, \text{ if } i + j \text{ is } 1 \text{ or } 5,$$

$$x_{03}^s + x_{12}^s + x_{21}^s + x_{30}^s = 0,$$

$$x_{00}^s = \alpha_s x_s^3 + \beta_s x_s^2 + \gamma_s x_s + \delta_s,$$

$$x_{02}^s + x_{11}^s + x_{20}^s = 3\alpha_s x_s^2 x_{s+1} + \beta_s \left(2x_s x_{s+1} + x_s^2\right)$$
$$+ \gamma_s(x_{s+1} + 2x_s) + 3\delta_s, \tag{10.12}$$

$$x_{13}^s + x_{22}^s + x_{31}^s = 3\alpha_s x_s x_{s+1}^2 + \beta_s \left(2x_s x_{s+1} + x_{s+1}^2\right)$$
$$+ \gamma_s(x_s + 2x_{s+1}) + 3\delta_s,$$

$$x_{33}^s = \alpha_s x_{s+1}^3 + \beta_s x_{s+1}^2 + \gamma_s x_{s+1} + \delta_s,$$

$$X^s \succeq 0.$$

Observe that the positive semidefiniteness of the matrix X^s implies that the first diagonal entry x_{00}^s is nonnegative, which corresponds to our earlier requirement $f_s(x_s) \geq 0$. In light of Corollary 10.1 , we see that introducing the additional variables X^s and the constraints (10.12), for $s = 1, \ldots, n_s$, into the earlier quadratic programming problem in Section 8.4, we obtain a new optimization problem which necessarily leads to a risk-neutral probability distribution function that is nonnegative in its entire domain. The new formulation has the following form:

$$\min_{y, X^1, \ldots, X^{n_s}} \quad E(y)$$
$$\text{s.t.} \quad (8.19), (8.20), (8.21), (8.22), (8.25), [(10.12), s = 1, \ldots, n_s].$$
$$\tag{10.13}$$

All constraints in (10.13), with the exception of the positive semidefiniteness constraints $X^s \succeq 0$, $s = 1, \ldots, n_s$, are linear in the optimization variables $(\alpha_s, \beta_s, \gamma_s, \delta_s)$ and $X^s, s = 1, \ldots, n_s$. The positive semidefiniteness constraints are convex constraints and thus the resulting problem can be reformulated as a convex semidefinite programming problem with a quadratic objective function.

For appropriate choices of the vectors c, f_i, g_k^s, and matrices Q and H_k^s, we can rewrite problem (10.13) in the following equivalent form:

$$\min_{y, X^1, \ldots, X^{n_s}} c^T y + \frac{1}{2} y^T Q y$$
$$\text{s.t.} \quad f_i^T y = b_i, \ i = 1, \ldots, 3n_s,$$
$$H_k^s \bullet X^s = 0, \ k = 1, 2, \ s = 1, \ldots, n_s, \tag{10.14}$$
$$\left(g_k^s\right)^T y + H_k^s \bullet X^s = 0, \ k = 3, 4, 5, 6, \ s = 1, \ldots, n_s,$$
$$X^s \succeq 0, \ s = 1, \ldots, n_s,$$

where $\bullet$ denotes the trace matrix inner product.

We should note that standard semidefinite optimization software such as SDPT3 [81] can solve only problems with linear objective functions. Since the objective function of (10.14) is quadratic in y a reformulation is necessary to solve this problem using SDPT3 or other SDP solvers. We can replace the objective function with min t, where t is a new artificial variable, and impose the constraint $t \geq c^T y + \frac{1}{2} y^T Q y$. This new constraint can be expressed as a second-order cone constraint after a simple change of variables; see, e.g., [53]. This final formulation is a standard form *conic optimization* problem – a class of problems that contain semidefinite programming and second-order cone programming as special classes.

Exercise 10.6   Express the constraint $t \geq c^T y + \frac{1}{2} y^T Q y$ using linear constraints and a second-order cone constraint.

10.4 Arbitrage bounds for forward start options

When pricing securities with complicated payoff structures, one of the strategies analysts use is to develop a portfolio of "related" securities in order to form a super (or sub) hedge and then use no-arbitrage arguments to bound the price of the complicated security. Finding the super or sub hedge that gives the sharpest no-arbitrage bounds is formulated as an optimization problem. We considered a similar approach in Section 4.2 when we used linear programming models for detecting arbitrage possibilities in prices of European options with a common underlying asset and same maturity.

In this section, we consider the problem of finding arbitrage bounds for prices of forward start options using prices of standard options expiring either at the activation or expiration date of the forward start option. As we will see this problem can be solved using semidefinite optimization. The tool we use to achieve this is the versatile result of Bertsimas and Popescu given in Proposition 10.1 .

A forward start option is an advance purchase, say at time T_0, of a put or call option that will become active at some specified future time, say T_1. These options are encountered frequently in employee incentive plans where an employee may be offered an option on the company stock that will be available after the employee remains with the company for a predetermined length of time. A premium is paid at T_0, and the underlying security and the expiration date (T_2) are specified at that time. Let S_1 and S_2 denote the spot price of the underlying security at times T_1 and T_2, respectively.

The strike price is described as a known function of S_1 but is unknown at T_0. It is determined at T_1 when the option becomes active. Typically, it is chosen to be the value of the underlying asset at that time, i.e., S_1, so that the option is at-the-money at time T_1. More generally, the strike can be chosen as γS_1 for some positive constant γ. We address the general case here. The payoff to the buyer of a forward start call option at time T_2 is $\max(0, S_2 - \gamma S_1) = (S_2 - \gamma S_1)^+$, and similarly it is $(\gamma S_1 - S_2)^+$ for puts.

Our primary objective is to find tightest possible no-arbitrage bounds (i.e., maximize the lower bound and minimize the upper bound) by finding the best possible sub- and super-replicating portfolios of European options of several strikes with exercise dates at T_1 and also others with exercise dates at T_2. We will also consider the possibility of trading the underlying asset at time T_1 in a self-financing manner (via risk-free borrowing/lending). For concreteness, we limit our attention to the forward start call option problem and only consider calls for replication purposes. Since we allow the shorting of calls, the omission of puts does not lose generality.

We show how to (approximately) solve the following problem: find the cheapest portfolio of the underlying (traded now and/or at T_1), cash, calls expiring at time T_1, and calls expiring at time T_2, such that the payoff from the portfolio always is at least $(S_2 - \gamma S_1)^+$, no matter what S_1 and S_2 turn out to be. There is a similar lower bound problem that can be solved identically.

For simplification, we assume throughout the rest of this discussion that the risk-free interest rate r is zero and that the underlying does not pay any dividends. We also assume throughout the discussion that the prices of options available for replication are arbitrage-free, which implies the existence of equivalent martingale measures consistent with these prices. Furthermore, we ignore trading costs.

For replication purposes, we assume that a number of options expiring at T_1 and T_2 are available for trading. Let $K_1^1 < K_2^1 < \cdots < K_m^1$ denote the strike prices of options expiring at T_1 and $K_1^2 < K_2^2 < \cdots < K_n^2$ denote the strike prices of the options expiring at T_2. Let $p^1 = (p_1^1, \ldots, p_m^1)$ and $p^2 = (p_1^2, \ldots, p_n^2)$ denote the (arbitrage-free) prices of these options at time T_0.

We assume that $K_1^1 = 0$, so that the first "call" is the underlying itself and $p_1^1 = S_0$, the price of the underlying at T_0. For our formulation, let $x = (x_1, x_2, \ldots, x_m)$ and $y = (y_1, y_2, \ldots, y_n)$ correspond to the positions in the T_1- and T_2-expiry options in our portfolio. Let B denote the cash position in the portfolio at time T_0. Then, the cost of this portfolio is

$$c(x, y, B) := \sum_{i=1}^{m} p_i^1 x_i + \sum_{j=1}^{n} p_j^2 y_j + B. \tag{10.15}$$

Holding only these call options and not trading until T_2, we would have a static hedge. To improve the bounds, we consider a semi-static hedge that is rebalanced at time T_1 through the purchase of underlying shares whose quantity is determined based on the price of the underlying at that time. If $f(S_1)$ shares of the underlying are purchased at time T_1 and if this purchase is financed by risk-free borrowing, our overall position would have the final payoff of:

$$g(S_1, S_2) := g_S(S_1, S_2) + f(S_1)(S_2 - S_1) \qquad (10.16)$$

$$= \sum_{i=1}^{m} (S_1 - K_i^1)^+ x_i + \sum_{j=1}^{n} (S_2 - K_j^2)^+ y_j + B + f(S_1)(S_2 - S_1).$$

Exercise 10.7   Verify equation (10.16).

Then, we would find the upper bound on the price of the forward start option by solving the following problem:

$$u := \min_{x, y, B, f} c(x, y, B)$$
$$\text{s.t.} \quad g(S_1, S_2) \geq (S_2 - \gamma S_1)^+, \quad \forall S_1, S_2 \geq 0. \qquad (10.17)$$

The inequalities in this optimization problem ensure the super-replication properties of the semi-static hedge we constructed. Unfortunately, there are infinitely many constraints indexed by the parameters S_1 and S_2. Therefore, (10.17) is a semi-infinite linear optimization problems and can be difficult.

Fortunately, however, the constraint functions are expressed using piecewise-linear functions of S_1 and S_2. The breakpoints for these functions are at the strike sets $\{K_1^1, \ldots, K_m^1\}$ and $\{K_1^2, \ldots, K_n^2\}$. The right-hand-side function $(S_2 - \gamma S_1)^+$ has breakpoints along the line $S_2 = \gamma S_1$. The remaining difficulty is about the specification of the function f. By limiting our attention to functions f that are piecewise linear we will obtain a conic optimization formulation.

A piecewise linear function $f(S_1)$ is determined by its values at the breakpoints: $z_i = f(K_i^1)$ for $i = 1, \ldots, m$ and its slope past K_m^1 (the last breakpoint) given by $\lambda_z = f(K_m^1 + 1) - f(K_m^1)$.

Thus, we approximate $f(S_1)$ as

$$f(S_1) = \begin{cases} z_i + (S_1 - K_i^1) \dfrac{z_{i+1} - z_i}{K_{i+1}^1 - K_i^1} & \text{if } S_1 \in [K_i^1, K_{i+1}^1), \\ z_m + (S_1 - K_m^1)\lambda_z & \text{if } S_1 \geq K_m^1. \end{cases}$$

Next, we consider a decomposition of the nonnegative orthant ($S_1, S_2 \geq 0$) into a grid with breakpoints at K_i^1's and K_j^2's such that the payoff function is **linear** in each *box* $B_{ij} = [K_i^1, K_{i+1}^1] \times [K_j^2, K_{j+1}^2]$:

$$g(S_1, S_2) = \sum_{k=1}^{n} (S_1 - K_k^1)^+ x_k + \sum_{l=1}^{n} (S_2 - K_l^2)^+ y_l + B + (S_2 - S_1)f(S_1)$$

$$= \sum_{k=1}^{i} (S_1 - K_k^1) x_k + \sum_{l=1}^{j} (S_2 - K_l^2) y_l + B$$

$$+ (S_2 - S_1)\left(z_i + (S_1 - K_i^1)\frac{z_{i+1} - z_i}{K_{i+1}^1 - K_i^1}\right)$$

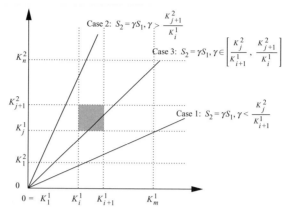

Figure 10.2 Three possible relative positions of the $S_2 = \gamma S_1$ line

Recall that we want to super-replicate the payoff $(S_2 - \gamma S_1)^+$. This means that the term $g(S_1, S_2)$ must exceed it for all S_1, S_2. When we consider the box $B_{ij} := [K_i^1, K_{i+1}^1] \times [K_j^2, K_{j+1}^2]$ there are three possibilities involving γ; see also Figure 10.2:

1. $S_2 > \gamma S_1$ for all $(S_1, S_2) \in B_{ij}$. Then, we replace $(S_2 - \gamma S_1)^+$ with $(S_2 - \gamma S_1)$.
2. $S_2 < \gamma S_1$ for all $(S_1, S_2) \in B_{ij}$. Then, we replace $(S_2 - \gamma S_1)^+$ with 0.
3. Otherwise, we replace $g(S_1, S_2) \geq (S_2 - \gamma S_1)^+$ with the two inequalities $g(S_1, S_2) \geq (S_2 - \gamma S_1)$ and $g(S_1, S_2) \geq 0$.

In all cases, we remove the nonlinearity on the right-hand side. Now, we can rewrite the super-replication inequality

$$g(S_1, S_2) \geq (S_2 - \gamma S_1)^+, \forall (S_1, S_2) \in B_{ij} \tag{10.18}$$

as

$$\alpha_{ij}(w)S_1^2 + \beta_{ij}(w)S_1 + \delta_{ij}(w)S_1 S_2 + \epsilon_{ij}(w)S_2 + \eta_{ij}(w)$$
$$\geq 0, \forall (S_1, S_2) \in B_{ij}, \tag{10.19}$$

where $w = (x, y, z, B)$ represents the variables of the problem collectively and the constants α_{ij}, etc., are easily obtained linear functions of these variables. In Case 3, we have two such inequalities rather than one.

Thus, the super-replication constraints in each box are polynomial inequalities that must hold within these boxes. This is very similar to the situation addressed by Proposition 10.1 with the important distinction that these polynomial inequalities are in two variables rather than one.

Next, observe that, for a fixed value of S_1, the function on the left-hand side of inequality (10.18) is linear in S_2. Let us denote this function with

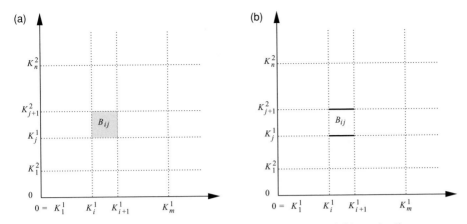

Figure 10.3 Super-replication constraints (a) in the box B_{ij} and (b) on the line segments

$h_{ij}(S_1, S_2)$. Since it is linear in S_2, for a fixed value of S_1, h_{ij} will assume its minimum value in the interval $[K_j^2, K_{j+1}^2]$ either at $S_2 = K_j^2$ or $S_2 = K_{j+1}^2$. Thus, if $h_{ij}(S_1, K_j^2) \geq 0$ and $h_{ij}(S_1, K_{j+1}^2) \geq 0$, then $h_{ij}(S_1, S_2) \geq 0, \forall S_2 \in [K_j^2, K_{j+1}^2]$. As a result, $h_{ij}(S_1, S_2) \geq 0, \forall (S_1, S_2) \in B_{ij}$ is equivalent to the following two constraints:

$$H_{ij}^l(S_1) := h_{ij}(S_1, K_j^2) \geq 0, \forall S_1 \in [K_i^1, K_{i+1}^1],$$
$$H_{ij}^u(S_1) := h_{ij}(S_1, K_{j+1}^2) \geq 0, \forall S_1 \in [K_i^1, K_{i+1}^1].$$

The situation is illustrated in Figure 10.3. Instead of satisfying the inequality on the whole box as in Figure 10.3(a), we only need to consider two line segments as in Figure 10.3(b).

The bivariate polynomial inequality is reduced to two univariate polynomial inequalities. Now, we can use the Bertsimas/Popescu result and represent this inequality efficiently. In summary, the super-replication constraints can be rewritten using a **finite** number of linear constraints and semidefiniteness constraints. Since H_{ij}^l and H_{ij}^u are quadratic polynomials in S_1, semidefiniteness constraints are on 3×3 matrices (see Proposition 10.1) and are easily handled with semidefinite programming software.

11

Integer programming: theory and algorithms

11.1 Introduction

A linear programming model for constructing a portfolio of assets might produce a solution with 3205.7 shares of stock XYZ and similarly complicated figures for the other assets. Most portfolio managers would have no trouble rounding the value 3205.7 to 3205 shares or even 3200 shares. In this case, a linear programming model would be appropriate. Its optimal solution can be used effectively by the decision maker, with minor modifications. On the other hand, suppose that the problem is to find the best among many alternatives (for example, a traveling salesman wants to find a shortest route going through ten specified cities). A model that suggests taking fractions of the roads between the various cities would be of little value. A 0,1 decision has to be made (a road between a pair of cities is either on the shortest route or it is not), and we would like the model to reflect this.

This integrality restriction on the variables is the central aspect of integer programming. From a modeling standpoint, integer programming has turned out to be useful in a wide variety of applications. With integer variables, one can model logical requirements, fixed costs, and many other problem aspects. Many software products can change a linear programming problem into an integer program with a single command.

The downside of this power, however, is that problems with more than a thousand variables are often not possible to solve unless they show a specific exploitable structure. Despite the possibility (or even likelihood) of enormous computing times, there are methods that can be applied to solving integer programs. The most widely used is "branch and bound" (it is used, for example, in SOLVER). More sophisticated commercial codes (CPLEX and XPRESS are currently two of the best) use a combination of "branch and bound" and another complementary approach called "cutting plane." Open source software codes in the COIN-OR library also implement a combination of branch and bound and cutting plane, called "branch

192

and cut" (such as cbc, which stands for COIN Branch and Cut, or bcp, which stands for Branch, Cut, and Price). The purpose of this chapter is to describe some of the solution techniques. For the reader interested in learning more about integer programming, we recommend Wolsey's introductory book [83]. The next chapter discusses problems in finance that can be modeled as integer programs: combinatorial auctions, constructing an index fund, portfolio optimization with minimum transaction levels.

First we introduce some terminology. An *integer linear program* is a linear program with the additional constraint that some of, or all, the variables are required to be integer. When all variables are required to be integer the problem is called a *pure integer linear program*. If some variables are restricted to be integer and some are not then the problem is a *mixed integer linear program,* denoted MILP. The case where the integer variables are restricted to be 0 or 1 comes up surprisingly often. Such problems are called *pure (mixed) 0–1 linear programs* or *pure (mixed) binary integer linear programs*. The case of an NLP with the additional constraint that some of the variables are required to be integer is called MINLP and is receiving an increasing amount of attention from researchers. In this chapter, we concentrate on MILP.

11.2 Modeling logical conditions

Suppose we wish to invest \$19 000. We have identified four investment opportunities. Investment 1 requires an investment of \$6700 and has a net present value of \$8000; investment 2 requires \$10 000 and has a value of \$11 000; investment 3 requires \$5500 and has a value of \$6000; and investment 4 requires \$3400 and has a value of \$4000. Into which investments should we place our money so as to maximize our total present value? Each project is a "take it or leave it" opportunity: we are not allowed to invest partially in any of the projects. Such problems are called *capital budgeting problems*.

As in linear programming, our first step is to decide on the variables. In this case, it is easy: we will use a 0–1 variable x_j for each investment. If x_j is 1 then we will make investment j. If it is 0, we will not make the investment.

This leads to the 0–1 programming problem:

$$\max\ 8x_1 + 11x_2 + 6x_3 + 4x_4$$

subject to

$$6.7x_1 + 10x_2 + 5.5x_3 + 3.4x_4 \le 19$$

$$x_j = 0 \text{ or } 1.$$

Now, a straightforward "bang for buck" suggests that investment 1 is the best choice. In fact, ignoring integrality constraints, the optimal linear programming solution is $x_1 = 1$, $x_2 = 0.89$, $x_3 = 0$, $x_4 = 1$ for a value of \$21 790. Unfortunately, this solution is not integral. Rounding x_2 down to 0 gives a feasible solution with a value of \$12 000. There is a better integer solution, however, of $x_1 = 0$, $x_2 = 1$, $x_3 = 1$, $x_4 = 1$ for a value of \$21 000. This example shows that rounding does not necessarily give an optimal solution.

There are a number of additional constraints we might want to add. For instance, consider the following constraints:

1. We can only make two investments.
2. If investment 2 is made, then investment 4 must also be made.
3. If investment 1 is made, then investment 3 cannot be made.

All of these, and many more *logical restrictions*, can be enforced using 0–1 variables. In these cases, the constraints are:

1. $x_1 + x_2 + x_3 + x_4 \leq 2$.
2. $x_2 - x_4 \leq 0$.
3. $x_1 + x_3 \leq 1$.

Solving the model with SOLVER

Modeling an integer program in SOLVER is almost the same as modeling a linear program. For example, if you placed binary variables x_1, x_2, x_3, x_4 in cells \$B\$5:\$B\$8, simply Add the constraint

   \$B\$5:\$B\$8 Bin

to your other constraints in the SOLVER dialog box. Note that the Bin option is found in the small box where you usually indicate the type of inequality: =, <=, or >=. Just click on Bin. That's all there is to it!

It is equally easy to model an integer program within other commercial codes. The formulation might look as follows:

```
! Capital budgeting example
VARIABLES
x(i=1:4)
OBJECTIVE
Max: 8*x(1)+11*x(2)+6*x(3)+4*x(4)
CONSTRAINTS
Budget: 6.7*x(1)+10*x(2)+5.5*x(3)+3.4*x(4) < 19
BOUNDS
x(i=1:4) Binary
END
```

Exercise 11.1   As the leader of an oil exploration drilling venture, you must determine the best selection of five out of ten possible sites. Label the sites $s_1, s_2, \ldots, s_{10}$ and the expected profits associated with each as $p_1, p_2, \ldots, p_{10}$.

(i) If site s_2 is explored, then site s_3 must also be explored.
(ii) Exploring sites s_1 *and* s_7 will prevent you from exploring site s_8.
(iii) Exploring sites s_3 *or* s_4 will prevent you from exploring site s_5.

Formulate an integer program to determine the best exploration scheme and solve with SOLVER.

Answer:

$$\max \ \sum_{j=1}^{10} p_j x_j$$

subject to

$$\sum_{j=1}^{10} x_j = 5$$
$$x_2 - x_3 \leq 0$$
$$x_1 + x_7 + x_8 \leq 2$$
$$x_3 + x_5 \leq 1$$
$$x_4 + x_5 \leq 1$$
$$x_j = 0 \text{ or } 1 \text{ for } j = 1, \ldots, 10.$$

Exercise 11.2   Consider the following investment projects where, for each project, you are given its NPV as well as the cash outflow required during each year (in million dollars).

	NPV	Year 1	Year 2	Year 3	Year 4
Project 1	30	12	4	4	0
Project 2	30	0	12	4	4
Project 3	20	3	4	4	4
Project 4	15	10	0	0	0
Project 5	15	0	11	0	0
Project 6	15	0	0	12	0
Project 7	15	0	0	0	13
Project 8	24	8	8	0	0
Project 9	18	0	0	10	0
Project 10	18	0	0	0	10

No partial investment is allowed in any of these projects. The firm has 18 million dollars available for investment each year.

(i) Formulate an integer linear program to determine the best investment plan and solve with SOLVER.

(ii) Formulate the following conditions as linear constraints.
- Exactly one of Projects 4, 5, 6, 7 must be invested in.
- If Project 1 is invested in, then Project 2 cannot be invested in.
- If Project 3 is invested in, then Project 4 must also be invested in.
- If Project 8 is invested in, then either Project 9 or Project 10 must also be invested in.
- If either Project 1 or Project 2 is invested in, then neither Project 8 nor Project 9 can be invested in.

11.3 Solving mixed integer linear programs

Historically, the first method developed for solving MILPs was based on cutting planes (adding constraints to the underlying linear program to cut off noninteger solutions). This idea was proposed by Gomory [32] in 1958. Branch and bound was proposed in 1960 by Land and Dong [50]. It is based on dividing the problem into a number of smaller problems (branching) and evaluating their quality based on solving the underlying linear programs (bounding). Branch and bound has been the most effective technique for solving MILPs in the following 40 years or so. However, in the last ten years, cutting planes have made a resurgence and are now efficiently combined with branch and bound into an overall procedure called branch and cut. This term was coined by Padberg and Rinaldi [62] in 1987. All these approaches involve solving a series of *linear* programs. So that is where we begin.

11.3.1 Linear programming relaxation

Given a mixed integer linear program

$$\text{(MILP)} \quad \min c^{\text{T}} x$$
$$Ax \geq b$$
$$x \geq 0$$
$$x_j \text{ integer for } j = 1, \ldots, p,$$

there is an associated linear program called the *relaxation* formed by dropping the integrality restrictions:

$$\text{(R)} \quad \min c^{\text{T}} x$$
$$Ax \geq b$$
$$x \geq 0.$$

Since R is less constrained than MILP, the following are immediate:

- The optimal objective value for R is less than or equal to the optimal objective for MILP.
- If R is infeasible, then so is MILP.
- If the optimal solution x^* of R satisfies x_j^* integer for $j = 1, \ldots, p$, then x^* is also optimal for MILP.

So solving R does give some information: it gives a bound on the optimal value, and, if we are lucky, it may give the optimal solution to MILP. However, rounding the solution of R will not in general give the optimal solution of MILP.

Exercise 11.3 Consider the problem

$$\max 20x_1 + 10x_2 + 10x_3$$
$$2x_1 + 20x_2 + 4x_3 \leq 15$$
$$6x_1 + 20x_2 + 4x_3 = 20$$
$$x_1, x_2, x_3 \geq 0 \text{ integer.}$$

Solve its linear programming relaxation. Then, show that it is impossible to obtain a feasible integral solution by rounding the values of the variables.

Exercise 11.4

(a) Compare the feasible solutions of the following three integer linear programs:

(i) $\max 14x_1 + 8x_2 + 6x_3 + 6x_4$
$28x_1 + 15x_2 + 13x_3 + 12x_4 \leq 39$
$x_1, x_2, x_3, x_4 \in \{0, 1\}$,

(ii) $\max 14x_1 + 8x_2 + 6x_3 + 6x_4$
$2x_1 + x_2 + x_3 + x_4 \leq 2$
$x_1, x_2, x_3, x_4 \in \{0, 1\}$,

(iii) $\max 14x_1 + 8x_2 + 6x_3 + 6x_4$
$x_2 + x_3 + x_4 \leq 2$
$x_1 + x_2 \leq 1$
$x_1 + x_3 \leq 1$
$x_1 + x_4 \leq 1$
$x_1, x_2, x_3, x_4 \in \{0, 1\}$.

(b) Compare the relaxations of the above integer programs obtained by replacing $x_1, x_2, x_3, x_4 \in \{0, 1\}$ by $0 \leq x_j \leq 1$ for $j = 1, \ldots, 4$. Which is the best formulation among (i), (ii), and (iii) for obtaining a tight bound from the linear programming relaxation?

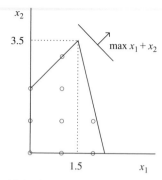

Figure 11.1 A two-variable integer program

11.3.2 Branch and bound

An example

We first explain branch and bound by solving the following pure integer linear program (see Figure 11.1):

$$\max\ x_1 + x_2$$
$$-x_1 + x_2 \le 2$$
$$8x_1 + 2x_2 \le 19$$
$$x_1, x_2 \ge 0$$
$$x_1, x_2 \text{ integer.}$$

The first step is to solve the linear programming relaxation obtained by ignoring the last constraint. The solution is $x_1 = 1.5$, $x_2 = 3.5$ with objective value 5. This is not a feasible solution to the integer program since the values of the variables are fractional. How can we exclude this solution while preserving the feasible integral solutions? One way is to *branch*, creating two linear programs, say one with $x_1 \le 1$, the other with $x_1 \ge 2$. Clearly, any solution to the integer program must be feasible to one or the other of these two problems. We will solve both of these linear programs. Let us start with

$$\max\ x_1 + x_2$$
$$-x_1 + x_2 \le 2$$
$$8x_1 + 2x_2 \le 19$$
$$x_1 \le 1$$
$$x_1, x_2 \ge 0.$$

The solution is $x_1 = 1$, $x_2 = 3$ with objective value 4. This is a feasible integral solution. So we now have an upper bound of 5 as well as a lower bound of 4 on the value of an optimum solution to the integer program.

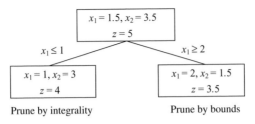

Figure 11.2 Branch-and-bound tree

Now we solve the second linear program

$$\max\ x_1 + x_2$$
$$-x_1 + x_2 \le 2$$
$$8x_1 + 2x_2 \le 19$$
$$x_1 \ge 2$$
$$x_1, x_2 \ge 0.$$

The solution is $x_1 = 2$, $x_2 = 1.5$ with objective value 3.5. Because this value is worse that the lower bound of 4 that we already have, we do not need any further branching. We conclude that the feasible integral solution of value 4 found earlier is optimum.

The solution of the above integer program by branch and bound required the solution of three linear programs. These problems can be arranged in a *branch-and-bound tree*, see Figure 11.2. Each *node* of the tree corresponds to one of the problems that were solved.

We can stop the enumeration at a node of the branch-and-bound tree for three different reasons (when they occur, the node is said to be *pruned*).

- *Pruning by integrality* occurs when the corresponding linear program has an optimum solution that is integral.
- *Pruning by bounds* occurs when the objective value of the linear program at that node is worse than the value of the best feasible solution found so far.
- *Pruning by infeasibility* occurs when the linear program at that node is infeasible.

To illustrate a larger tree, let us solve the same integer program as above, with a different objective function:

$$\max\ 3x_1 + x_2$$
$$-x_1 + x_2 \le 2$$
$$8x_1 + 2x_2 \le 19$$
$$x_1, x_2 \ge 0$$
$$x_1, x_2\ \text{integer}.$$

The solution of the linear programming relaxation is $x_1 = 1.5$, $x_2 = 3.5$ with objective value 8. Branching on variable x_1, we create two linear programs. The one with the additional constraint $x_1 \leq 1$ has solution $x_1 = 1$, $x_2 = 3$ with value 6 (so now we have an upper bound of 8 and a lower bound of 6 on the value of an optimal solution of the integer program). The linear program with the additional constraint $x_2 \geq 2$ has solution $x_1 = 2$, $x_2 = 1.5$ and objective value 7.5. Note that the value of x_2 is fractional, so this solution is not feasible to the integer program. Since its objective value is higher than 6 (the value of the best integer solution found so far), we need to continue the search. Therefore we branch on variable x_2. We create two linear programs, one with the additional constraint $x_2 \geq 2$, the other with $x_2 \leq 1$, and solve both. The first of these linear programs is infeasible. The second is

$$\max\ 3x_1 + x_2$$
$$-x_1 + x_2 \leq 2$$
$$8x_1 + 2x_2 \leq 19$$
$$x_1 \geq 2$$
$$x_2 \leq 1$$
$$x_1, x_2 \geq 0.$$

The solution is $x_1 = 2.125$, $x_2 = 1$ with objective value 7.375. Because this value is greater than 6 and the solution is not integral, we need to branch again on x_1. The linear program with $x_1 \geq 3$ is infeasible. The one with $x_1 \leq 2$ is

$$\max\ 3x_1 + x_2$$
$$-x_1 + x_2 \leq 2$$
$$8x_1 + 2x_2 \leq 19$$
$$x_1 \geq 2$$
$$x_2 \leq 1$$
$$x_1 \leq 2$$
$$x_1, x_2 \geq 0.$$

The solution is $x_1 = 2$, $x_2 = 1$ with objective value 7. This node is pruned by integrality and the enumeration is complete. The optimal solution is the one with value 7. See Figure 11.3.

The branch-and-bound algorithm

Consider a mixed integer linear program:

$$(\text{MILP})\ \ z_I = \ \min\ c^T x$$
$$Ax \geq b$$
$$x \geq 0$$
$$x_j\ \text{integer for } j = 1, \ldots, p.$$

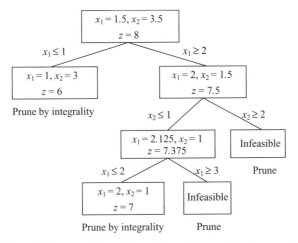

Figure 11.3 Branch-and-bound tree for modified example

The data are an n-vector c, an $m \times n$ matrix A, an m-vector b and an integer p such that $1 \le p \le n$. The set $I = \{1, \ldots, p\}$ indexes the integer variables whereas the set $C = \{p+1, \ldots, n\}$ indexes the continuous variables. The branch-and-bound algorithm keeps a list of linear programming problems obtained by relaxing the integrality requirements on the variables and imposing constraints such as $x_j \le u_j$ or $x_j \ge l_j$. Each such linear program corresponds to a *node* of the branch-and-bound tree. For a node N_i, let z_i denote the value of the corresponding linear program (it will be convenient to denote this linear program by N_i as well). Let $\mathcal{L}$ denote the list of nodes that must still be solved (i.e., that have not been pruned nor branched on). Let z_U denote an upper bound on the optimum value z_I (initially, the bound z_U can be derived from a heuristic solution of (MILP), or it can be set to $+\infty$).

0. Initialize
$\mathcal{L} = \{MILP\}$, $z_U = +\infty$, $x^* = \emptyset$.

1. Terminate?
If $\mathcal{L} = \emptyset$, the solution x^* is optimal.

2. Select node
Choose and delete a problem N_i from $\mathcal{L}$.

3. Bound
Solve N_i. If it is infeasible, go to Step 1. Else, let x^i be its solution and z_i its objective value.

4. Prune
If $z_i \ge z_U$, go to Step 1.
If x^i is not feasible to (MILP), go to Step 5.
If x^i is feasible to (MILP), let $z_U = z_i$, $x^* = x^i$ and delete from $\mathcal{L}$ all problems with $z_j \ge z_U$. Go to Step 1.

5. Branch

From N_i, construct linear programs $N_i^1, \ldots, N_i^k$ with smaller feasible regions whose union contains all the feasible solutions of (MILP) in N_i. Add $N_i^1, \ldots, N_i^k$ to $\mathcal{L}$ and go to Step 1.

Various choices are left open by the algorithm, such as the node selection criterion and the branching strategy. We will discuss some options for these choices. Even more important to the success of branch-and-bound is the ability to prune the tree (Step 4). This will occur when z_U is a good upper bound on z_I and when z_i is a good lower bound. For this reason, it is crucial to have a formulation of (MILP) such that the value of its linear programming relaxation z_{LP} is as close as possible to z_I. To summarize, four issues need attention when solving MILPs by branch and bound:

- formulation (so that the gap $z_I - z_{LP}$ is small);
- heuristics (to find a good upper bound z_U);
- branching;
- node selection.

We defer the formulation issue to Section 11.3.3 on cutting planes. This issue will also be addressed in Chapter 12. Heuristics can be designed either as stand alone (an example will be given in Section 12.3) or as part of the branch-and-bound algorithm (by choosing branching and node selection strategies that are more likely to produce feasible solutions x^i to (MILP) in Step 4). We discuss branching strategies first, followed by node selection strategies and heuristics.

Branching

Problem N_i is a linear program. A way of dividing its feasible region is to impose bounds on a variable. Let x_j^i be one of the fractional values for $j = 1, \ldots, p$, in the optimal solution x^i of N_i (we know that there is such a j, since otherwise N_i would have been pruned in Step 4 on account of x^i being feasible to (MILP)). From problem N_i, we can construct two linear programs N_{ij}^- and N_{ij}^+ that satisfy the requirements of Step 5 by adding the constraints $x_j \leq \lfloor x_j^i \rfloor$ and $x_j \geq \lceil x_j^i \rceil$ respectively to N^i. The notation $\lfloor a \rfloor$ and $\lceil a \rceil$ means a rounded down and up to the nearest integer respectively. This is called *branching on a variable*. The advantage of branching on a variable is that the number of constraints in the linear programs does not increase, since linear programming solvers treat bounds on variables implicitly.

An important question is: on which variable x_j should we branch, among the $j = 1, \ldots, p$ such that x_j^i is fractional? To answer this question, it would be very helpful to know the increase D_{ij}^- in objective value between N_i and N_{ij}^-, and D_{ij}^+ between N_i and N_{ij}^+. A good branching variable x_j at node N^i is one for which both D_{ij}^- and D_{ij}^+ are relatively large (thus tightening the lower bound z_i, which is useful for pruning). For example, researchers have proposed to choose $j = 1, \ldots, p$

such that $\min(D_{ij}^-, D_{ij}^+)$ is the largest. Others have proposed to choose j such that $D_{ij}^- + D_{ij}^+$ is the largest. Combining these two criteria is even better, with more weight on the first.

The strategy which consists in computing D_{ij}^- and D_{ij}^+ explicitly for each j is called *strong branching*. It involves solving linear programs that are small variations of N_i by performing dual simplex pivots (recall Section 2.4.5), for each $j = 1, \ldots, p$ such that x_j^i is fractional and each of the two bounds. Experiments indicate that strong branching reduces the size of the enumeration tree by a factor of 20 or more in most cases, relative to a simple branching rule such as branching on the most fractional variable. Thus there is a clear benefit to spending time on strong branching. But the computing time of doing it at each node N_i, for every fractional variable x_j^i, may be too high. A reasonable strategy is to restrict the j's that are evaluated to those for which the fractional part of x_j^i is closest to 0.5 so that the amount of computing time spent performing these evaluations is limited. Significantly more time should be spent on these evaluations towards the top of the tree. This leads to the notion of *pseudocosts* that are initialized at the root node and then updated throughout the branch-and-bound tree.

Let $f_j^i = x_j^i - \lfloor x_j^i \rfloor$ be the fractional part of x_j^i, for $j = 1, \ldots p$. For an index j such that $f_j^i > 0$, define the *down pseudocost* and *up pseudocost* as

$$P_j^- = \frac{D_{ij}^-}{f_j^i} \text{ and } P_j^+ = \frac{D_{ij}^+}{1 - f_j^i}$$

respectively. Benichou *et al.* [10] observed that the pseudocosts tend to remain fairly constant throughout the branch-and-bound tree. Therefore the pseudocosts need not be computed at each node of the tree. They are estimated instead. How are they initialized and how are they updated in the tree? A good way of initializing the pseudocosts is through strong branching at the root node or other nodes of the tree when a variable becomes fractional for the first time. The down pseudocost P_j^- is updated by averaging the observations D_{ij}^-/f_j^i over all the nodes of the tree where x_j was branched on. Similarly for the up pseudocost P_j^+. The decision of which variable to branch on at a node N_i of the tree is done as follows. The estimated pseudocosts P_j^- and P_j^+ are used to compute estimates of D_{ij}^- and D_{ij}^+ at node N_i, namely $D_{ij}^- = P_j^- f_j^i$ and $D_{ij}^+ = P_j^+(1 - f_j^i)$ for each $j = 1, \ldots, p$ such that $f_j^i > 0$. Among these candidates, the branching variable x_j is chosen to be the one with largest $\min(D_{ij}^-, D_{ij}^+)$ (or other criteria such as those mentioned earlier).

Node selection

How does one choose among the different problems N_i available in Step 2 of the algorithm? Two goals need to be considered: finding good feasible solutions (thus decreasing the upper bound z_U) and proving optimality of the current best feasible solution (by increasing the lower bound as quickly as possible).

For the first goal, we estimate the value of the best feasible solution in each node N_i. For example, we could use the following estimate:

$$E_i = z_i + \sum_{j=1}^{p} \min \left(P_j^- f_j^i, P_j^+ (1 - f_j^i) \right),$$

based on the pseudocosts defined above. This corresponds to rounding the noninteger solution x^i to a nearby integer solution and using the pseudocosts to estimate the degradation in objective value. We then select a node N_i with the smallest E_i. This is the so-called "best estimate criterion" node selection strategy.

For the second goal, the best strategy depends on whether the first goal has been achieved already. If we have a very good upper bound z_U, it is reasonable to adopt a depth-first search strategy. This is because the linear programs encountered in a depth-first search are small variations of one another. As a result they can be solved faster in sequence, using the dual simplex method initialized with the optimal solution of the father node (about ten times faster, based on empirical evidence). On the other hand, if no good upper bound is available, depth-first search is wasteful: it may explore many nodes with a value z_i that is larger than the optimum z_I. This can be avoided by using the "best bound" node selection strategy, which consists in picking a node N_i with the smallest bound z_i. Indeed, no matter how good a solution of (MILP) is found in other nodes of the branch-and-bound tree, the node with the smallest bound z_i cannot be pruned by bounds (assuming no ties) and therefore it will have to be explored eventually. So we might as well explore it first. This strategy minimizes the total number of nodes in the branch-and-bound tree.

The most successful node selection strategy may differ depending on the application. For this reason, most MILP solvers have several node selection strategies available as options. The default strategy is usually a combination of the "best estimate criterion" (or a variation) and depth-first search. Specifically, the algorithm may dive using depth-first search until it reaches an infeasible node N_i or it finds a feasible solution of (MILP). At this point, the next node might be chosen using the "best estimate criterion" strategy, and so on, alternating between dives in a depth-first search fashion to get feasible solutions at the bottom of the tree and the "best estimate criterion" to select the next most promising node.

Heuristics

Heuristics are useful for improving the bound z_U, which helps in Step 4 for pruning by bounds. Of course, heuristics are even more important when the branch-and-bound algorithm is too time consuming and has to be terminated before completion, returning a solution of value z_U without a proof of its optimality.

We have already presented all the ingredients needed for a diving heuristic: solve the linear programming relaxation, use strong branching or pseudocosts to determine a branching variable; then compute the estimate E_i at each of the two sons and move down the branch corresponding to the smallest of the two estimates. Solve the new linear programming relaxation with this variable fixed and repeat until infeasibility is reached or a solution of (MILP) is found. The diving heuristic can be repeated from a variety of starting points (corresponding to different sets of variables being fixed) to improve the chance of getting good solutions.

An interesting idea that has been proposed recently to improve a feasible solution of (MILP) is called *local branching* [28]. This heuristic is particularly suited for MILPs that are too large to solve to optimality, but where the linear programming relaxation can be solved in reasonable time. For simplicity, assume that all the integer variables are 0,1 valued. Let $\bar{x}$ be a feasible solution of (MILP) (found by a diving heuristic, for example). The idea is to define a neighborhood of $\bar{x}$ as follows:

$$\sum_{j=1}^{p} |x_j - \bar{x}_j| \le k,$$

where k is an integer chosen by the user (for example $k = 20$ seems to work well), to add this constraint to (MILP) and apply your favorite MILP solver. Instead of getting lost in a huge enumeration tree, the search is restricted to the neighborhood of $\bar{x}$ by this constraint. Note that the constraint should be linearized before adding it to the formulation, which is easy to do:

$$\sum_{j \in I: \bar{x}_j=0} x_j + \sum_{j \in I: \bar{x}_j=1} (1 - x_j) \le k.$$

If a better solution than $\bar{x}$ is found, the neighborhood is redefined relatively to this new solution, and the procedure is repeated until no better solution can be found.

Exercise 11.5 Consider an investment problem as in Section 11.2. We have $14\,000 to invest among four different investment opportunities. Investment 1 requires an investment of $7000 and has a net present value of $11\,000; investment 2 requires $5000 and has a value of $8000; investment 3 requires $4000 and has a value of $6000; and investment 4 requires $3000 and has a value of $4000. As in Section 11.2, these are "take it or leave it" opportunities and we are not allowed to invest partially in any of the projects. The objective is to maximize our total value given the budget constraint. We do not have any other (logical) constraints.

We formulate this problem as an integer program using 0–1 variables x_j for each investment. As before, x_j is 1 if we make investment j and 0 if we do not. This leads to the following formulation:

$$\max \ 11x_1 + 8x_2 + 6x_3 + 4x_4$$
$$7x_1 + 5x_2 + 4x_3 + 3x_4 \le 14$$
$$x_j = 0 \ or \ 1.$$

The linear relaxation solution is $x_1 = 1$, $x_2 = 1$, $x_3 = 0.5$, $x_4 = 0$ with a value of 22. Since x_3 is not integer, we do not have an integer solution yet. Solve this problem using the branch-and-bound technique.

Exercise 11.6 Solve the three integer linear programs of Exercise 11.4 using your favorite solver. In each case, report the number of nodes in the enumeration tree. Is it related to the tightness of the linear programming relaxation studied in Exercise 11.4 (b)?

Exercise 11.7 Modify the branch-and-bound algorithm so that it stops as soon as it has a feasible solution that is guaranteed to be within $p\%$ of the optimum.

11.3.3 Cutting planes

In order to solve the mixed integer linear program

$$(MILP) \ \min \ c^T x$$
$$Ax \ge b$$
$$x \ge 0$$
$$x_j \ integer \ for \ j = 1, \ldots, p,$$

a possible approach is to strengthen the linear programming relaxation

$$(R) \ \min c^T x$$
$$Ax \ge b$$
$$x \ge 0,$$

by adding valid inequalities for (MILP). When the optimal solution x^* of the strengthened linear program is valid for (MILP), then x^* is also an optimal solution of (MILP). Even when this does not occur, the strengthened linear program may provide better lower bounds in the context of a branch-and-bound algorithm. How do we generate valid inequalities for (MILP)?

Gomory [33] proposed the following approach. Consider nonnegative variables x_j for $j \in I \cup C$, where x_j must be integer valued for $j \in I$. We allow the possibility that $C = \emptyset$. Let

$$\sum_{j \in I} a_j x_j + \sum_{j \in C} a_j x_j = b \tag{11.1}$$

be an equation satisfied by these variables. Assume that b is not an integer and let f_0 be its fractional part, i.e., $b = \lfloor b \rfloor + f_0$ where $0 < f_0 < 1$. For $j \in I$, let $a_j = \lfloor a_j \rfloor + f_j$ where $0 \le f_j < 1$. Replacing in (11.1) and moving sums of integer products to the right, we get:

$$\sum_{j \in I:\, f_j \le f_0} f_j x_j + \sum_{j \in I:\, f_j > f_0} (f_j - 1) x_j + \sum_{j \in C} a_j x_j = k + f_0,$$

where k is some integer.

Using the fact that $k \le -1$ or $k \ge 0$, we get the disjunction

$$\sum_{j \in I:\, f_j \le f_0} \frac{f_j}{f_0} x_j - \sum_{j \in I:\, f_j > f_0} \frac{1 - f_j}{f_0} x_j + \sum_{j \in C} \frac{a_j}{f_0} x_j \ge 1$$

or

$$- \sum_{j \in I:\, f_j \le f_0} \frac{f_j}{1 - f_0} x_j + \sum_{j \in I:\, f_j > f_0} \frac{1 - f_j}{1 - f_0} x_j - \sum_{j \in C} \frac{a_j}{1 - f_0} x_j \ge 1.$$

This is of the form $\sum_j a_j^1 x_j \ge 1$ or $\sum_j a_j^2 x_j \ge 1$, which implies

$$\sum \max \left(a_j^1, a_j^2 \right) x_j \ge 1$$

for $x \ge 0$.

Which is the largest of the two coefficients in our case? The answer is easy since one coefficient is positive and the other is negative for each variable:

$$\sum_{j \in I:\, f_j \le f_0} \frac{f_j}{f_0} x_j + \sum_{j \in I:\, f_j > f_0} \frac{1 - f_j}{1 - f_0} x_j + \sum_{j \in C:\, a_j > 0} \frac{a_j}{f_0} x_j - \sum_{j \in C:\, a_j < 0} \frac{a_j}{1 - f_0} x_j \ge 1. \tag{11.2}$$

Inequality (11.2) is valid for all $x \ge 0$ that satisfy (11.1) with x_j integer for all $j \in I$. It is called the *Gomory mixed integer cut* (GMI cut).

Let us illustrate the use of Gomory's mixed integer cuts on the two-variable example of Figure 11.1. Recall that the corresponding integer program is

$$\max z = x_1 + x_2$$
$$- x_1 + x_2 \le 2$$
$$8x_1 + 2x_2 \le 19$$
$$x_1, x_2 \ge 0$$
$$x_1, x_2 \text{ integer.}$$

We first add slack variables x_3 and x_4 to turn the inequality constraints into equalities. The problem becomes:

$$z - x_1 - x_2 = 0$$
$$- x_1 + x_2 + x_3 = 2$$
$$8x_1 + 2x_2 + x_4 = 19$$
$$x_1, x_2, x_3, x_4 \geq 0$$
$$x_1, x_2, x_3, x_4 \text{ integer.}$$

Solving the linear programming relaxation by the simplex method (Section 2.4), we get the optimal tableau:

$$z + 0.6x_3 + 0.2x_4 = 5$$
$$x_2 + 0.8x_3 + 0.1x_4 = 3.5$$
$$x_1 - 0.2x_3 + 0.1x_4 = 1.5$$
$$x_1, x_2, x_3, x_4 \geq 0.$$

The corresponding basic solution is $x_3 = x_4 = 0$, $x_1 = 1.5$, $x_2 = 3.5$, and $z = 5$. This solution is not integer. Let us generate the Gomory mixed integer cut corresponding to the equation

$$x_2 + 0.8x_3 + 0.1x_4 = 3.5$$

found in the final tableau. We have $f_0 = 0.5$, $f_1 = f_2 = 0$, $f_3 = 0.8$, and $f_4 = 0.1$. Applying formula (11.2), we get the GMI cut

$$\frac{1 - 0.8}{1 - 0.5}x_3 + \frac{0.1}{0.5}x_4 \geq 1, \quad \text{i.e.,} \quad 2x_3 + x_4 \geq 5.$$

We could also generate a GMI cut from the other equation in the final tableau $x_1 - 0.2x_3 + 0.1x_4 = 1.5$. It turns out that, in this case, we get exactly the same GMI cut. We leave it to the reader to verify this.

Since $x_3 = 2 + x_1 - x_2$ and $x_4 = 19 - x_1 - 2x_2$, we can express the above GMI cut in the space (x_1, x_2). This yields

$$3x_1 + 2x_2 \leq 9.$$

Adding this cut to the linear programming relaxation, we get the following formulation (see Figure 11.4):

$$\max x_1 + x_2$$
$$- x_1 + x_2 \leq 2$$
$$8x_1 + 2x_2 \leq 19$$
$$3x_1 + 2x_2 \leq 9$$
$$x_1, x_2 \geq 0.$$

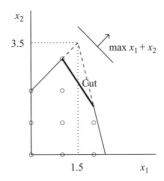

Figure 11.4 Formulation strengthened by a cut

Solving this linear program by the simplex method, we find the basic solution $x_1 = 1$, $x_2 = 3$, and $z = 4$. Since x_1 and x_2 are integer, this is the optimal solution to the integer program.

Exercise 11.8 Consider the integer program

$$\max 10x_1 + 13x_2$$
$$10x_1 + 14x_2 \leq 43$$
$$x_1, x_2 \geq 0$$
$$x_1, x_2 \text{ integer.}$$

(i) Introduce slack variables and solve the linear programming relaxation by the simplex method. (Hint: You should find the following optimal tableau:

$$\min x_2 + x_3$$
$$x_1 + 1.4x_2 + 0.1x_3 = 4.3$$
$$x_1, x_2 \geq 0.$$

with basic solution $x_1 = 4.3$, $x_2 = x_3 = 0$.)
(ii) Generate a GMI cut that cuts off this solution.
(iii) Multiply both sides of the equation $x_1 + 1.4x_2 + 0.1x_3 = 4.3$ by the constant $k = 2$ and generate the corresponding GMI cut. Repeat for $k = 3, 4$, and 5. Compare the five GMI cuts that you found.
(iv) Add the GMI cut generated for $k = 3$ to the linear programming relaxation. Solve the resulting linear program by the simplex method. What is the optimum solution of the integer program?

Exercise 11.9

(i) Consider the two-variable mixed integer set

$$S := \{(x, y) \in I\!N \times I\!R_+ : x - y \leq b\},$$

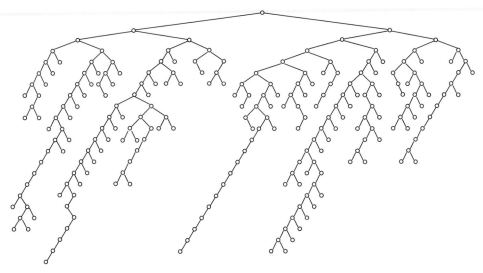

Figure 11.5 A branch-and-cut enumeration tree

where $b \in \mathbb{R}$. Let $f_0 = b - \lfloor b \rfloor$. Show that

$$x - \frac{1}{1 - f_0} y \le \lfloor b \rfloor$$

is a valid inequality for S.

(ii) Consider the mixed integer set

$$S := \{(x, y) \in \mathbb{N}_+^n \times \mathbb{R}_+^p : a^\mathsf{T} x + g^\mathsf{T} y \le b\}$$

where $a \in \mathbb{R}^n$, $g \in \mathbb{R}^p$, and $b \in \mathbb{R}$. Let $f_0 = b - \lfloor b \rfloor$ and $f_j = a_j - \lfloor a_j \rfloor$. Show that

$$\sum_{j=1}^n \left(\lfloor a_j \rfloor + \frac{(f_j - f_0)^+}{1 - f_0} \right) x_j + \frac{1}{1 - f_0} \sum_{j: g_j < 0} g_j y_j \le \lfloor b \rfloor$$

is a valid inequality for S.

11.3.4 Branch and cut

The best software packages for solving MILPs use neither pure branch-and-bound nor pure cutting plane algorithms. Instead they combine the two approaches in a method called branch and cut. The basic structure is essentially the same as branch and bound. The main difference is that, when a node N_i is explored, cuts may be generated to strengthen the formulation, thus improving the bound z_i. Some cuts may be local (i.e., valid only at node N_i and its descendants) or global (valid at all the nodes of the branch-and-bound tree). Cplex and Xpress are two excellent commercial branch-and-cut codes. cbc (COIN branch and cut) and bcp (branch, cut and price) are open source codes in the COIN-OR library.

In Figure 11.5 we give an example of an enumeration tree obtained when running the branch-and-cut algorithm of a commercial code on an instance with 89 binary variables and 28 constraints. Nodes of degree two (other than the root) occur when one of the sons can be pruned immediately by bounds or infeasibility.

12

Integer programming models: constructing an index fund

This chapter presents several applications of integer linear programming: combinatorial auctions, the lockbox problem, and index funds. We also present a model of integer quadratic programming: portfolio optimization with minimum transaction levels.

12.1 Combinatorial auctions

In many auctions, the value that a bidder has for acquiring a set of items may not be the sum of the values that he has for acquiring the individual items in the set. It may be more or it may be less. Examples are equity trading, electricity markets, pollution right auctions, and auctions for airport landing slots. To take this into account, combinatorial auctions allow the bidders to submit bids on combinations of items.

Specifically, let $M = \{1, 2, \ldots, m\}$ be the set of items that the auctioneer has to sell. A *bid* is a pair $B_j = (S_j, p_j)$ where $S_j \subseteq M$ is a nonempty set of items and p_j is the price offer for this set. Suppose that the auctioneer has received n bids $B_1, B_2, \ldots, B_n$. How should the auctioneer determine the winners in order to maximize his revenue? This can be done by solving an integer program. Let x_j be a 0,1 variable that takes the value 1 if bid B_j wins, and 0 if it looses. The auctioneer maximizes his revenue by solving the integer program:

$$\max \quad \sum_{i=1}^{n} p_j x_j$$

$$\text{subject to} \quad \sum_{j:\, i \in S_j} x_j \leq 1 \quad \text{for } i = 1, \ldots, m$$

$$x_j = 0 \text{ or } 1 \quad \text{for } j = 1, \ldots, n.$$

The constraints impose that each item i is sold at most once.

For example, if there are four items for sale and the following bids have been received: $B_1 = (\{1\}, 6)$, $B_2 = (\{2\}, 3)$, $B_3 = (\{3, 4\}, 12)$, $B_4 = (\{1, 3\}, 12)$, $B_5 =$

$(\{2, 4\}, 8)$, $B_6 = (\{1, 3, 4\}, 16)$, the winners can be determined by the following integer program:

$$\max \quad 6x_1 + 3x_2 + 12x_3 + 12x_4 + 8x_5 + 16x_6$$

$$\text{subject to} \quad x_1 + x_4 + x_6 \leq 1$$

$$x_2 + x_5 \leq 1$$

$$x_3 + x_4 + x_6 \leq 1$$

$$x_3 + x_5 + x_6 \leq 1$$

$$x_j = 0 \text{ or } 1 \quad \text{for } j = 1, \ldots, 6.$$

In some auctions, there are multiple indistinguishable units of each item for sale. A bid in this setting is defined as $B_j = (\lambda_1^j, \lambda_2^j, \ldots, \lambda_m^j; p_j)$ where λ_i^j is the desired number of units of item i and p_j is the price offer. The auctioneer maximizes his revenue by solving the integer program:

$$\max \quad \sum_{i=1}^{n} p_j x_j$$

$$\text{subject to} \quad \sum_{j: i \in S_j} \lambda_i^j x_j \leq u_i \quad \text{for } i = 1, \ldots, m$$

$$x_j = 0 \text{ or } 1 \quad \text{for } j = 1, \ldots, n.$$

where u_i is the number of units of item i for sale.

Exercise 12.1 In a combinatorial exchange, both buyers and sellers can submit combinatorial bids. Bids are like in the multiple item case, except that the λ_i^j values can be negative, as can the prices p_j, representing selling instead of buying. Note that a single bid can be buying some items while selling other items. Write an integer linear program that will maximize the surplus generated by the combinatorial exchange.

12.2 The lockbox problem

Consider a national firm that receives checks from all over the United States. Due to the vagaries of the US Postal Service, as well as the banking system, there is a variable delay from when the check is postmarked (and hence the customer has met his/her obligation) and when the check clears (and when the firm can use the money). For instance, a check mailed in Pittsburgh sent to a Pittsburgh address might clear in just two days. A similar check sent to Los Angeles might take four days to clear. It is in the firm's interest to have the check clear as quickly as possible since then the firm can use the money. In order to speed up this clearing process, firms open offices (called lockboxes) in different cities to handle the checks.

Table 12.1 *Clearing times*

From	L.A.	Pittsburgh	Boston	Houston
West	2	4	6	6
Midwest	4	2	5	5
East	6	5	2	5
South	7	5	6	3

Table 12.2 *Lost interest ('000)*

From	L.A.	Pittsburgh	Boston	Houston
West	60	120	180	180
Midwest	48	24	60	60
East	216	180	72	180
South	126	90	108	54

For example, suppose we receive payments from four regions (West, Midwest, East, and South). The average daily value from each region is as follows: $600 000 from the West, $240 000 from the Midwest, $720 000 from the East, and $360 000 from the South. We are considering opening lockboxes in Los Angeles, Pittsburgh, Boston, and/or Houston. Operating a lockbox costs $90 000 per year. The average days from mailing to clearing is given in Table 12.1. Which lockboxes should we open?

First we must calculate the lost interest for each possible assignment. For example, if West sends its checks to a lockbox in Boston, then on average there will be $3 600 000 (= 6 × $600 000) in process on any given day. Assuming an investment rate of 5%, this corresponds to a yearly loss of $180 000. We can calculate the losses for the other combinations in a similar fashion to get Table 12.2.

To formulate the problem as an integer linear program, we will use the following variables. Let y_j be a 0–1 variable that is 1 if lockbox j is opened and 0 if it is not. Let x_{ij} be 1 if region i sends its checks to lockbox j.

The objective is to minimize total yearly costs:

$$60x_{11} + 120x_{12} + 180x_{13} + 180x_{14} + 48x_{21} + \cdots + 90y_1 + 90y_2 + 90y_3 + 90y_4.$$

Each region must be assigned to one lockbox:

$$\sum_j x_{ij} = 1 \quad \text{for all } i.$$

The regions cannot send checks to closed lockboxes. For lockbox 1 (Los Angeles), this can be written as:

$$x_{11} + x_{21} + x_{31} + x_{41} \leq 4y_1.$$

Indeed, suppose that we do not open a lockbox in L.A. Then y_1 is 0, so all of x_{11}, x_{21}, x_{31}, and x_{41} must also be. On the other hand, if we open a lockbox in L.A., then y_1 is 1 and there is no restriction on the x values.

We can create constraints for the other lockboxes to finish off the integer program. For this problem, we would have 20 variables (4 y variables, 16 x variables) and 8 constraints. This gives the following integer program:

```
MIN      60 X11 +  120 X12 +  180 X13 +  180 X14 +  48 X21
       +  24 X22 +   60 X23 +   60 X24 +  216 X31 +  180 X32
       +  72 X33 +  180 X34 +  126 X41 +   90 X42 +  108 X43
       +  54 X44 +   90 Y1  +   90 Y2  +   90 Y3  +   90 Y4
SUBJECT TO
              X11 + X12 + X13 + X14 =      1
              X21 + X22 + X23 + X24 =      1
              X31 + X32 + X33 + X34 =      1
              X41 + X42 + X43 + X44 =      1
              X11 + X21 + X31 + X41 - 4 Y1 <=   0
              X12 + X22 + X32 + X42 - 4 Y2 <=   0
              X13 + X23 + X33 + X43 - 4 Y3 <=   0
              X14 + X24 + X34 + X44 - 4 Y4 <=   0
ALL VARIABLES BINARY
```

If we ignore integrality, we get the solution $x_{11} = x_{22} = x_{33} = x_{44} = 1, y_1 = y_2 = y_3 = y_4 = 0.250$, and the rest equals 0. Note that we get no useful information out of this linear programming solution: all four regions look the same.

The above is a perfectly reasonable 0–1 programming formulation of the lockbox problem. There are other formulations, however. For instance, consider the 16 constraints of the form

$$x_{ij} \leq y_j.$$

These constraints also force a region to only use open lockboxes. It might seem that a larger formulation is less efficient and therefore should be avoided. This is not the case! If we solve the linear program with the above constraints, we get the solution $x_{11} = x_{21} = x_{33} = x_{43} = y_1 = y_3 = 1$ with the rest equal to zero. In fact, we have an integer solution, which must therefore be optimal! Different integer programming formulations can have very different properties with respect to their linear programming relaxations. As a general rule, one prefers an integer programming formulation whose linear programming relaxation provides a tight bound.

Exercise 12.2   Consider a lockbox problem where c_{ij} is the cost of assigning region i to a lockbox in region j, for $j = 1, \ldots, n$. Suppose that we wish to open exactly q lockboxes where q is a given integer, $1 \leq q \leq n$.

  (i) Formulate as an integer linear program the problem of opening q lockboxes so as to minimize the total cost of assigning each region to an open lockbox.
 (ii) Formulate in two different ways the constraint that regions cannot send checks to closed lockboxes.
(iii) For the following data,

$$q = 2 \quad \text{and} \quad (c_{ij}) = \begin{pmatrix} 0 & 4 & 5 & 8 & 2 \\ 4 & 0 & 3 & 4 & 6 \\ 5 & 3 & 0 & 1 & 7 \\ 8 & 4 & 1 & 0 & 4 \\ 2 & 6 & 7 & 4 & 0 \end{pmatrix},$$

compare the linear programming relaxations of your two formulations in question (ii).

12.3 Constructing an index fund

An old and recurring debate about investing lies in the merits of active versus passive management of a portfolio. Active portfolio management tries to achieve superior performance by using technical and fundamental analysis as well as fore-casting techniques. On the other hand, passive portfolio management avoids any forecasting techniques and rather relies on diversification to achieve a desired per-formance. There are two types of passive management strategies: "buy and hold" or "indexing". In the first one, assets are selected on the basis of some fundamental criteria and there is no active selling or buying of these stocks afterwards (see the sections on Dedication in Chapter 3 and Portfolio optimization in Chapter 8). In the second approach, absolutely no attempt is made to identify mispriced securities. The goal is to choose a portfolio that mirrors the movements of a broad market population or a market index. Such a portfolio is called an index fund. Given a target population of n stocks, one selects q stocks (and their weights in the index fund), to represent the target population as closely as possible.

In the last 20 years, an increasing number of investors, both large and small, have established index funds. Simply defined, an index fund is a portfolio designed to track the movement of the market as a whole or some selected broad market segment. The rising popularity of index funds can be explained both theoretically and empirically.

- **Market efficiency:** If the market is efficient, no superior risk-adjusted returns can be achieved by stock picking strategies since the prices reflect all the information available in the marketplace. Additionally, since the market portfolio provides the best possible return per unit of risk, to the extent that it captures the efficiency of the market via diversification, one may argue that the best theoretical approach to fund management is to invest in an index fund. On the other hand, there is some empirical evidence refuting market efficiency.
- **Empirical performance:** Empirical studies provide evidence that, on average, money managers have consistently underperformed the major indexes. In addition, studies show that, in most cases, top performing funds for a year are no longer amongst the top performers in the following years, leaving room for the intervention of luck as an explanation for good performance.
- **Transaction cost:** Actively managed funds incur transaction costs, which reduce the overall performance of these funds. In addition, active management implies significant research costs. Finally, some fund managers may have costly compensation packages that can be avoided to a large extent with index funds.

Here we take the point of view of a fund manager who wants to construct an index fund. Strategies for forming index funds involve choosing a broad market index as a proxy for an entire market, e.g., the Standard and Poor list of 500 stocks (S&P 500). A pure indexing approach consists in purchasing all the issues in the index, with the same exact weights as in the index. In most instances, this approach is impractical (many small positions) and expensive (rebalancing costs may be incurred frequently). An index fund with q stocks, where q is substantially smaller than the size n of the target population, seems desirable. We propose a large-scale deterministic model for aggregating a broad market index of stocks into a smaller more manageable index fund. This approach will not necessarily yield mean/variance efficient portfolios but will produce a portfolio that closely replicates the underlying market population.

12.3.1 A large-scale deterministic model

We present a model that clusters the assets into groups of similar assets and selects one representative asset from each group to be included in the index fund portfolio. The model is based on the following data, which we will discuss in more detail later:

$$\rho_{ij} = \text{similarity between stock } i \text{ and stock } j.$$

For example, $\rho_{ii} = 1$, $\rho_{ij} \leq 1$ for $i \neq j$ and ρ_{ij} is larger for more similar stocks. An example of this is the correlation between the returns of stocks i and j. But one could choose other similarity indices ρ_{ij}.

(M) $\qquad$ $Z = \max \sum_{i=1}^{n} \sum_{j=1}^{n} \rho_{ij} x_{ij}$

$\qquad$ subject to $\qquad \sum_{j=1}^{n} y_j = q$

$$\sum_{j=1}^{n} x_{ij} = 1 \qquad \text{for } i = 1, \ldots, n$$

$$x_{ij} \leq y_j \qquad \text{for } i = 1, \ldots, n; \ j = 1, \ldots, n$$

$$x_{ij}, y_j = 0 \text{ or } 1 \qquad \text{for } i = 1, \ldots, n; \ j = 1, \ldots, n.$$

The variables y_j describe which stocks j are in the index fund ($y_j = 1$ if j is selected in the fund, 0 otherwise). For each stock $i = 1, \ldots, n$, the variable x_{ij} indicates which stock j in the index fund is most similar to i ($x_{ij} = 1$ if j is the most similar stock in the index fund, 0 otherwise).

The first constraint selects q stocks in the fund. The second constraint imposes that each stock i has exactly one representative stock j in the fund. The third constraint guarantees that stock i can be represented by stock j only if j is in the fund. The objective of the model maximizes the similarity between the n stocks and their representatives in the fund.

Once the model has been solved and a set of q stocks has been selected for the index fund, a weight w_j is calculated for each j in the fund:

$$w_j = \sum_{i=1}^{n} V_i x_{ij},$$

where V_i is the market value of stock i. So w_j is the total market value of the stocks "represented" by stock j in the fund. The fraction of the index fund to be invested in stock j is proportional to the stock's weight w_j, i.e.,

$$\frac{w_j}{\sum_{f=1}^{n} w_f}.$$

Note that, instead of the objective function used in (M), one could have used an objective function that takes the weights w_j directly into account, such as $\sum_{i=1}^{n} \sum_{j=1}^{n} V_i \rho_{ij} x_{ij}$. The q stocks in the index fund found by this variation of Model (M) would still need to be weighted as explained in the previous paragraph.

Data requirements

We need a coefficient ρ_{ij} that measures the similarity between stocks i and j. There are several ways of constructing meaningful coefficients ρ_{ij}. One approach is to

Table 12.3 *Performance of a 25 stock index fund*

Length	Ratio
1 QTR	1.006
2 QTR	0.99
1 YR	0.985
3 YR	0.982

consider the time series of stock prices over a calibration period T and to compute the correlation between each pair of assets.

Testing the model

Stocks comprising the S&P 500 were chosen as the target population to test the model. A calibration period of 60 months was used. Then a portfolio of 25 stocks was constructed using model (M) and held for periods ranging from three months to three years. Table 12.3 gives the ratio of the population's market value (normalized) to the index fund's market value. A perfect index fund would have a ratio equal unity.

Solution strategy

Branch and bound is a natural candidate for solving model (M). Note however that the formulation is very large. Indeed, for the S&P 500, there are 250 000 variables x_{ij} and 250 000 constraints $x_{ij} \leq y_j$. So the linear programming relaxation needed to get upper bounds in the branch-and-bound algorithm is a very large linear program to solve. It turns out, however, that one does not need to solve this large linear program to obtain good upper bounds. Cornuéjols *et al.* [24] proposed using the following Lagrangian relaxation, which is defined for any vector $u = (u_1, \ldots, u_n)$:

$$L(u) = \max \sum_{i=1}^{n} \sum_{j=1}^{n} \rho_{ij} x_{ij} + \sum_{i=1}^{n} u_i \left(1 - \sum_{j=1}^{n} x_{ij} \right)$$

subject to
$$\sum_{j=1}^{n} y_j = q$$

$$x_{ij} \leq y_j \qquad \text{for } i = 1, \ldots, n; \, j = 1, \ldots, n$$
$$x_{ij}, y_j = 0 \text{ or } 1 \quad \text{for } i = 1, \ldots, n; \, j = 1, \ldots, n.$$

Property 1: $L(u) \geq Z$, where Z is the maximum for model (M).

Exercise 12.3 Prove Property 1.

The objective function $L(u)$ may be equivalently stated as

$$L(u) = \max \sum_{i=1}^{n} \sum_{j=1}^{n} (\rho_{ij} - u_i) x_{ij} + \sum_{i=1}^{n} u_i.$$

Let

$$(\rho_{ij} - u_i)^+ = \begin{cases} \rho_{ij} - u_i & \text{if } \rho_{ij} - u_i > 0 \\ 0 & \text{otherwise} \end{cases}$$

and

$$C_j = \sum_{i=1}^{n} (\rho_{ij} - u_i)^+.$$

Then

Property 2:

$$L(u) = \max \sum_{j=1}^{n} C_j y_j + \sum_{i=1}^{n} u_i$$

$$\text{subject to} \quad \sum_{j=1}^{n} y_j = q$$

$$y_j = 0 \text{ or } 1 \text{ for } j = 1, \dots, n.$$

Exercise 12.4 Prove Property 2.

Property 3: In an optimal solution of the Lagrangian relaxation, y_j is equal to 1 for the q largest values of C_j, and the remaining y_j are equal to 0. Furthermore, if $\rho_{ij} - u_i > 0$, then $x_{ij} = y_j$ and otherwise $x_{ij} = 0$.

Exercise 12.5 Prove Property 3.

Interestingly, the set of q stocks corresponding to the q largest values of C_j can also be used as a heuristic solution for model (M). Specifically, construct an index fund containing these q stocks and assign each stock $i = 1, \dots, n$ to the most similar stock in this fund. This solution is feasible to model (M), although not necessarily optimal. This heuristic solution provides a lower bound on the optimum value Z of model (M). As previously shown, $L(u)$ provides an upper bound on Z. So for any vector u, we can compute quickly both a lower bound and an upper bound on the optimum value of (M). To improve the upper bound $L(u)$, we would like to solve the nonlinear problem

$$\min \; L(u).$$

How does one minimize $L(u)$? Since $L(u)$ is nondifferentiable and convex, one can use the subgradient method (see Section 5.6). At each iteration, a revised set of Lagrange multipliers u and an accompanying lower bound and upper bound to model (M) are computed. The algorithm terminates when these two bounds match or when a maximum number of iterations is reached. (It is proved in [24] that min $L(u)$ is equal to the value of the linear programming relaxation of (M). In general, this value is not equal to Z, and therefore it is not possible to match the upper and lower bounds.) If one wants to solve the integer program (M) to optimality, one can use a branch-and-bound algorithm, using the upper bound min $L(u)$ for pruning the nodes.

12.3.2 A linear programming model

In this section, we consider a different approach to constructing an index fund. It can be particularly useful as one tries to rebalance the portfolio at minimum cost. This approach assumes that we have identified important characteristics of the market index to be tracked. Such characteristics might be the fraction f_i of the index in each sector i, the fraction of companies with market capitalization in various ranges (small, medium, large), the fraction of companies that pay no dividends, the fraction in each region, etc. Let us assume that there are m such characteristics that we would like our index fund to track as well as possible. Let $a_{ij} = 1$ if company j has characteristic i and 0 if it does not.

Let x_j denote the optimum weight of asset j in the portfolio. Assume that, initially, the portfolio has weights x_j^0. Let y_j denote the fraction of asset j bought and z_j the fraction sold. The problem of rebalancing the portfolio at minimum cost is the following:

$$\min \sum_{j=1}^{n} (y_j + z_j)$$

subject to

$$\sum_{j=1}^{n} a_{ij}x_j = f_i \quad \text{for } i = 1, \ldots, m$$

$$\sum_{j=1}^{n} x_j = 1$$

$$x_j - x_j^0 \le y_j \quad \text{for } j = 1, \ldots, n$$
$$x_j^0 - x_j \le z_j \quad \text{for } j = 1, \ldots, n$$
$$y_j \ge 0 \quad \quad \text{for } j = 1, \ldots, n$$
$$z_j \ge 0 \quad \quad \text{for } j = 1, \ldots, n$$
$$x_j \ge 0 \quad \quad \text{for } j = 1, \ldots, n.$$

12.4 Portfolio optimization with minimum transaction levels

When solving the classical Markowitz model, the optimal portfolio often contains positions x_i that are too small to execute. In practice, one would like a solution of

$$\min_x \tfrac{1}{2} x^T Q x$$
$$\mu^T x \geq R$$
$$Ax = b \tag{12.1}$$
$$Cx \geq d,$$

with the additional property that

$$x_j > 0 \Rightarrow x_j \geq l_j, \tag{12.2}$$

where l_j are given minimum transaction levels. This constraint states that, if an investment is made in a stock, then it must be "large enough," for example, at least 100 shares. Because the constraint (12.2) is not a simple linear constraint, it cannot be handled directly by quadratic programming.

This problem is considered by Bienstock [12]. He also considers the portfolio optimization problem where there is an upper bound on the number of positive variables, that is,

$$x_j > 0 \text{ for at most } K \text{ distinct } j = 1, \ldots, n. \tag{12.3}$$

Requirement (12.2) can easily be incorporated within a branch-and-bound algorithm: first solve the basic Markowitz model (12.1) using the usual algorithm (see Chapter 7). Let x^* be the optimal solution found. If no minimum transaction level constraint (12.2) is violated by x^*, then x^* is also optimum to (12.1) and (12.2) and we can stop. Otherwise, let j be an index for which (12.2) is violated by x^*. Form two subproblems, one obtained from (12.1) by adding the constraint $x_j = 0$, and the other obtained from (12.1) by adding the constraint $x_j \geq l_j$. Both are quadratic programs that can be solved using the usual algorithms of Chapter 7. Now we check whether the optimum solutions to these two problems satisfy the transaction level constraint (12.2). If a solution violates (12.2) for index k, the corresponding problem is further divided by adding the constraint $x_k = 0$ on one side and $x_k \geq l_k$ on the other. A branch-and-bound tree is expanded in this way.

The constraint (12.3) is a little more tricky to handle. Assume that there is a given upper bound u_j on how much can be invested in stock j. That is, we assume that constraints $x_j \leq u_j$ are part of the formulation (12.1). Then, clearly, constraint (12.3) implies the weaker constraint

$$\sum_j \frac{x_j}{u_j} \leq K. \tag{12.4}$$

We add this constraint to (12.1) and solve the resulting quadratic program. Let x^* be the optimal solution found. If x^* satisfies (12.3), it is optimum to (12.1)–(12.3) and we can stop. Otherwise, let k be an index for which $x_k > 0$. Form two subproblems, one obtained from (12.1) by adding the constraint $x_k = 0$ (down branch), and the other obtained from (12.1) by adding the constraint $\sum_{j \neq k}(x_j/u_j) \leq K - 1$ (up branch). The branch-and-bound tree is developped recursively. When a set T of variables has been branched up, the constraint added to the basic model (12.1) becomes

$$\sum_{j \notin T} \frac{x_j}{u_j} \leq K - |T|.$$

12.5 Additional exercises

Exercise 12.6 You have $250\,000$ to invest in the following possible investments. The cash inflows/outflows are as follows:

	Year 1	Year 2	Year 3	Year 4
Investment 1	-1.00		1.18	
Investment 2		-1.00		1.22
Investment 3			-1.00	1.10
Investment 4	-1.00	0.14	0.14	1.00
Investment 5		-1.00	0.20	1.00

For example, if you invest one dollar in Investment 1 at the beginning of Year 1, you receive $1.18 at the beginning of Year 3. If you invest in any of these investments, the required minimum level is $100\,000$ in each case. Any or all the available funds at the beginning of a year can be placed in a money market account that yields 3% per year. Formulate a mixed integer linear program to maximize the amount of money available at the beginning of Year 4. Solve the integer program using your favorite solver.

Exercise 12.7 You currently own a portfolio of eight stocks. Using the Markowitz model, you computed the optimal mean/variance portfolio. The weights of these two portfolios are shown in the following table:

Stock	A	B	C	D	E	F	G	H
Your portfolio	0.12	0.15	0.13	0.10	0.20	0.10	0.12	0.08
M/V portfolio	0.02	0.05	0.25	0.06	0.18	0.10	0.22	0.12

You would like to rebalance your portfolio in order to be closer to the M/V portfolio. To avoid excessively high transaction costs, you decide to rebalance only three stocks from your portfolio. Let x_i denote the weight of stock i in your rebalanced portfolio. The objective is to minimize the quantity

$$|x_1 - 0.02| + |x_2 - 0.05| + |x_3 - 0.25| + \cdots + |x_8 - 0.12|$$

which measures how closely the rebalanced portfolio matches the M/V portfolio.

Formulate this problem as a mixed integer linear program. Note that you will need to introduce new continuous variables in order to linearize the absolute values and new binary variables in order to impose the constraint that only three stocks are traded.

12.6 Case study: constructing an index fund

The purpose of this project is to construct an index fund that will track a given segment of the market. First, choose a segment of the market and discuss the collection of data. Then, compare different approaches for computing an index fund: model (M) solved as a large integer program, Lagrangian relaxations and the subgradient approach, the linear programming approach of Section 12.3.2, or others. The index fund should be computed using an in-sample period and evaluated on an out-of-sample period.

13

Dynamic programming methods

13.1 Introduction

Decisions must often be made in a sequential manner over time. Earlier decisions may affect the feasibility and performance of later decisions. In such environments, myopic decisions that optimize only the immediate impact are usually suboptimal for the overall process. To find optimal strategies one must consider current and future decisions simultaneously. These types of *multi-stage* decision problems are the typical settings where one employs *dynamic programming*, or DP. Dynamic programming is a term used both for the modeling methodology and the solution approaches developed to solve sequential decision problems. In some cases the sequential nature of the decision process is obvious and natural, in other cases one reinterprets the original problem as a sequential decision problem. We will consider examples of both types below.

Dynamic programming models and methods are based on Bellman's *Principle of Optimality*, namely that for overall optimality in a sequential decision process, all the remaining decisions after reaching a particular state must be optimal with respect to that state. In other words, if a strategy for a sequential decision problem makes a sub-optimal decision in any one of the intermediate stages, it cannot be optimal for the overall problem. This principle allows one to formulate *recursive relationships* between the optimal strategies of successive decision stages and these relationships form the backbone of DP algorithms.

Common elements of DP models include decision *stages*, a set of possible *states* in each stage, *transitions* from states in one stage to states in the next, *value functions* that measure the best possible objective values that can be achieved starting from each state, and finally the *recursive relationships* between value functions of different states. For each state in each stage, the decision-maker needs to specify the *decision* he/she would make in order to reach that state and the collection of all decisions associated with all states forms the *policy* or *strategy* of the decision-maker.

Table 13.1 *Project costs and profits ($ million)*

Project	Region 1		Region 2		Region 3	
	c_1	p_1	c_2	p_2	c_3	p_3
1	0	0	0	0	0	0
2	1	2	1	3	1	2
3	2	4	3	9	2	5
4	4	10	—	—	—	—

Transitions from the states of a given stage to those of the next may happen as a result of the actions of the decision-maker, as a result of random external events, or a combination of the two. If a decision at a particular state uniquely determines the transition state, the DP is a *deterministic DP*. If probabilistic events also affect the transition state, then one has a *stochastic DP*. We will discuss each one of these terms below.

Dynamic programming models are pervasive in the financial literature. The best-known and most common examples are the tree or lattice models (binomial, trino-mial, etc.) used to describe the evolution of security prices, interest rates, volatilities, etc., and the corresponding pricing and hedging schemes. We will discuss several such examples in the next chapter. Here, we focus on the fundamentals of the dynamic programming approach and, for this purpose, it is best to start with an example.

Consider a capital budgeting problem. A manager has $4 million to allocate to different projects in three different regions where her company operates. In each region, there are a number of possible projects to consider with estimated costs and projected profits. Let us denote the costs with c_j's and profits with p_j's. Table 13.1 lists the information for possible project options; both the costs and the profits are given in millions of dollars.

Note that the projects in the first row with zero costs and profits correspond to the option of doing nothing in that particular region. The manager's objective is to maximize the total profits from projects financed in all regions. She will choose only one project from each region.

One may be tempted to approach this problem using integer programming tech-niques we discussed in the previous two chapters. Indeed, since there is a one-to-one correspondence between the projects available at each region and their costs, by letting x_i denote the investment amount in region i, we can formulate an integer programming problem with the following constraints:

$$x_1 + x_2 + x_3 \leq 4$$
$$x_1 \in \{0, 1, 2, 4\}, x_2 \in \{0, 1, 3\}, x_3 \in \{0, 1, 2\}.$$

The problem with this approach is the profits are not *linear* functions of the variables x_i. For example, for region 3, while the last project costs twice as much as the the second one, the expected profit from this last project is two and half times that of the second project. To avoid formulating a *nonlinear integer programming* problem, which can be quite difficult to solve, one might consider a formulation that uses a binary variable for each project in each region. For example, we can use binary decision variables x_{ij} to represent whether project j in region i is to be financed. This results in an integer linear program but with many more variables.

Exercise 13.1 Formulate an integer linear program for the capital budgeting problem with project costs and profits given in Table 13.1.

Another strategy we can consider is total enumeration of all investment possibilities. We have four choices for the first region, and three choices for each of the second and third regions. Therefore, we would end up with $4 \times 3 \times 3 = 36$ possibilities to consider. We can denote these possibilities with (x_1, x_2, x_3) where, for example, $(2, 3, 1)$ corresponds to the choices of the second, the third and the first projects in regions 1, 2, and 3, respectively. We could evaluate each of these possibilities and then pick the best one. There are obvious problems with this approach, as well.

First of all, for larger problems with many regions and/or many options in each region, the total number of options we need to consider will grow very quickly and become computationally prohibitive. Furthermore, many of the combinations are not feasible with respect to the constraints of the problem. In our example, choosing the third project in each region would require $2 + 3 + 2 = 7$ million dollars, which is above the $4 million budget, and therefore is an infeasible option. In fact, only 21 of the 36 possibilities are feasible in our example. In an enumeration scheme, such infeasibilities will not be detected in advance leading to inefficiencies. Finally, an enumeration scheme does not take advantage of the information generated during the investigation of other alternatives. For example, after discovering that $(3, 3, 1)$ is an infeasible option, we should no longer consider the more expensive $(3, 3, 2)$ or $(3, 3, 3)$. Unfortunately, the total enumeration scheme will not take advantage of such simple deductions.

We will approach this problem using the dynamic programming methodology. For this purpose, we will represent our problem in a graph. The construction of this graph representation is not necessary for the solution procedure; it is provided here for didactic purposes. We will use the *root* node of the graph to correspond to stage 0 with $4 million to invest and use the pair $(0, 4)$ to denote this node. In stage 1 we will consider investment possibilities in region 1. In stage 2, we will consider investment possibilities in regions 1 and 2, and finally in stage 3 we will consider all three regions. Throughout the graph, nodes will be denoted by pairs

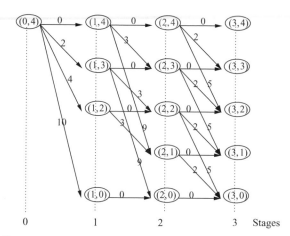

Figure 13.1 Graphical representation of the three-region capital budgeting problem

(i, j) where i represents the stage and j represents the particular state of that stage. States in stage i will correspond to the different amounts of money left after some projects are already funded in regions 1 through i. For example, the node $(2, 3)$ in stage 2 of the graph represents the *state* of having \$3 million left for investment after funding projects in regions 1 and 2.

The branches in the graphical representation correspond to the projects undertaken in a particular region. To be at node (i, j) means that we have already considered regions 1 to i and have j million dollars left for investment. Then, the branch corresponding to project k in the next region will take us to the node $(i + 1, j')$ where j' equals j minus the cost of project k. For example, starting from node $(1, 3)$, the branch corresponding to project 2 in the second region will take us to node $(2, 2)$. For each one of these branches, we will use the expected profit from the corresponding project as the *weight* of the branch. The resulting graph is shown in Figure 13.1. Now the manager's problem is to find the largest weight *path* from node $(0, 4)$ to a third stage node.

At this point, we can proceed in two alternative ways: using either a backward or a forward progression on the graph. In the backward mode, we first identify the largest weight path from each one of the nodes in stage 2 to a third stage node. Then using this information and the Principle of Optimality, we will determine the largest weight paths from each of the nodes in stage 1 to a third stage node, and finally from node $(0, 4)$ to a third stage node. In contrast, the forward mode will first determine the largest weight path from $(0, 4)$ to all first stage nodes, then to all second stage nodes and finally to all third stage nodes. We illustrate the backward method first and then the forward method.

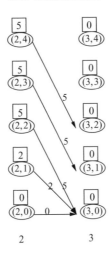

Figure 13.2 Optimal allocations from stage 2 nodes

13.1.1 Backward recursion

For each state, or node, we keep track of the largest profit that can be collected starting from that state. These quantities form what we will call the *value function* associated with each state. For the backward approach, we start with stage 3 nodes. Since we are assuming that any money that is not invested in regions 1 through 3 will generate no profits, the value function for each one of the stage 3 states is zero and there are no decisions associated with these states.

Next, we identify the largest weight paths from each one of the second stage nodes to the third stage nodes. It is clear that for nodes $(2, 4)$, $(2, 3)$, and $(2, 2)$ the best alternative is to choose project 3 of the third region and collect an expected profit of $5 million. Since node $(2, 1)$ corresponds to the state where there is only $1 million left for investment, the best alternative from the third region is project 2, with the expected profit of $2 million. For node $(2, 0)$, the only alternative is project 1 ("do nothing") with no profit. We illustrate these choices in Figure 13.2.

For each node, we indicated the *value function* associated with that node in a box on top of the node label in Figure 13.2. Next, we determine the value function and optimal decisions for each one of the first stage nodes. These computations are slightly more involved, but still straightforward. Let us start with node $(1, 4)$. From Figure 13.1 we see that one can reach the third stage nodes via one of $(2, 4)$, $(2, 3)$, and $(2, 1)$. The maximum expected profit on the paths through $(2, 4)$ is $0 + 5 = 5$, the sum of the profit on the arc from $(1, 4)$ to $(2, 4)$, which is zero, and the largest profit from $(2, 4)$ to a period 3 node. Similarly, we compute the maximum expected profit on the paths through $(2, 3)$ and $(2, 1)$ to be $3 + 5 = 8$, and $9 + 2 = 11$. The

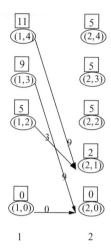

Figure 13.3 Optimal allocations from stage 1 nodes

maximum profit from $(1, 4)$ to a stage 3 node is then

$$\max\{0 + v(2, 4),\, 3 + v(2, 3),\, 9 + v(2, 1)\} = \{0 + 5,\, 3 + 5,\, 9 + 2\} = 11,$$

which is achieved by following the path $(1, 4) \rightarrow (2, 1) \rightarrow (3, 0)$. After performing similar computations for all period 1 nodes we obtain the node *values* and optimal branches given in Figure 13.3.

Finally, we need to compute the best allocations from node $(0, 4)$ by comparing the profits along the branches to first stage nodes and the best possible profits starting from those first period nodes. To be exact, we compute

$$\max\{0 + v(1, 4),\, 2 + v(1, 3),\, 4 + v(1, 2),\, 10 + v(1, 0)\}$$
$$= \{0 + 11,\, 2 + 9,\, 4 + 5,\, 10 + 0\} = 11.$$

Therefore, the optimal expected profit is \$11 million and is achieved on either of the two alternative paths $(0, 4) \rightarrow (1, 4) \rightarrow (2, 1) \rightarrow (3, 0)$ and $(0, 4) \rightarrow (1, 3) \rightarrow (2, 0) \rightarrow (3, 0)$. These paths correspond to the selections of project 1 in region 1, project 3 in region 2, and project 2 in region 3 in the first case, and project 2 in region 1, project 3 in region 2, and project 1 in region 3 in the second case. Figure 13.4 summarizes the whole process. The optimal paths are shown using thicker lines.

Exercise 13.2 Construct a graphical representation of a five-region capital budgeting problem with the project costs and profits given in Table 13.2. Exactly one project must be chosen in each region and there is a total budget of 10. Solve by backward recursion.

Table 13.2 *Project costs and profits*

Project	Region 1		Region 2		Region 3		Region 4		Region 5	
	c_1	p_1	c_2	p_2	c_3	p_3	c_4	p_4	c_5	p_5
1	1	8	3	20	2	15	0	3	1	6
2	2	15	2	14	4	26	1	10	2	15
3	3	25	1	7	5	40	3	25	3	22

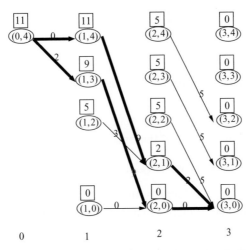

Figure 13.4 Optimal paths from $(0, 4)$ to $(3, 0)$

13.1.2 Forward recursion

Next, we explore the "forward" method. In this case, in the first step we will identify the best paths from $(0, 4)$ to all nodes in stage 1, then best paths from $(0, 4)$ to all stage 2 nodes, and finally to stage 3 nodes. The first step is easy since there is only one way to get from node $(0, 4)$ to each one of the stage 1 nodes, and hence all these paths are optimal. Similar to the backward method, we will keep track of a *value function* for each node. For node (i, j), its value function will represent the highest total expected profit we can collect from investments in regions 1 through i if we want to have $\$j$ million left for future investment. For $(0, 4)$ the value function is zero and for all stage 1 nodes, they are equal to the weight of the tree branch that connects $(0, 4)$ and the corresponding node.

For most of the second stage nodes, there are multiple paths from $(0, 4)$ to that corresponding node and we need to determine the best option. For example, let us consider the node $(2, 2)$. One can reach $(2, 2)$ from $(0, 4)$ either via $(1, 3)$ or $(1, 2)$. The value function at $(2, 2)$ is the maximum of the sum of the value function at

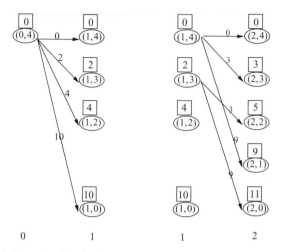

Figure 13.5 Optimal paths between stage 0, stage 1, and stage 2 nodes

(1, 3) and the weight of the branch from (1, 3) to (2, 2), and the sum of the value function at (1, 2) and the weight of the branch from (1, 2) to (2, 2):

$$v(2, 2) = \max\{v(1, 3) + 3, v(1, 2) + 0\} = \max\{2 + 3, 4 + 0\} = 5.$$

After similar calculations we identify the value function at all stage 2 nodes and the corresponding optimal branches one must follow. The results are shown on the right side of Figure 13.5.

Finally, we perform similar calculations for stage 3 nodes. For example, we can calculate the value function at (3, 0) as follows:

$$v(3, 0) = \max\{v(2, 2) + 5, v(2, 1) + 2, v(2, 0) + 0\} = \{5 + 5, 9 + 2, 11 + 0\} = 11.$$

Optimal paths for all nodes are depicted in Figure 13.6. Note that there are three alternative optimal ways to reach node (3, 2) from (0, 4).

Clearly, both the forward and the backward method identified the two alternative optimal paths between (0, 4) and (3, 0). However, the additional information generated by these two methods differ. In particular, studying Figures 13.4 and 13.6, we observe that while the backward method produces the optimal paths **from** each node in the graph to the final stage nodes, in contrast, the forward method produces the optimal paths from the initial stage node **to** all nodes in the graph. There may be situations where one prefers to have one set of information above the other and this preference dictates which method to use. For example, if for some reason the actual transition state happens to be different from the one intended by an optimal decision, it would be important to know what to do when in a state that is not on the optimal path. In that case, the paths generated by the backward method would give the answer.

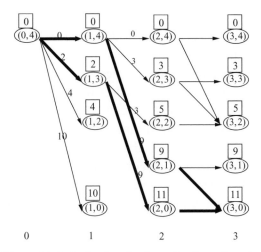

Figure 13.6 Optimal paths from (0, 4) to all nodes

Exercise 13.3 Solve the capital budgeting problem of Exercise 13.2 by forward recursion.

13.2 Abstraction of the dynamic programming approach

Before proceeding with additional examples, we study the common characteristics of dynamic programming models and methods. In particular, we will identify the aspects of the example considered in the previous section that qualified our approach as dynamic programming.

We already mentioned the sequential nature of the decision-making process as the most important ingredient of a DP problem. Every DP model starts with the identification of **stages** that correspond to the order of the decisions to be made. There is an initial stage (for a forward recursion) or final stage (for a backward recursion) for which the optimal decisions are immediately or easily available and do not depend on decisions of other stages. In our example in Section 13.1, the number of regions considered for different project options constituted the stages of our formulation. Stage 0 was the initial stage and stage 3 the final stage.

Each stage consists of a number of possible **states**. In allocation problems, states are typically used to represent the possible levels of availability for scarce resources in each stage. In financial binomial lattice models, states may correspond to spot prices of assets.

In many cases, the set of states in each particular stage is finite or, at least, discrete. Such DPs are categorized as *discrete DPs*, in contrast to *continuous DPs*, which may have a continuum of states in each stage. In the example of Section 13.1, the states represented the amount of money still available for investment at the end of that particular stage. For consistency with our earlier example, we continue to denote states of a DP formulation with the pair (i, j), where i specifies the stage and j specifies the particular state in that stage.

A DP formulation must also specify a **decision set** for each one of the states. As with states, decision sets may be discrete or continuous. In our example in Section 13.1, the decision sets were formed from the set of possible projects in each stage. Because of feasibility considerations, decision sets are not necessarily identical for all states in a given stage. For example, while the decision set consists of region 2 projects 1, 2, and 3 for state $(1, 4)$, the decision set for state $(1, 0)$ is the singleton corresponding to project 1 (do nothing). We denote the decision set associated with state (i, j) with $S(i, j)$.

In a deterministic DP, a choice d made from the decision set $S(i, j)$ uniquely determines what state one transitions to. We call this state the **transition state** associated with the particular state (i, j) and decision $d \in S(i, j)$ and use the notation $T((i, j), d)$ to denote this state. Furthermore, there is a cost (or benefit, for a maximization problem) associated with each transition that we indicate with $c((i, j), d)$. In our example in the previous section, from state $(2, 1)$, we can either transition to state $(3, 1)$ by choosing project 1 with an associated profit of 0, or to state $(3, 0)$ by choosing project 2 with an associated profit of 2.

In our example above, all the transition states from a given state were among the states of the next stage. Although this is common, it is not required. All that is necessary for the DP method to function is that all the transition states from a given state are in the later stages whose computations are already completed. So, for example, in a five-stage formulation, transition states of a state in stage 2 can be in any one of stages 3, 4, and 5.

A **value function** keeps track of the costs (or benefits) accumulated *optimally* from the initial stage up to a particular state (in the forward method) or from a particular state to the final stage (in the backward method). Each such quantity will be called the **value** of the corresponding state. We use the notation $v(i, j)$ to denote the value of the state (i, j).

The Principle of Optimality implies a recursive relationship between the values of states in consecutive stages. For example, in the backward method, to compute the optimal decision at and the value of a particular state, all we need to do is to compare the following quantity for each transition state of that state: the value of the transition state plus the cost of transitioning to that state. Namely, we do the following computation:

$$v(i, j) = \min_{d \in S(i,j)} \{v(T((i, j), d)) + c((i, j), d)\}. \tag{13.1}$$

In a benefit maximization problem, as in our example in the previous section, the values would be the benefits rather than costs and the *min* in (13.1) would be replaced by a *max*. Equation (13.1) is known as the Bellman equation and is a discrete-time deterministic special case of the Hamilton–Jacobi–Bellman (HJB) equation often encountered in optimal control texts.

To illustrate the definitions above and equation (13.1), let us explicitly perform one of the calculations of the example in the previous section. Say, in the backward method we have already calculated the values of the states in stage 2 (5, 5, 5, 2, and 0, for states (2, 4), (2, 3), (2, 2), (2, 1), and (2, 0), respectively) and we intend to compute the value of the state (1, 3). We first identify the decision set for (1, 3): $S(1, 3) = \{1, 2, 3\}$, i.e., projects 1, 2, and 3. The corresponding transition states are easily determined:

$$T((1, 3), 1) = (2, 3), \quad T((1, 3), 2) = (2, 2), \quad T((1, 3), 3) = (2, 0).$$

The associated benefits (or expected profits, in this case) are

$$c((1, 3), 1) = 0, \quad c((1, 3), 2) = 3, \quad c((1, 3), 3) = 9.$$

Now we can derive the value of state (1, 3):

$$
\begin{aligned}
v(1, 3) &= \max_{d \in S(1,3)} \{v(T((1, 3), d)) + c((1, 3), d)\} \\
&= \max\{v(T((1, 3), 1)) + c((1, 3), 1), \, v(T((1, 3), 2)) + c((1, 3), 2), \\
&\qquad v(T((1, 3), 3)) + c((1, 3), 3)\} \\
&= \max\{v(2, 3) + 0, \, v(2, 2) + 3, \, v(2, 0) + 9\} \\
&= \max\{5 + 0, 5 + 3, 0 + 9\} = 9,
\end{aligned}
$$

and the corresponding optimal decision at (1, 3) is project 3. Note that for us to be able to compute the values recursively as above, we must be able to compute the values at the final stage without any recursion.

If a given optimization problem can be formulated with the ingredients and properties outlined above, we can solve it using dynamic programming methods. Most often, finding the right formulation of a given problem, and specifying the stages, states, transitions, and recursions in a way that fits the framework above is the most challenging task in the dynamic programming approach. Even when a problem admits a DP formulation, there may be several alternative ways to do this (see, for example, Section 13.3) and it may not be clear which of these formulations would produce the quickest computational scheme. Developing the best formulations for a given optimization problem must be regarded as a form of art and, in our opinion, is best learned through examples. We continue in the next section with a canonical example of both integer and dynamic programming.

13.3 The knapsack problem

A traveler has a knapsack that she plans to take along for an expedition. Each item she would like to take with her in the knapsack has a given size and a value associated with the benefit the traveler receives by carrying that item. Given that the knapsack has a fixed and finite capacity, how many of each of these items should she put in the knapsack to maximize the total value of the items in the knapsack? This is the well-known and well-studied integer program called the *knapsack problem*. It has the special property that it only has a single constraint other than the nonnegative integrality condition on the variables.

We recall the investment problem considered in Exercise 11.5 in Chapter 11 which is an instance of the knapsack problem. We have $14\,000 to invest among four different investment opportunities. Investment 1 requires an investment of $7000 and has a net present value of $11\,000; investment 2 requires $5000 and has a value of $8000; investment 3 requires $4000 and has a value of $6000; and investment 4 requires $3000 and has a value of $4000.

As we discussed in Chapter 11, this problem can be formulated and solved as an integer program, say using the branch-and-bound method. Here, we will formulate it using the DP approach. To make things a bit more interesting, we will allow the possibility of multiple investments in the same investment opportunity. The effect of this modification is that the variables are now general integer variables rather than 0–1 binary variables and therefore the problem

$$\text{max} = 11x_1 + 8x_2 + 6x_3 + 4x_4$$
$$7x_1 + 5x_2 + 4x_3 + 3x_4 \leq 14$$
$$x_j \geq 0 \text{ integer}, \forall j$$

is an instance of the knapsack problem. We will consider two alternative DP formulations of this problem. For future reference, let y_j and p_j denote the cost and the net present value of investment j (in thousands of dollars), respectively, for $j = 1$ to 4.

13.3.1 Dynamic programming formulation

One way to approach this problem using the dynamic programming methodology is by considering the following question that already suggests a recursion: if I already know how to allocate i thousand dollars to the investment options optimally for all $i = 1, \ldots, k - 1$, can I determine how to optimally allocate k thousand dollars to these investment option? The answer to this question is yes, and building the recursion equation is straightforward.

The first element of our DP construction is the determination of the stages. The question in the previous paragraph suggests the use of stages $0, 1, \ldots,$ up to 14, where stage i corresponds to i thousand dollars left to invest. Note that we need only one state per stage and therefore can denote stages/states using the single index i. The decision set in state i is the set of investments we can afford with the i thousand dollars we have left for investment. That is, $S(i) = \{d : y_d \leq i\}$. The transition state is given by $T(i, d) = i - y_d$ and the benefit associated with the transition is $c(i, d) = p_d$. Therefore, the recursion for the value function is given by the following equation:

$$v(i) = \max_{d:y_d \leq i} \{v(i - y_d) + p_d\}.$$

Note that $S(i) = \emptyset$ and $v(i) = 0$ for $i = 0$, 1, and 2 in our example.

Exercise 13.4   Using the recursion given above, determine $v(i)$ for all i from 0 to 14 and the corresponding optimal decisions.

13.3.2 An alternative formulation

As we discussed in Section 13.2, dynamic programming formulation of a given optimization problem need not be unique. Often, there exist alternative ways of defining the stages and states, and obtaining recursions. Here we develop an alternative formulation of our investment problem by choosing stages to correspond to each one of the investment possibilities.

So, we will have four stages, $i = 1$, 2, 3, and 4. For each stage i, we will have states j corresponding to the total investment in opportunities i through 4. So, for example, in the fourth stage we will have states $(4, 0)$, $(4, 3)$, $(4, 6)$, $(4, 9)$, and $(4, 12)$, corresponding to 0, 1, 2, 3, and 4 investments in the fourth opportunity.

The decision to be made at stage i is the number of times one invests in the investment opportunity i. Therefore, for state (i, j), the decision set is given by

$$S(i, j) = \left\{d: \frac{j}{y_i} \geq d, d \text{ non-negative integer}\right\}.$$

The transition states are given by $T((i, j), d) = (i + 1, j - y_i d)$ and the value function recursion is:

$$v(i, j) = \max_{d \in S(i,j)} \{v(i + 1, j - y_i d) + p_i d\}.$$

Finally, note that $v(4, 3k) = 4k$ for $k = 0$, 1, 2, 3, and 4.

Exercise 13.5   Using the DP formulation given above, determine $v(0, 14)$ and the corresponding optimal decisions. Compare your results with the optimal decisions from Exercise 13.4 .

Exercise 13.6   Formulate a dynamic programming recursion for the following shortest path problem. City O (the origin) is in stage 0, one can go from any city i in stage $k-1$ to any city j in stage k for $k = 1, \ldots N$. The distance between such cities i and j is denoted by d_{ij}. City D (the destination) is in stage N. The goal is to find a shortest path from the origin O to the destination D.

13.4 Stochastic dynamic programming

So far, we have only considered dynamic programming models that are deterministic, meaning that given a particular state and a decision from its decision set, the transition state is known and unique. This is not always the case for optimization problems involving uncertainty. Consider a blackjack player trying to maximize his earnings by choosing a strategy or a commuter trying to minimize her commute time by picking the roads to take. Suppose the blackjack player currently holds 12 (his current "state") and asks for another card (his "decision"). His next state may be a "win" if he gets a 9, a "lose" if he gets a 10, or "15 (and keep playing)" if he gets a 3. The state he ends up in depends on the card he receives, which is beyond his control. Similarly, the commuter may choose road 1 over road 2, but her actual commute time will depend on the current level of congestion on the road she picks, a quantity beyond her control.

Stochastic dynamic programming addresses optimization problems with uncertainty. The DP methodology we discussed above must be modified to incorporate uncertainty. This is done by allowing multiple transition states for a given state and decision. Each one of the possible transition states is assigned a probability associated with the likelihood of the corresponding state being reached when a certain decision is made. Since the costs are not certain anymore, the value function calculations and optimal decisions will be based on expected values.

We have the following formalization: stages and states are defined as before, and a decision set associated with each state. Given a state (i, j) and $d \in \mathcal{S}(i, j)$, a random event will determine the transition state. We denote with $\mathcal{R}((i, j), d)$ the set of possible outcomes of the random event when we make decision d at state (i, j). For each possible outcome $r \in \mathcal{R}((i, j), d)$ we denote the likelihood of that outcome by $p((i, j), d, r)$. We observe that the probabilities $p((i, j), d, r)$ must be nonnegative and satisfy

$$\sum_{r \in \mathcal{R}((i,j),d)} p((i, j), d, r) = 1, \quad \forall (i, j) \text{ and } \forall d \in \mathcal{S}(i, j).$$

When we make decision d at state (i, j) and when the random outcome r is realized, we transition to the state $T((i, j), d, r)$ and the cost (or benefit) associated with this transition is denoted by $c((i, j), d, r)$. The value function $v(i, j)$ computes expected value of the costs accumulated and must satisfy the following recursion:

$$v(i, j) = \min_{d \in S(i,j)} \left\{ \sum_{r \in R((i,j),d)} p((i, j), d, r) [v(T((i, j), d, r)) + c((i, j), d, r)] \right\}.$$

(13.2)

As before, in a benefit maximization problem, the *min* in (13.2) must be replaced by a *max*.

In some problems, the uncertainty is only in the transition costs and not in the transition states. Such problems can be handled in our notation above by letting $R((i, j), d)$ correspond to the possible outcomes for the cost of the transition. The transition state is independent of the random event, that is, $T((i, j), d, r_1) = T((i, j), d, r_2)$ for all $r_1, r_2 \in R((i, j), d)$. The cost function $c((i, j), d, r)$ reflects the uncertainty in the problem.

Exercise 13.7 Recall the investment problem we discussed in Section 13.3. We have $14\,000$ to invest in four different options which cost y_j thousand dollars for $j = 1$ to 4. Here we introduce the element of uncertainty to the problem. While the cost of investment j is fixed at y_j (all quantities in thousands of dollars), its net present value is uncertain because of the uncertainty of future cash flows and interest rates. We believe that the net present value of investment j has a discrete uniform distribution in the set $\{p_j - 2, p_j - 1, p_j, p_j + 1, p_j + 2\}$. We want to invest in these investment options in order to maximize the expected net present value of our investments. Develop a stochastic DP formulation of this problem and solve it using the recursion (13.2).

14

DP models: option pricing

The most common use of dynamic programming models and principles in financial mathematics is through the lattice models. The binomial lattice has become an indispensable tool for pricing and hedging of derivative securities. We study the binomial lattice in Section 14.2 below. Before we do that, however, we will show how the dynamic programming principles lead to optimal exercise decisions in a more general model than the binomial lattice.

14.1 A model for American options

For a given stock, let S_k denote its price on day k. We can write

$$S_k = S_{k-1} + X_k,$$

where X_k is the change in price from day $k-1$ to day k. The *random-walk model* for stock prices assumes that the random variables X_k are independent and identically distributed, and are also independent of the known initial price S_0. We will also assume that the distribution F of X_k has a finite mean μ.

Now consider an American call option on this stock. Purchasing such an option entitles us to buy the stock at a fixed price c on any day between today (let us call it day 0) and day N, when the option expires. We do not have to ever exercise the option, but if we do at a time when the stock price is S, then our profit is $S - c$. What exercise strategy maximizes our expected profit? We assume that the interest rate is zero throughout the life of the option for simplicity.

Let $v(k, S)$ denote the maximum expected profit when the stock price is S and the option has k additional days before expiration. In our dynamic programming terminology, the stages are $k = 0, 1, 2, \ldots, N$ and the state in each stage is S, the current stock price. Note that stage 0 corresponds to day N and vice versa. In contrast to the DP examples we considered in the previous chapter, we do not assume that the state space is finite in this model. That is, we are considering a continuous DP here, not a discrete DP. The decision set for each state has two elements, namely "exercise" or "do not exercise." The "exercise" decision takes one to the transition state "option exercised," which should be placed at stage N for convenience. The immediate benefit from the "exercise" decision is $S - c$. If we "do not exercise" the option in stage k, we hold the option for at least one more period and observe the random shock x to the stock price that takes us to state $S + x$ in stage $k - 1$.

Given this formulation, our value function $v(k, S)$ satisfies the following recursion:

$$v(k, S) = \max \left\{ S - c, \int v(k - 1, S + x) \mathrm{d}F(x) \right\},$$

with the boundary condition

$$v(0, S) = \max\{S - c, 0\}.$$

For the case that we are considering (American call options), there is no closed form formula for $v(k, S)$. However, dynamic programming can be used to compute a numerical solution. In the remainder of this section, we use the recursion formula to derive the structure of the optimal policy.

Exercise 14.1   Using induction on k, show that $v(k, S) - S$ is a nonincreasing function of S.

Solution: The fact that $v(0, S) - S$ is a nonincreasing function of S follows from the definition of $v(0, S)$. Assume now $v(k - 1, S) - S$ is a nonincreasing function of S. Using the recursion equation, we get

$$v(k, S) - S = \max \left\{ -c, \int (v(k - 1, S + x) - S) \mathrm{d}F(x) \right\}$$

$$= \max \left\{ -c, \int (v(k - 1, S + x) - (S + x)) \mathrm{d}F(x) + \int x \mathrm{d}F(x) \right\}$$

$$= \max \left\{ -c, \mu + \int (v(k - 1, S + x) - (S + x)) \mathrm{d}F(x) \right\},$$

recalling that $\mu = \int x \mathrm{d}F(x)$ denotes the expected value of the random variable x representing daily shocks to the stock price.

For any x, the function $v(k - 1, S + x) - (S + x)$ is a nonincreasing function of S, by the induction hypothesis. It follows that $v(k, S) - S$ is a nonincreasing function of S.   □

Theorem 14.1   *The optimal policy for an American call option has the following form:*

There are nondecreasing numbers $s_1 \leq s_2 \leq \cdots \leq s_k \leq \cdots s_N$ such that, if the current stock price is S and there are k days until expiration, then one should exercise the option if and only if $S \geq s_k$.

Proof: It follows from the recursion equation that if $v(k, S) \leq S - c$, then it is optimal to exercise the option when the stock price is S and there remain k days until expiration. Indeed this yields $v(k, S) = S - c$, which is the maximum possible under the above assumption. Define

$$s_k = \min\{S : v(k, S) = S - c\}.$$

If no S satisfies $v(k, S) = S - c$, then s_k is defined as $+\infty$. From the exercise above, it follows that

$$v(k, S) - S \leq v(k, s_k) - s_k = -c$$

for any $s \geq s_k$ since $v(k, S) - S$ is nonincreasing. Therefore it is optimal to exercise the option with k days to expiration whenever $S \geq s_k$. Since $v(k, S)$ is nondecreasing in k, it immediately follows that s_k is also nondecreasing in k, i.e., $s_1 \leq s_2 \leq \cdots \leq s_k \leq \cdots s_N$. □

A consequence of the above result is that, when $\mu > 0$, it is always optimal to wait until the maturity date to exercise an American call option. The optimal policy described above becomes nontrivial when $\mu < 0$ however.

Exercise 14.2   A *put* option is an agreement to sell an asset for a fixed price c (the *strike price*). An American put option can be exercised at any time up to the maturity date. Prove a theorem similar to Theorem 14.1 for American put options. Can you deduce that it is optimal to wait until maturity to exercise a put option when $\mu > 0$?

14.2 Binomial lattice

If we want to buy or sell an option on an asset (whether a call or a put, an American, European, or another type of option), it is important to determine the fair value of the option today. Determining this fair value is called *option pricing*. The option price depends on the structure of the movements in the price of the underlying asset using information such as the volatility of the underlying asset, the current value of the asset, the dividends (if any) the strike price, the time to maturity, and the riskless interest rate. Several approaches can be used to determine the option price. One popular approach uses dynamic programming on a *binomial lattice* that models the

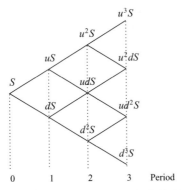

Figure 14.1 Asset price in the binomial lattice model

price movements of the underlying asset. Our discussion here is based on the work of Cox *et al.* [25].

In the binomial lattice model, a basic period length is used, such as a day or a week. If the price of the asset is S in a period, the asset price can only take two values in the next period. Usually, these two possibilities are represented as uS and dS where $u > 1$ and $d < 1$ are multiplicative factors (u stands for up and d for down). The probabilities assigned to these possibilities are p and $1 - p$ respectively, where $0 < p < 1$. This can be represented on a lattice (see Figure 14.1).

After several periods, the asset price can take many different values. Starting from price S_0 in period 0, the price in period k is $u^j d^{k-j} S_0$ if there are j up moves and $k - j$ down moves. The probability of an up move is p whereas that of a down move is $1 - p$ and there are $\binom{k}{j}$ possible paths to reach the corresponding node. Therefore, the probability that the price is $u^j d^{k-j} S_0$ in period k is $\binom{k}{j} p^j (1 - p)^{k-j}$. This is the binomial distribution. As k increases, this distribution converges to the normal distribution.

14.2.1 Specifying the parameters

To specify the model completely, one needs to choose values for u, d, and p. This is done by matching the mean and volatility of the asset price to the mean and volatility of the above binomial distribution. Because the model is multiplicative (the price S of the asset being either uS or dS in the next period), it is convenient to work with logarithms.

Let S_k denote the asset price in periods $k = 0, \ldots, n$. Let μ and σ be the mean and volatility of $\ln(S_n/S_0)$ (we assume that this information about the asset is known). Let $\Delta = 1/n$ denote the length between consecutive periods. Then, the mean and volatility of $\ln(S_1/S_0)$ are $\mu\Delta$ and $\sigma\sqrt{\Delta}$, respectively. In the binomial lattice, we get by direct computation that the mean and variance of $\ln(S_1/S_0)$ are $p \ln u + (1 - p) \ln d$ and $p(1 - p)(\ln u - \ln d)^2$ respectively. Matching these values we get two equations:

$$p \ln u + (1 - p) \ln d = \mu \Delta,$$
$$p(1 - p)(\ln u - \ln d)^2 = \sigma^2 \Delta.$$

Note that there are three parameters but only two equations, so we can set $d = 1/u$ as in [25]. Then the equations simplify to

$$(2p - 1) \ln u = \mu \Delta,$$
$$4p(1 - p)(\ln u)^2 = \sigma^2 \Delta.$$

Squaring the first and adding it to the second, we get $(\ln u)^2 = \sigma^2 \Delta + (\mu \Delta)^2$. This yields

$$u = e^{\sqrt{\sigma^2 \Delta + (\mu \Delta)^2}},$$
$$d = e^{-\sqrt{\sigma^2 \Delta + (\mu \Delta)^2}},$$
$$p = \frac{1}{2} \left(1 + \frac{1}{\sqrt{1 + (\sigma^2/\mu^2 \Delta)}} \right).$$

When Δ is small, these values can be approximated as

$$u = e^{\sigma \sqrt{\Delta}},$$
$$d = e^{-\sigma \sqrt{\Delta}},$$
$$p = \frac{1}{2} \left(1 + \frac{\mu}{\sigma} \sqrt{\Delta} \right).$$

As an example, consider a binomial model with 52 periods of a week each. Consider a stock with current known price S_0 and random price S_{52} a year from today. We are given the mean μ and volatility σ of $\ln(S_{52}/S_0)$, say $\mu = 10\%$ and $\sigma = 30\%$. What are the parameters u, d, and p of the binomial lattice? Since $\Delta = 1/52$ is small, we can use the second set of formulas:

$$u = e^{0.30/\sqrt{52}} = 1.0425 \quad \text{and} \quad d = e^{-0.30/\sqrt{52}} = 0.9592,$$

$$p = \frac{1}{2} \left(1 + \frac{0.10}{0.30\sqrt{52}} \right) = 0.523.$$

14.2.2 Option pricing

Using the binomial lattice described above for the price process of the underlying asset, the value of an option on this asset can be computed by dynamic programming, using backward recursion, working from the maturity date T (period n) back to period 0 (the current period). The stages of the dynamic program are the periods $k = 0, \ldots, N$ and the states are the nodes of the lattice in a given period. Thus there are $k + 1$ states in stage k, which we label $j = 0, \ldots, k$. The nodes in stage N are called the *terminal nodes*.

From a nonterminal node j, we can go either to node $j + 1$ (up move) or to node j (down move) in the next stage. So, to reach node j at stage k we must make exactly j up moves, and $k - j$ down moves between stage 0 and stage k.

We denote by $v(k, j)$ the value of the option in node j of stage k. The value of the option at time 0 is then given by $v(0, 0)$. This is the quantity we have to compute in order to solve the option pricing problem.

The option values at maturity are simply given by the payoff formulas, i.e., $\max(S - c, 0)$ for call options and $\max(c - S, 0)$ for put options, where c denotes the strike price and S is the asset price at maturity. Recall that, in our binomial lattice after N time steps, the asset price in node j is $u^j d^{N-j} S_0$. Therefore the option values in the terminal nodes are:

$$v(N, j) = \max(u^j d^{N-j} S_0 - c, 0) \quad \text{for call options,}$$
$$v(N, j) = \max(c - u^j d^{N-j} S_0, 0) \quad \text{for put options.}$$

We can compute $v(k, j)$ knowing $v(k + 1, j)$ and $v(k + 1, j + 1)$. Recall (Section 4.1.1) that this is done using the risk-neutral probabilities

$$p_u = \frac{R - d}{u - d} \quad \text{and} \quad p_d = \frac{u - R}{u - d},$$

where $R = 1 + r$ and r is the one-period return on the risk-free asset. For European options, the value of $f_k(j)$ is

$$v(k, j) = \frac{1}{R} \left(p_u v(k + 1, j + 1) + p_d v(k + 1, j) \right).$$

For an American call option, we have

$$v(k, j) = \max \left\{ \frac{1}{R} \left(p_u v(k + 1, j + 1) + p_d v(k + 1, j) \right), u^j d^{k-j} S_0 - c \right\},$$

and for an American put option, we have

$$v(k, j) = \max \left\{ \frac{1}{R} \left(p_u v(k + 1, j + 1) + p_d v(k + 1, j) \right), c - u^j d^{k-j} S_0 \right\}.$$

Let us illustrate the approach. We wish to compute the value of an American put option on a stock. The current stock price is \$100. The strike price is \$98 and the expiration date is four weeks from today. The yearly volatility of the logarithm of the stock return is $\sigma = 0.30$. The risk-free interest rate is 4%.

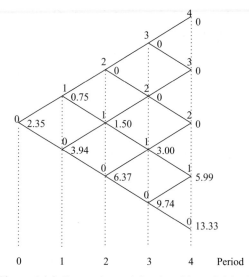

Figure 14.2 Put option pricing in a binomial lattice

We consider a binomial lattice with $N = 4$; see Figure 14.2. To get an accurate answer one would need to take a much larger value of N. Here the purpose is just to illustrate the dynamic programming recursion and $N = 4$ will suffice for this purpose. We recall the values of u and d computed in the previous section:

$$u = 1.0425 \quad \text{and} \quad d = 0.9592.$$

In period $N = 4$, the stock price in node j is given by $u^j d^{4-j} S_0 = 1.0425^j \times 0.9592^{4-j} \times 100$ and therefore the put option payoff is given by:

$$v(4, j) = \max(98 - 1.0425^j \times 0.9592^{4-j} \times 100, 0).$$

That is, $v(4, 0) = 13.33$, $v(4, 1) = 5.99$ and $v(4, 2) = v(4, 3) = v(4, 4) = 0$. Next, we compute the stock price in period $k = 3$. The one-period return on the risk-free asset is $r = 0.04/52 = 0.00077$ and thus $R = 1.00077$.

Accordingly, the risk-neutral probabilities are

$$p_u = \frac{1.00077 - 0.9592}{1.0425 - 0.9592} = 0.499, \quad \text{and} \quad p_d = \frac{1.0425 - 1.00077}{1.0425 - 0.9592} = 0.501.$$

We deduce that, in period 3, the stock price in node j is

$$v(3, j) = \max \left\{ \frac{1}{1.00077} (0.499 v(4, j+1) + 0.501 v(4, j)), \right.$$

$$\left. 98 - 1.0425^j \times 0.9592^{3-j} \times 100 \right\}.$$

That is, $v(3, 0) = \max\{9.67, 9.74\} = 9.74$ (as a side remark, note that it is optimal to exercise the American option before its expiration in this case), $v(3, 1) =$

max{3.00, 2.08} = \$3.00 and $v(3, 2) = v(3, 3) = 0$. Continuing the computations going backward, we compute $v(2, j)$ for $j = 0, 1, 2$, then $v(1, j)$ for $j = 0, 1$, and finally $v(0, 0)$. See Figure 14.2. The option price is $v(0, 0) = \$2.35$.

Note that the approach we outlined above can be used with various types of derivative securities with payoff functions that may make other types of analysis difficult.

Exercise 14.3 Compute the value of an American put option on a stock with current price equal to \$100, strike price equal to \$98, and expiration date five weeks from today. The yearly volatility of the logarithm of the stock return is $\sigma = 0.30$. The risk-free interest rate is 4%. Use a binomial lattice with $N = 5$.

Exercise 14.4 Compute the value of an American call option on a stock with current price equal to \$100, strike price equal to \$102 and expiration date four weeks from today. The yearly volatility of the logarithm of the stock return is $\sigma = 0.30$. The risk-free interest rate is 4%. Use a binomial lattice with $N = 4$.

Exercise 14.5 Computational exercise. Repeat Exercises 14.3 and 14.4 using a binomial lattice with $N = 1000$.

15

DP models: structuring asset-backed securities

The structuring of collateralized mortgage obligations will give us an opportunity to apply the dynamic programming approach studied in Chapter 13.

Mortgages represent the largest single sector of the US debt market, surpassing even the federal government. In 2000, there were over \$5 trillion in outstanding mortgages. Because of the enormous volume of mortgages and the importance of housing in the US economy, numerous mechanisms have been developed to facilitate the provision of credit to this sector. The predominant method by which this has been accomplished since 1970 is securitization, the bundling of individual mortgage loans into capital market instruments. In 2000, \$2.3 trillion of mortgage-backed securities were outstanding, an amount comparable to the \$2.1 trillion corporate bond market and \$3.4 trillion market in federal government securities.

A mortgage-backed security (MBS) is a bond backed by a pool of mortgage loans. Principal and interest payments received from the underlying loans are passed through to the bondholders. These securities contain at least one type of embedded option due to the right of the home buyer to prepay the mortgage loan before maturity. Mortgage payers may prepay for a variety of reasons. By far the most important factor is the level of interest rates. As interest rates fall, those who have fixed rate mortgages tend to repay their mortgages faster.

MBSs were first packaged using the pass-through structure. The pass-through's essential characteristic is that investors receive a pro rata share of the cash flows that are generated by the pool of mortgages – interest, scheduled amortization, and principal prepayments. Exercise of mortgage prepayment options has pro rata effects on all investors. The pass-through allows banks that initiate mortgages to take their fees up front, and sell the mortgages to investors. One troublesome feature of the pass-through for investors is that the timing and level of the cash flows are uncertain. Depending on the interest rate environment, mortgage holders may prepay substantial portions of their mortgage in order to refinance at lower interest rates.

A collateralized mortgage obligation (CMO) is a more sophisticated MBS. The CMO rearranges the cash flows to make them more predictable. This feature makes CMOs more desirable to investors. The basic idea behind a CMO is to restructure the cash flows from an underlying mortgage collateral (pool of mortgage loans) into a set of bonds with different maturities. These two or more series of bonds (called "tranches") receive sequential, rather than pro rata, principal pay down. Interest payments are made on all tranches (except possibly the last tranche, called Z tranche or "accrual" tranche). A two-tranche CMO is a simple example. Assume that there is $100 in mortgage loans backing two $50 tranches, say tranche A and tranche B. Initially, both tranches receive interest, but principal payments are used to pay down only the A tranche. For example, if $1 in mortgage scheduled amortization and pre-payments is collected in the first month, the balance of the A tranche is reduced (paid down) by $1. No principal is paid on the B tranche until the A tranche is fully retired, i.e., $50 in principal payments have been made. Then the remaining $50 in mortgage principal pays down the $50 B tranche. In effect, the A or "fast-pay" tranche has been assigned all of the early mortgage principal payments (amortization and prepayments) and reaches its maturity sooner than would an ordinary pass-through security. The B or "slow-pay" tranche has only the later principal payments and it begins paying down much later than an ordinary pass-through security.

By repackaging the collateral cash flow in this manner, the life and risk characteristics of the collateral are restructured. The fast-pay tranches are guaranteed to be retired first, implying that their lives will be less uncertain, although not completely fixed. Even the slow-pay tranches will have less cash-flow uncertainty than the underlying collateral. Therefore the CMO allows the issuer to target different investor groups more directly than when issuing pass-through securities. The low maturity (fast-pay) tranches may be appealing to investors with short horizons while the long maturity bonds (slow-pay) may be attractive to pension funds and life insurance companies. Each group can find a bond that is better customized to their particular needs.

A by-product of improving the predictability of the cash flows is being able to structure tranches of different credit quality from the same mortgage pool. With the payments of a very large pool of mortgages dedicated to the "fast-pay" tranche, it can be structured to receive a AAA credit rating even if there is a significant default risk on part of the mortgage pool. This high credit rating lowers the interest rate that must be paid on this slice of the CMO. While the credit rating for the early tranches can be very high, the credit quality for later tranches will necessarily be lower because there is less principal left to be repaid and therefore there is increased default risk on slow-pay tranches.

We will take the perspective of an issuer of CMOs. How many tranches should be issued? Which sizes? Which coupon rates? Issuers make money by issuing CMOs because they can pay interest on the tranches that is lower than the interest payments being made by mortgage holders in the pool. The mortgage holders pay 10- or 30-year interest rates on the entire outstanding principal, while some tranches only pay two, four, six, and eight-year interest rates plus an appropriate spread.

The convention in mortgage markets is to price bonds with respect to their weighted average life (WAL), which is much like duration, i.e.,

$$\text{WAL} = \frac{\sum_{t=1}^{T} t P_t}{\sum_{t=1}^{T} P_t},$$

where P_t is the principal payment in period t $(t = 1, \ldots, T)$.

A bond with a WAL of 3 years will be priced at the 3-year Treasury rate plus a spread, while a bond with a WAL of 7 years will be priced at the 7-year Treasury rate plus a spread. The WAL of the CMO collateral is typically high, implying a high rate for (normal) upward sloping rate curves. By splitting the collateral into several tranches, some with a low WAL and some with a high WAL, lower rates are obtained on the fast-pay tranches while higher rates result for the slow-pay. Overall, the issuer ends up with a better (lower) average rate on the CMO than on the collateral.

15.1 Data

When issuing a CMO, several restrictions apply. First it must be demonstrated that the collateral can service the payments on the issued CMO tranches under several scenarios. These scenarios are well defined and standardized, and cover conditional prepayment models (see below) as well as the two extreme cases of full immediate prepayment and no prepayment at all. Second, the tranches are priced using their expected WAL. For example, a tranche with a WAL between 2.95 and 3.44 will be priced at the 3-year Treasury rate plus a spread that depends on the tranche's rating. For an AAA rating the spread might be 1%, whereas for a BB rating the spread might be 2%.

Table 15.1 contains the payment schedule for a $100 million pool of ten-year mortgages with 10% interest, assuming the same total payment (interest + scheduled amortization) each year. It may be useful to remember that, if the outstanding principal is Q, interest is r and amortization occurs over k years, then the scheduled amortization in the first year is

$$\frac{Qr}{(1 + r)^k - 1}.$$

Table 15.1 *Payment schedule*

Period (t)	Interest (I_t)	Scheduled amortization (P_t)	Outstanding principal (Q_t)
1	10.00	6.27	93.73
2	9.37	6.90	86.83
3	8.68	7.59	79.24
4	7.92	8.35	70.89
5	7.09	9.19	61.70
6	6.17	10.11	51.59
7	5.16	11.12	40.47
8	4.05	12.22	28.25
9	2.83	13.45	14.80
10	1.48	14.80	0
Total		100.00	

Exercise 15.1 Derive this formula, using the fact that the total payment (interest + scheduled amortization) is the same for years 1 through k.

For the mortgage pool described above, $Q = 100, r = 0.10$, and $k = 10$, thus the scheduled amortization in the first year is 6.27. Adding the 10% interest payment on Q, the total payments (interest + scheduled amortization) are \$16.27 million per year.

Table 15.1 assumes no prepayment. Next we want to analyze the following scenario: a conditional prepayment model reflecting the 100% PSA (Public Securities Association) industry-standard benchmark. For simplicity, we present a yearly PSA model, even though the actual PSA model is defined monthly. The rate of mortgage prepayments is 1% of the outstanding principal at the end of the first year. At the end of the second year, prepayment is 3% of the outstanding principal at that time. At the end of the third year, it is 5% of the outstanding principal. For each later year $t \geq 3$, prepayment is 6% of the outstanding principal at the end of year t. Let us denote by PP_t the prepayment in year t. For example, in year 1, in addition to the interest payment $I_1 = 10$ and the amortization payment $A_1 = 6.27$, there is a 1% prepayment on the $100 - 6.27 = 93.73$ principal remaining after amortization. That is, there is a prepayment $PP_1 = 0.9373$ collected at the end of year 1. Thus the principal pay down is $P_1 = A_1 + PP_1 = 6.27 + 0.9373 = 7.2073$ in year 1. The outstanding principal at the end of year 1 is $Q_1 = 100 - 7.2073 = 92.7927$. In year 2, the interest paid is $I_2 = 9.279$ (that is 10% of Q_1), the amortization payment is $A_2 = \frac{Q_1 \times 0.10}{(1.10)^9 - 1} = 6.8333$, the prepayment is $PP_2 = 2.5788$ (that is, 3% of $Q_1 - A_2$), and the principal pay down is $P_2 = A_2 + PP_2 = 9.412$, etc.

Exercise 15.2   Construct the table containing I_t, P_t, and Q_t to reflect the above scenario.

Loss multiple and required buffer

In order to achieve a high quality rating, tranches should be able to sustain higher than expected default rates without compromising payments to the tranche holders. For this reason, credit ratings are assigned based on how much money is "behind" the current tranche. That is, how much outstanding principal is left after the current tranche is retired, as a percentage of the total amount of principal. This is called the "buffer." Early tranches receive higher credit ratings since they have greater buffers, which means that the CMO would have to experience very large default rates before their payments would be compromised. A tranche with AAA rating must have a buffer equal to six times the expected default rate. This is referred to as the "loss multiple." The loss multiples are as follows:

Credit rating	AAA	AA	A	BBB	BB	B	CCC
Loss multiple	6	5	4	3	2	1.5	0

The required buffer is computed by the following formula:

$$\text{Required buffer} = \text{WAL} \times \text{expected default rate} \times \text{loss multiple}.$$

Let us assume a 0.9% expected default rate, based on foreclosure rates reported by the M&T Mortgage Corporation in 2004. With this assumption, the required buffer to get an AAA rating for a tranche with a WAL of 4 years is $4 \times 0.009 \times 6 = 21.6\%$.

Exercise 15.3   Construct the table containing the required buffer as a function of rating and WAL, assuming a 0.9% expected default rate.

Coupon yields and spreads

Each tranche is priced based on a credit spread to the current treasury rate for a risk-free bond of that approximate duration. These rates appear in Table 15.2, based on the yields on US Treasuries as of 10/12/04. The reader can get more current figures from online sources. Spreads on corporate bonds with similar credit ratings would provide reasonable figures.

15.2 Enumerating possible tranches

We are going to consider every possible tranche: since there are ten possible maturities t and t possible starting dates j with $j \leq t$ for each t, there are 55 possible tranches. Specifically, tranche (j, t) starts amortizing at the beginning of year j and ends at the end of year t.

Table 15.2 *Yields and spreads*

Period (t)	Risk-free spot (%)	Credit spread in basis points					
		AAA	AA	A	BBB	BB	B
1	2.18	13	43	68	92	175	300
2	2.53	17	45	85	109	195	320
3	2.80	20	47	87	114	205	330
4	3.06	26	56	90	123	220	343
5	3.31	31	65	92	131	235	355
6	3.52	42	73	96	137	245	373
7	3.72	53	81	99	143	255	390
8	3.84	59	85	106	151	262	398
9	3.95	65	90	112	158	268	407
10	4.07	71	94	119	166	275	415

Exercise 15.4 From the principal payments P_t that you computed in Exercise 15.2, construct a table containing WAL_{jt} for each possible combination (j, t).
For each of the 55 possible tranches (j, t), compute the buffer $\sum_{k=t+1}^{10} P_k / \sum_{k=1}^{10} P_k$. If there is no buffer, the corresponding tranche is a Z-tranche. When there is a buffer, calculate the loss multiple from the formula:

$$\text{Required buffer} = WAL \times \text{expected default rate} \times \text{loss multiple}.$$

Finally, construct a table containing the credit rating for each tranche that is not a Z-tranche.
For each of the 55 tranches, construct a table containing the appropriate coupon rate c_{jt} (no coupon rate on a Z-tranche). As described earlier, these rates depend on the WAL and credit rating just computed.

Define T_{jt} to be the present value of the payments on a tranche (j, t). Armed with the proper coupon rate c_{jt} and a full curve of spot rates r_t, T_{jt} is computed as follows. In each year k, the payment C_k for tranche (j, t) is equal to the coupon rate c_{jt} times the remaining principal, plus the principal payment made to tranche (j, t) if it is amortizing in year k. The present value of C_k is simply equal to $C_k/(1 + r_k)^k$. Now T_{jt} is obtained by summing the present values of all the payments going to tranche (j, t).

15.3 A dynamic programming approach

Based on the above data, we would like to structure a CMO with four sequential tranches A, B, C, Z. The objective is to maximize the profits from the issuance by choosing the size of each tranche. In this section, we present a dynamic programming recursion for solving the problem.

Let $t = 1, \ldots, 10$ index the years. The states of the dynamic program will be the years t and the stages will be the number k of tranches up to year t.

Now that we have the matrix T_{jt}, we are ready to describe the dynamic programming recursion. Let

$v(k, t) = $ Minimum present value of total payments to bondholders in years

1 through t when the CMO has k tranches up to year t.

Obviously, $v(1, t)$ is simply T_{1t}. For $k \geq 2$, the value $v(k, t)$ is computed recursively by the formula:

$$v(k, t) = \min_{j=k-1,\ldots,t-1} (v(k-1, j) + T_{j+1,t}).$$

For example, for $k = 2$ and $t = 4$, we compute $v(1, j) + T_{j+1,4}$ for each $j = 1, 2, 3$ and we take the minimum. The power of dynamic programming becomes clear as k increases. For example, when $k = 4$, there is no need to compute the minimum of thousands of possible combinations of four tranches. Instead, we use the optimal structure $v(3, j)$ already computed in the previous stage. So the only enumeration is over the size of the last tranche.

Exercise 15.5   Compute $v(4, 10)$ using the above recursion. Recall that $v(4, 10)$ is the least-cost solution of structuring the CMO into four tranches. What are the sizes of the tranches in this optimal solution? To answer this question, you will need to backtrack from the last stage and identify how the minimum leading to $v(4, 10)$ was achieved at each stage.

Exercise 15.6   The dynamic programming approach presented in this section is based on a single prepayment model. How would you deal with several scenarios for prepayment and default rates, each occuring with a given probability?

15.4 Case study: structuring CMOs

Repeat the above steps for a pool of mortgages using current data. Study the influence of the expected default rate on the profitability of structuring your CMO. What other factors have a significant impact on profitability?

16

Stochastic programming: theory and algorithms

16.1 Introduction

In the introductory chapter and elsewhere, we argued that many optimization problems are described by uncertain parameters. There are different ways of incorporating this uncertainty. We consider two approaches: stochastic programming in the present chapter and robust optimization in Chapter 19. *Stochastic programming* assumes that the uncertain parameters are random variables with known probability distributions. This information is then used to transform the stochastic program into a so-called *deterministic equivalent*, which might be a linear program, a nonlinear program, or an integer program (see Chapters 2, 5, and 11 respectively).

While stochastic programming models have existed for several decades, computational technology has only recently allowed the solution of realistic size problems. The field continues to develop with the advancement of available algorithms and computing power. It is a popular modeling tool for problems in a variety of disciplines including financial engineering.

The uncertainty is described by a certain sample space Ω, a σ-field of random events, and a probability measure P (see Appendix C). In stochastic programming, Ω is often a finite set $\{\omega_1, \ldots, \omega_S\}$. The corresponding probabilities $p(\omega_k) \geq 0$ satisfy $\sum_{k=1}^{S} p(\omega_k) = 1$. For example, to represent the outcomes of flipping a coin twice in a row, we would use four random events $\Omega = \{HH, HT, TH, TT\}$, each with probability $1/4$, where H stands for heads and T stands for tails.

Stochastic programming models can include *anticipative* and/or *adaptive* decision variables. Anticipative variables correspond to those decisions that must be made *here-and-now* and cannot depend on the future observations/partial realizations of the random parameters. Adaptive variables correspond to *wait-and-see* decisions that can be made after some (or, sometimes all) of the random parameters are observed.

Stochastic programming models that include both anticipative and adaptive variables are called *recourse* models. Using a multi-stage stochastic programming formulation, with recourse variables at each stage, one can model a decision environment where information is revealed progressively and the decisions are adapted to each new piece of information.

In investment planning, each new trading opportunity represents a new decision to be made. Therefore, trading dates where investment portfolios can be rebalanced become natural choices for decision stages, and these problems can be formulated conveniently as multi-stage stochastic programming problems with recourse.

16.2 Two-stage problems with recourse

In Chapter 1, we have already seen a generic form of a *two-stage stochastic linear program with recourse*:

$$\max_x \quad a^T x + E[\max_{y(\omega)} c(\omega)^T y(\omega)]$$
$$Ax \qquad\qquad\qquad\qquad = b$$
$$B(\omega)x + \qquad C(\omega)y(\omega) = d(\omega) \qquad (16.1)$$
$$x \geq 0, \qquad\qquad y(\omega) \geq 0.$$

In this formulation, the first-stage decisions are represented by vector x. These decisions are made *before* the random event ω is observed. The second-stage decisions are represented by vector $y(\omega)$. These decisions are made *after* the random event ω has been observed, and therefore the vector y is a function of ω. A and b define deterministic constraints on the first-stage decisions x, whereas $B(\omega)$, $C(\omega)$, and $d(\omega)$ define stochastic constraints linking the recourse decisions $y(\omega)$ to the first-stage decisions x. The objective function contains a deterministic term $a^T x$ and the expectation of the second-stage objective $c(\omega)^T y(\omega)$ taken over all realizations of the random event ω.

Notice that the first-stage decisions will not necessarily satisfy the linking constraints $B(\omega)x + C(\omega)y(\omega) = d(\omega)$, if no recourse action is taken. Therefore, recourse allows one to make sure that the initial decisions can be "corrected" with respect to this second set of feasibility equations.

In Section 1.2.1, we also argued that problem (16.1) can be represented in an alternative manner by considering the *second-stage* or *recourse* problem that is defined as follows, given x, the first-stage decisions:

$$f(x, \omega) = \max c(\omega)^T y(\omega)$$
$$C(\omega)y(\omega) = d(\omega) - B(\omega)x \qquad (16.2)$$
$$y(\omega) \geq 0.$$

Let $f(x) = E[f(x, \omega)]$ denote the expected value of this optimum. If the function $f(x)$ is available, the two-stage stochastic linear program (16.1) reduces to a deterministic nonlinear program:

$$\begin{aligned}
\max \ & a^T x + f(x) \\
& Ax = b \\
& x \geq 0.
\end{aligned} \qquad (16.3)$$

Unfortunately, computing $f(x)$ is often very hard, especially when the sample space Ω is infinite. Next, we consider the case where Ω is a finite set.

Assume that $\Omega = \{\omega_1, \ldots, \omega_S\}$ and let $p = (p_1, \ldots, p_S)$ denote the probability distribution on this sample space. The S possibilities ω_k, for $k = 1, \ldots, S$ are also called *scenarios*. The expectation of the second-stage objective becomes:

$$E[\max_{y(\omega)} c(\omega)^T y(\omega)] \ = \ \sum_{k=1}^{S} p_k \max_{y(\omega_k)} c(\omega_k)^T y(\omega_k).$$

For brevity, we write c_k instead of $c(\omega_k)$, etc. Under this *scenario approach* the two-stage stochastic linear programming problem (16.1) takes the following form:

$$\begin{aligned}
\max_x \ & a^T x + \sum_{k=1}^{S} p_k \max_{y_k} c_k^T y_k \\
& Ax = b \\
& B_k x + C_k y_k = d_k \quad \text{for } k = 1, \ldots S \\
& x \geq 0 \\
& y_k \geq 0 \quad \text{for } k = 1, \ldots, S.
\end{aligned} \qquad (16.4)$$

Note that there is a different second stage decision vector y_k for each scenario k. The maximum in the objective is achieved by optimizing over all variables x and y_k simultaneously. Therefore, this optimization problem is:

$$\begin{aligned}
\max_{x, y_1, \ldots, y_S} \ & a^T x + p_1 c_1^T y_1 + \ldots + p_S c_S^T y_S \\
& Ax && = b \\
& B_1 x + \ C_1 y_1 && = d_1 \\
& \quad \vdots \qquad\qquad \ddots && \quad \vdots \\
& B_S x && + \ C_S y_S = d_S \\
& x, \qquad y_1, \quad \ldots && \quad y_S \geq 0.
\end{aligned} \qquad (16.5)$$

This is a deterministic linear programming problem called the *deterministic equivalent* of the original uncertain problem. This problem has S copies of the second-stage decision variables and therefore, can be significantly larger than the original problem before we considered the uncertainty of the parameters. Fortunately, however, the constraint matrix has a very special sparsity structure that can be exploited by modern decomposition based solution methods (see Section 16.4).

Exercise 16.1   Consider an investor with an initial wealth W_0. At time 0, the investor constructs a portfolio comprising one riskless asset with return R_1 in the first period and one risky asset with return R_1^+ with probability 0.5 and R_1^- with probability 0.5. At the end of the first period, the investor can rebalance her portfolio. The return in the second period is R_2 for the riskless asset, while it is R_2^+ with probability 0.5 and R_2^- with probability 0.5 for the risky asset. The objective is to meet a liability $L_2 = 0.9$ at the end of period 2 and to maximize the expected remaining wealth W_2. Formulate a two-stage stochastic linear program that solves the investor's problem.

Exercise 16.2   In Exercise 3.2, the cash requirement in quarter Q1 is known to be 100 but, for the remaining quarters, the company considers three equally likely scenarios:

	Q2	Q3	Q4	Q5	Q6	Q7	Q8
Scenario 1	450	100	−650	−550	200	650	−850
Scenario 2	500	100	−600	−500	200	600	−900
Scenario 3	550	150	−600	−450	250	600	−800

Formulate a linear program that maximizes the expected wealth of the company at the end of quarter Q8.

16.3 Multi-stage problems

In a *multi-stage stochastic program with recourse*, the recourse decisions can be made at several points in time, called stages. Let $n \geq 2$ be the number of stages. The random event ω is a vector $(o_1, \ldots, o_{n-1})$ that gets revealed progressively over time. The first-stage decisions are taken before any component of ω is revealed. Then o_1 is revealed. With this knowledge, one takes the second-stage decisions. After that, o_2 is revealed, and so on, alternating between a new component of ω being revealed and new recourse decisions being implemented. We assume that $\Omega = \{\omega_1, \ldots, \omega_S\}$ is a finite set. Let p_k be the probability of scenario ω_k, for $k = 1, \ldots, S$.

Some scenarios ω_k may be identical in their first components and only become differentiated in the later stages. Therefore it is convenient to introduce the *scenario tree*, which illustrates how the scenarios branch off at each stage. The nodes are labeled 1 through N, where node 1 is the root. Each node is in one stage, where the root is the unique node in stage 1. Each node i in stage $k \geq 2$ is adjacent to a unique node $a(i)$ in stage $k - 1$. Node $a(i)$ is called the *father* of node i. The paths from the root to the leaves (in stage n) represent the scenarios. Thus the last stage has as many nodes as scenarios. These nodes are called the *terminal nodes*. The collection of scenarios passing through node i in stage k have identical components $o_1, \ldots, o_{k-1}$.

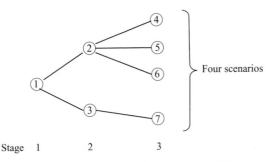

Figure 16.1 A scenario tree with three stages and four scenarios

In Figure 16.1, node 1 is the root, nodes 4, 5, 6, and 7 are the terminal nodes. The father of node 6 is node 2, in other words $a(6) = 2$.

Associated with each node i is a recourse decision vector x_i. For a node i is stage k, the decisions x_i are taken based on the information that has been revealed up to stage k. Let q_i be the sum of the probabilities p_k over all the scenarios ω_k that go through node i. Therefore q_i is the probability of node i, conditional on being in stage k. The multi-stage stochastic program with recourse can be formulated as follows:

$$\max_{x_1,\dots,x_N} \quad \sum_{i=1}^{N} q_i c_i^{\mathrm{T}} x_i$$
$$\begin{aligned} Ax_1 &= b \\ B_i x_{a(i)} + \quad C_i x_i &= d_i \quad \text{for } i = 2,\dots,N \\ x_i &\geq 0. \end{aligned} \tag{16.6}$$

In this formulation, A and b define deterministic constraints on the first-stage decisions x_1, whereas B_i, C_i, and d_i define stochastic constraints linking the recourse decisions x_i in node i to the recourse decisions $x_{a(i)}$ in its father node. The objective function contains a term $c_i^{\mathrm{T}} x_i$ for each node.

To illustrate, we present formulation (16.6) for the example of Figure 16.1. The terminal nodes 4 to 7 correspond to scenarios 1 to 4 respectively. Thus we have $q_4 = p_1, q_5 = p_2, q_6 = p_3$, and $q_7 = p_4$, where p_k is the probability of scenario k. We also have $q_2 = p_1 + p_2 + p_3$, $q_3 = p_4$, and $q_2 + q_3 = 1$.

$$\max \ c_1^{\mathrm{T}} x_1 + q_2 c_2^{\mathrm{T}} x_2 + q_3 c_3^{\mathrm{T}} x_3 + p_1 c_4^{\mathrm{T}} x_4 + p_2 c_5^{\mathrm{T}} x_5 + p_3 c_6^{\mathrm{T}} x_6 + p_4 c_7^{\mathrm{T}} x_7$$
$$\begin{aligned} Ax_1 &= b \\ B_2 x_1 + C_2 x_2 &= d_2 \\ B_3 x_1 \qquad + C_3 x_3 &= d_3 \\ B_4 x_2 \qquad + C_4 x_4 &= d_4 \\ B_5 x_2 \qquad\qquad + C_5 x_5 &= d_5 \\ B_6 x_2 \qquad\qquad\qquad + C_6 x_6 &= d_6 \\ B_7 x_3 \qquad\qquad\qquad\qquad + C_7 x_7 &= d_7 \\ x_i &\geq 0. \end{aligned}$$

Note that the size of the linear program (16.6) increases rapidly with the number of stages. For example, for a problem with ten stages and a binary tree, there are 1024 scenarios and therefore the linear program (16.6) may have several thousand constraints and variables, depending on the number of variables and constraints at each node. Modern commercial codes can handle such large linear programs, but a moderate increase in the number of stages or in the number of branches at each stage could make (16.6) too large to solve by standard linear programming solvers. When this happens, one may try to exploit the special structure of (16.6) to solve the model (see Section 16.4).

Exercise 16.3 In Exercise 3.2, the cash requirements in quarters Q1, Q2, Q3, Q6, and Q7 are known. On the other hand, the company considers two equally likely (and independent) possibilities for each of the quarters Q4, Q5, and Q8, giving rise to eight equally likely scenarios. In quarter Q4, the cash inflow will be either 600 or 650. In quarter Q5, it will be either 500 or 550. In quarter Q8, it will be either 850 or 900. Formulate a linear program that maximizes the expected wealth of the company at the end of quarter Q8.

Exercise 16.4 Develop the linear program (16.6) for the following scenario tree.

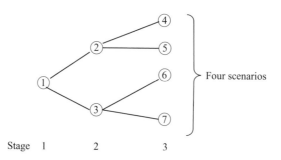

16.4 Decomposition

The size of the linear program (16.6) depends on the number of decision stages and the branching factor at each node of the scenario tree. For example, a four-stage model with 25 branches at each node has $25 \times 25 \times 25 \times 25 = 390\,625$ scenarios. Increasing the number of stages and branches quickly results in an explosion of dimensionality. Obviously, the size of (16.6) can be a limiting factor in solving realistic problems. When this occurs, it becomes essential to take advantage of the special structure of the linear program (16.6). In this section, we present a decomposition algorithm for exploiting this structure. It is called *Benders decomposition* or, in the stochastic programming literature, the *L-shaped method*.

The structure that we really want to exploit is that of the two-stage problem (16.5). So we start with (16.5). We will explain subsequently how to deal with the general multi-stage model (16.6). The constraint matrix of (16.5) has the following form:

$$\begin{pmatrix} A & & & \\ B_1 & C_1 & & \\ \vdots & & \ddots & \\ B_S & & & C_S \end{pmatrix}.$$

Note that the blocks $C_1, \ldots, C_S$ of the constraint matrix are only interrelated through the blocks $B_1, \ldots, B_S$, which correspond to the first-stage decisions. In other words, once the first-stage decisions x have been fixed, (16.5) decomposes into S independent linear programs. The idea of Benders decomposition is to solve a "master problem" involving only the variables x and a series of independent "recourse problems" each involving a different vector of variables y_k. The master problem and recourse problems are linear programs. The size of these linear programs is much smaller than the size of full model (16.5). The recourse problems are solved for a given vector x and their solutions are used to generate inequalities that are added to the master problem. Solving the new master problem produces a new x and the process is repeated. More specifically, let us write (16.5) as

$$\begin{aligned} \max_x\ a^\mathsf{T}x\ +\ P_1(x)\ +\ \cdots\ +\ P_S(x) \\ Ax = b \\ x \geq 0, \end{aligned} \tag{16.7}$$

where, for $k = 1, \ldots S$,

$$\begin{aligned} P_k(x)\ =\ \max_{y_k}\ p_k c_k^\mathsf{T} y_k \\ C_k y_k = d_k - B_k x \\ y_k \geq 0. \end{aligned} \tag{16.8}$$

The dual linear program of the recourse problem (16.8) is

$$\begin{aligned} P_k(x)\ =\ \min_{u_k}\ u_k^\mathsf{T}(d_k - B_k x) \\ C_k^\mathsf{T} u_k \geq p_k c_k. \end{aligned} \tag{16.9}$$

For simplicity, we assume that the dual (16.9) is feasible, which is the case of interest in applications. The recourse linear program (16.8) will be solved for a sequence of vectors x^i, for $i = 0, \ldots$. The initial vector x^0 might be obtained by solving

$$\begin{aligned} \max_x\ a^\mathsf{T}x \\ Ax = b \\ x \geq 0. \end{aligned} \tag{16.10}$$

For a given vector x^i, two possibilities can occur for the recourse linear program (16.8): either (16.8) has an optimal solution or it is infeasible.

If (16.8) has an optimal solution y_k^i, and u_k^i is the corresponding optimal dual solution, then (16.9) implies that

$$P_k(x^i) = \left(u_k^i\right)^{\mathrm{T}}(d_k - B_k x^i)$$

and, since

$$P_k(x) \le \left(u_k^i\right)^{\mathrm{T}}(d_k - B_k x),$$

we get that

$$P_k(x) \le \left(u_k^i\right)^{\mathrm{T}}(B_k x^i - B_k x) + P_k(x^i).$$

This inequality, which is called an *optimality cut*, can be added to the current master linear program. Initially, the master linear program is just (16.10).

If (16.8) is infeasible, then the dual problem is unbounded. Let u_k^i denote a direction where (16.9) is unbounded, i.e., $(u_k^i)^{\mathrm{T}}(d_k - B_k x^i) < 0$ and $C_k^{\mathrm{T}} u_k^i \ge p_k c_k$. Since we are only interested in first-stage decisions x that lead to feasible second-stage decisions y_k, the following *feasibility cut* can be added to the current master linear program:

$$\left(u_k^i\right)^{\mathrm{T}}(d_k - B_k x) \ge 0.$$

After solving the recourse problems (16.8) for each k, we have the following lower bound on the optimal value of (16.5):

$$\mathrm{LB} = a^{\mathrm{T}} x^i + P_1(x^i) + \cdots + P_S(x^i),$$

where we set $P_k(x^i) = -\infty$ if the corresponding recourse problem is infeasible.

Adding all the optimality and feasibility cuts found so far (for $j = 0, \ldots, i$) to the master linear program, we obtain:

$$\begin{aligned}
\max_{x, z_1, \ldots, z_S}\ & a^{\mathrm{T}} x + \sum_{k=1}^{S} z_k \\
& Ax = b \\
& z_k \le \left(u_k^j\right)^{\mathrm{T}}(B_k x^j - B_k x) + P_k(x^j) \quad \text{for some pairs } (j, k) \\
& 0 \le \left(u_k^j\right)^{\mathrm{T}}(d_k - B_k x) \quad \text{for the remaining pairs } (j, k) \\
& x \ge 0.
\end{aligned}$$

Denoting by $x^{i+1}, z_1^{i+1}, \ldots, z_S^{i+1}$ an optimal solution to this linear program we get an upper bound on the optimal value of (16.5):

$$\mathrm{UB} = a^{\mathrm{T}} x^{i+1} + z_1^{i+1} + \cdots + z_S^{i+1}.$$

Benders decomposition alternately solves the recourse problems (16.8) and the master linear program with new optimality and feasibility cuts added at each iteration until the gap between the upper bound UB and the lower bound LB falls below a given threshold. One can show that UB − LB converges to zero in a finite number of iterations. See, for instance, the book of Birge and Louveaux [13], pages 159–162.

Benders decomposition can also be used for multi-stage problems (16.6) in a straightforward way: the stages are partitioned into a first set that gives rise to the "master problem" and a second set that gives rise to the "recourse problems." For example in a six-stage problem, the variables of the first two stages could define the master problem. When these variables are fixed, (16.6) decomposes into separate linear programs each involving variables of the last four stages. The solutions of these recourse linear programs provide optimality or feasibility cuts that can be added to the master problem. As before, upper and lower bounds are computed at each iteration and the algorithm stops when the difference drops below a given tolerance. Using this approach, Gondzio and Kouwenberg [34] were able to solve an asset liability management problem with over 4 million scenarios, whose linear programming formulation (16.6) had 12 million constraints and 24 million variables. This linear program was so large that storage space on the computer became an issue. The scenario tree had 6 levels and 13 branches at each node. In order to apply two-stage Benders decomposition, Gondzio and Kouwenberg divided the six-stage problem into a first-stage problem containing the first three periods and a second stage containing periods 4 to 6. This resulted in 2197 recourse linear programs, each involving 2197 scenarios. These recourse linear programs were solved by an interior-point algorithm. Note that Benders decomposition is ideally suited for parallel computations since the recourse linear programs can be solved simultaneously. When the solution of all the recourse linear programs is completed (which takes the bulk of the time), the master problem is then solved on one processor while the other processors remain idle temporarily. Gondzio and Kouwenberg tested a parallel implementation on a computer with 16 processors and they obtained an almost perfect speedup, that is a speedup factor of almost k when using k processors.

16.5 Scenario generation

How should one generate scenarios in order to formulate a deterministic equivalent formulation (16.6) that accurately represents the underlying stochastic program? There are two separate issues. First, one needs to model the correlation over time among the random parameters. For a pension fund, such a model might relate wage inflation (which influences the liability side) to interest rates and stock prices (which influence the asset side). Mulvey [59] describes the system developed by Towers Perrin, based on a cascading set of stochastic differential equations. Simpler autoregressive models can also be used. This is discussed below. The second issue is the construction of a scenario tree from these models: a finite number of scenarios must reflect as accurately as possible the random processes modeled in the previous step, suggesting the need for a large number of scenarios. On the other hand, the linear program (16.6) can only be solved if the size of the scenario tree is reasonably small, suggesting a rather limited number of scenarios. To reconcile these two conflicting objectives, it might be crucial to use variance reduction techniques. We address these issues in this section.

16.5.1 Autoregressive model

In order to generate the random parameters underlying the stochastic program, one needs to construct an economic model reflecting the correlation between the parameters. Historical data may be available. The goal is to generate meaningful time series for constructing the scenarios. One approach is to use an autoregressive model.

Specifically, if r_t denotes the random vector of parameters in period t, an *autoregressive model* is defined by:

$$r_t = D_0 + D_1 r_{t-1} + \cdots + D_p r_{t-p} + \epsilon_t,$$

where p is the number of lags used in the regression, $D_0, D_1, \ldots, D_p$ are time-independent constant matrices which are estimated through statistical methods such as maximum likelihood, and ϵ_t is a vector of i.i.d. random disturbances with mean zero.

To illustrate this, consider the example of Section 8.1.1. Let s_t, b_t, and m_t denote the rates of return of stocks, bonds, and the money market, respectively, in year t. An autoregressive model with $p = 1$ has the form:

$$\begin{pmatrix} s_t \\ b_t \\ m_t \end{pmatrix} = \begin{pmatrix} d_1 \\ d_2 \\ d_3 \end{pmatrix} + \begin{pmatrix} d_{11} & d_{12} & d_{13} \\ d_{21} & d_{22} & d_{23} \\ d_{31} & d_{32} & d_{33} \end{pmatrix} \begin{pmatrix} s_{t-1} \\ b_{t-1} \\ m_{t-1} \end{pmatrix} + \begin{pmatrix} \epsilon_t^s \\ \epsilon_t^b \\ \epsilon_t^m \end{pmatrix} \quad t = 2, \ldots, T.$$

In particular, to find the parameters $d_1, d_{11}, d_{12}, d_{13}$ in the first equation:

$$s_t = d_1 + d_{11} s_{t-1} + d_{12} b_{t-1} + d_{13} m_{t-1} + \epsilon_t^s,$$

one can use standard linear regression tools that minimize the sum of the squared errors ϵ_t^s. Within an Excel spreadsheet, for instance, one can use the function LINEST. Suppose that the rates of return on the stocks are stored in cells B2 to B44 and that, for bonds and the money market, the rates are stored in columns C and D, rows 2 to 44 as well. LINEST is an array formula. Its first argument contains the known data for the left-hand side of the equation (here the column s_t), the second argument contains the known data in the right-hand side (here the columns s_{t-1}, b_{t-1}, and m_{t-1}). Typing LINEST(B3:B44, B2:D43,,) one obtains the following values of the parameters:

$$d_1 = 0.077, \quad d_{11} = -0.058, \quad d_{12} = 0.219, \quad d_{13} = 0.448.$$

Using the same approach for the other two equations we get the following autoregressive model:

$$
\begin{aligned}
s_t &= 0.077 - 0.058s_{t-1} + 0.219b_{t-1} + 0.448m_{t-1} + \epsilon_t^s, \\
b_t &= 0.047 - 0.053s_{t-1} - 0.078b_{t-1} + 0.707m_{t-1} + \epsilon_t^b, \\
m_t &= 0.016 + 0.033s_{t-1} - 0.044b_{t-1} + 0.746m_{t-1} + \epsilon_t^m.
\end{aligned}
$$

The option LINEST(B3:B44, B2:D43,,TRUE) provides some useful statistics, such as the standard error of the estimate s_t. Here we get a standard error of $\sigma_s = 0.173$. Similarly, the standard error for b_t and m_t are $\sigma_b = 0.108$ and $\sigma_m = 0.022$ respectively.

Exercise 16.5 Instead of an autoregressive model relating the rates of returns r_t, b_t, and m_t, construct an autoregressive model relating the logarithms of the returns $g_t = log(1 + r_t)$, $h_t = log(1 + b_t)$, and $k_t = log(1 + m_t)$. Use one lag, i.e., $p = 1$. Solve using LINEST or your prefered linear regression tool.

Exercise 16.6 In the above autoregressive model, the coefficients of m_{t-1} are significantly larger than those of s_{t-1} and b_{t-1}. This suggests that these two variables might not be useful in the regression. Resolve the example, assuming the following autoregressive model:

$$
\begin{aligned}
s_t &= d_1 + d_{13}m_{t-1} + \epsilon_t^s, \\
b_t &= d_2 + d_{23}m_{t-1} + \epsilon_t^b, \\
m_t &= d_3 + d_{33}m_{t-1} + \epsilon_t^m.
\end{aligned}
$$

16.5.2 Constructing scenario trees

The random distributions relating the various parameters of a stochastic program must be discretized to generate a set of scenarios that is adequate for its deterministic equivalent. Too few scenarios may lead to approximation errors. On the other hand, too many scenarios will lead to an explosion in the size of the scenario tree, leading to an excessive computational burden. In this section, we discuss a simple random sampling approach and two variance reduction techniques: adjusted random sampling and tree fitting. Unfortunately, scenario trees constructed by these methods could contain spurious arbitrage opportunities. We end this section with a procedure to test that this does not occur.

Random sampling

One can generate scenarios directly from the autoregressive model introduced in the previous section:

$$r_t = D_0 + D_1 r_{t-1} + \cdots + D_p r_{t-p} + \epsilon_t,$$

where $\epsilon_t \sim N(0, \Sigma)$ are independently distributed multivariate normal distributions with mean 0 and covariance matrix Σ.

In our example, Σ is a 3×3 diagonal matrix, with diagonal entries σ_s, σ_b, and σ_m. Using the parameters $\sigma_s = 0.173$, $\sigma_b = 0.108$, $\sigma_m = 0.022$ computed earlier, and a random number generator, we obtained $\epsilon_t^s = -0.186$, $\epsilon_t^b = 0.052$, and $\epsilon_t^m = 0.007$. We use the autoregressive model to get rates of return for 2004 based on the known rates of returns for 2003 (see Table 8.3 in Section 8.1.1):

$$s_{2004} = 0.077 - 0.058 \times 0.2868 + 0.219 \times 0.0054 + 0.448 \times 0.0098 - 0.186$$
$$= -0.087,$$
$$b_{2004} = 0.047 - 0.053 \times 0.2868 - 0.078 \times 0.0054 + 0.707 \times 0.0098 + 0.052$$
$$= 0.091,$$
$$m_{2004} = 0.016 + 0.033 \times 0.2868 - 0.044 \times 0.0054 + 0.746 \times 0.0098 + 0.007$$
$$= 0.040.$$

These are the rates of return for one of the branches from node 1. For each of the other branches from node 1, one generates random values of ϵ_t^s, ϵ_t^b, and ϵ_t^m and computes the corresponding values of s_{2004}, b_{2004}, and m_{2004}. Thirty branches or so may be needed to get a reasonable approximation of the distribution of the rates of return in stage 1. For a problem with three stages, 30 branches at each stage represent 27 000 scenarios. With more stages, the size of the linear program (16.6) explodes. Kouwenberg [48] performed tests on scenario trees with fewer branches at each node (such as a five-stage problem with branching structure 10-6-6-4-4, meaning 10 branches at the root, then 6 branches at each node in the next stage, and so on) and he concluded that random sampling on such trees leads to unstable investment strategies. This occurs because the approximation error made by representing parameter distributions by random samples can be significant in a small scenario tree. As a result the optimal solution of (16.6) is not optimal for the actual parameter distributions. How can one construct a scenario tree that more accurately represents these distributions, without blowing up the size of (16.6)?

Adjusted random sampling

An easy way of improving upon random sampling is as follows. Assume that each node of the scenario tree has an even number $K = 2k$ of branches. Instead of generating $2k$ random samples from the autoregressive model, generate k random samples only and use the negative of their error terms to compute the values on the remaining k branches. This will fit all the odd moments of the distributions correctly. In order to fit the variance of the distributions as well, one can scale the sampled values. The sampled values are all scaled by a multiplicative factor until their variance fits that of the corresponding parameter.

As an example, corresponding to the branch with $\epsilon_t^s = -0.186$, $\epsilon_t^b = 0.052$, and $\epsilon_t^m = 0.007$ at node 1, one would also generate another branch with $\epsilon_t^s = 0.186$, $\epsilon_t^b = -0.052$, and $\epsilon_t^m = -0.007$. For this branch the autoregressive model gives the following rates of return for 2004:

$$s_{2004} = 0.077 - 0.058 \times 0.2868 + 0.219 \times 0.0054 + 0.448 \times 0.0098 + 0.186$$
$$= 0.285,$$
$$b_{2004} = 0.047 - 0.053 \times 0.2868 - 0.078 \times 0.0054 + 0.707 \times 0.0098 - 0.052$$
$$= -0.013,$$
$$m_{2004} = 0.016 + 0.033 \times 0.2868 - 0.044 \times 0.0054 + 0.746 \times 0.0098 - 0.007$$
$$= 0.026.$$

Suppose that the set of ϵ_t^s generated on the branches leaving from node 1 has standard deviation 0.228 but the corresponding parameter should have standard deviation 0.165. Then the ϵ_t^s would be scaled down by $0.165/0.228$ on all the branches from node 1. For example, instead of $\epsilon_t^s = -0.186$ on the branch discussed earlier, one would use $\epsilon_t^s = -0.186(0.165/0.228) = -0.135$. This corresponds to the following rate of return:

$$s_{2004} = 0.077 - 0.058 \times 0.2868 + 0.219 \times 0.0054 + 0.448 \times 0.0098 - 0.135$$
$$= -0.036.$$

The rates of returns on all the branches from node 1 would be modified in the same way.

Tree fitting

How can one best approximate a continuous distribution by a discrete distribution with K values? In other words, how should one choose values v_k and their probabilities p_k, for $k = 1, \ldots, K$, in order to approximate the given distribution as accurately as possible? A natural answer is to match as many of the moments as possible. In the context of a scenario tree, the problem is somewhat more complicated since there are several correlated parameters at each node and there is interdependence between periods as well. Hoyland and Wallace [41] propose to formulate this fitting problem as a nonlinear program. The fitting problem can be solved either at each node separately or on the overall tree. We explain the fitting problem at a node. Let S_l be the values of the statistical properties of the distributions that one desires to fit, for $l = 1, \ldots, s$. These might be the expected values of the distributions, the correlation matrix, and the skewness and kurtosis. Let v_k and p_k denote the vector of values on branch k and its probability, respectively, for $k = 1, \ldots, K$. Let $f_l(v, p)$ be the mathematical expression of property l for the discrete distribution (for example, the mean of the vectors v_k, and their correlation, skewness, and kurtosis). Each property has a positive weight w_l indicating its importance in the desired fit. Hoyland and Wallace formulate the fitting problem as

$$\min_{v,p} \sum_l w_l(f_l(v, p) - S_l)^2$$
$$\sum_k p_k = 1 \qquad\qquad (16.11)$$
$$p \geq 0.$$

One might want some statistical properties to match exactly. As an example, consider again the autoregressive model:

$$r_t = D_0 + D_1 r_{t-1} + \cdots + D_p r_{t-p} + \epsilon_t,$$

where $\epsilon_t \sim N(0, \Sigma)$ are independently distributed multivariate normal distributions with mean 0 and covariance matrix Σ. To simplify notation, let us write ϵ instead of ϵ_t. The random vector ϵ has distribution $N(0, \Sigma)$ and we would like to approximate this continuous distribution by a finite number of disturbance vectors ϵ^k occuring with probability p_k, for $k = 1, \ldots, K$. Let ϵ_q^k denote the qth component of vector ϵ^k. One might want to fit the mean of ϵ exactly and its covariance matrix as well as possible. In this case, the fitting problem is:

$$\min_{\epsilon^1, \ldots, \epsilon^K, p} \sum_{q=1}^l \sum_{r=1}^l \left(\sum_{k=1}^K p_k \epsilon_q^k \epsilon_r^k - \Sigma_{qr} \right)^2$$
$$\sum_{k=1}^K p_k \epsilon^k = 0$$
$$\sum_k p_k = 1$$
$$p \geq 0.$$

Arbitrage-free scenario trees

Approximating the continuous distributions of the uncertain parameters by a finite number of scenarios in the linear programming (16.6) typically creates modeling errors. In fact, if the scenarios are not chosen properly or if their number is too small, the supposedly "linear programming equivalent" could be far from being equivalent to the original stochastic program. One of the most disturbing aspects of this phenomenon is the possibility of creating arbitrage opportunities when constructing the scenario tree. When this occurs, model (16.6) might produce unrealistic solutions that exploit these arbitrage opportunities. Klaassen [45] was the first to address this issue. In particular, he shows how arbitrage opportunities can be detected *ex post* in a scenario tree. When such arbitrage opportunities exist, a simple solution is to discard the scenario tree and to construct a new one with more branches. Klaassen [45] also discusses what constraints to add to the nonlinear program (16.11) in order to preclude arbitrage opportunities *ex ante*. The additional constraints are nonlinear, thus increasing the difficulty of solving (16.11). We present below Klassen's *ex-post* check.

Recall that there are two types of arbitrage (Definition 4.1). We start we Type A. An arbitrage of Type A is a trading strategy with an initial positive cash flow and no risk of loss later. Let us express this at a node i of the scenario tree. Let r^k denote the vectors of rates of return on the branches connecting node i to its sons in the next stage, for $k = 1, \ldots, K$. There exists an arbitrage of Type A if there exists an asset allocation $x = (x_1, \ldots, x_Q)$ at node i such that

$$\sum_{q=1}^{Q} x_q < 0$$

$$\text{and } \sum_{q=1}^{Q} x_q r_q^k \geq 0 \quad \text{for all } k = 1, \ldots, K.$$

To check whether such an allocation x exists, it suffices to solve the linear program

$$\min_x \sum_{q=1}^{Q} x_q \tag{16.12}$$
$$\sum_{q=1}^{Q} x_q r_q^k \geq 0 \quad \text{for all } k = 1, \ldots, K.$$

There is an arbitrage opportunity of Type A at node i if and only if this linear program is unbounded.

Next we turn to Type B. An arbitrage of Type B requires no initial cash input, has no risk of a loss and a positive probability of making profits in the future. At node i of the scenario tree, this is expressed by the conditions:

$$\sum_{q=1}^{Q} x_q = 0,$$

$$\sum_{q=1}^{Q} x_q r_q^k \geq 0 \quad \text{for all } k = 1, \ldots, K,$$

$$\text{and } \sum_{q=1}^{Q} x_q r_q^k > 0 \quad \text{for at least one } k = 1, \ldots, K.$$

These conditions can be checked by solving the linear program

$$\max_x \sum_{q=1}^{Q} x_q r_q^k$$

$$\sum_{q=1}^{Q} x_q = 0 \tag{16.13}$$

$$\sum_{q=1}^{Q} x_q r_q^k \geq 0 \quad \text{for all } k = 1, \ldots, K.$$

There is an arbitrage opportunity of Type B at node i if and only if this linear program is unbounded.

Exercise 16.7 Show that the linear program (16.12) is always feasible.

Write the dual linear program of (16.12). Let u_k be the dual variable associated with the kth constraint of (16.12).

Recall that a feasible linear program is unbounded if and only if its dual is infeasible. Show that there is no arbitrage of Type A at node i if and only if there exists $u_k \geq 0$, for $k = 1, \ldots, K$, such that

$$\sum_{k=1}^{K} u_k r_q^k = 1 \quad \text{for all } q = 1, \ldots, Q.$$

Similarly, write the dual of (16.13). Let v_0, v_k, for $k = 1, \ldots, K$, be the dual variables. Write necessary and sufficient conditions for the nonexistence of arbitrage of Type B at node i, in terms of v_k, for $k = 0, \ldots, K$.

Modify the nonlinear program (16.11) in order to formulate a fitting problem at node i that contains no arbitrage opportunities.

17

Stochastic programming models: Value-at-Risk and Conditional Value-at-Risk

In this chapter, we discuss Value-at-Risk, a widely used measure of risk in finance, and its relative, Conditional Value-at-Risk. We then present an optimization model that optimizes a portfolio when the risk measure is the Conditional Value-at-Risk instead of the variance of the portfolio as in the Markowitz model. This is acheived through stochastic programming. In this case, the variables are anticipative. The random events are modeled by a large but finite set of scenarios, leading to a linear programming equivalent of the original stochastic program.

17.1 Risk measures

Financial activities involve risk. Our stock or mutual fund holdings carry the risk of losing value due to market conditions. Even money invested in a bank carries a risk – that of the bank going bankrupt and never returning the money let alone some interest. While individuals generally just have to live with such risks, financial and other institutions can and very often must manage risk using sophisticated mathematical techniques. Managing risk requires a good understanding of quantitative risk measures that adequately reflect the vulnerabilities of a company.

Perhaps the best-known risk measure is Value-at-Risk (VaR) developed by financial engineers at J.P. Morgan. VaR is a measure related to percentiles of loss distributions and represents the predicted maximum loss with a specified probability level (e.g., 95%) over a certain period of time (e.g., one day). Consider, for example, a random variable X that represents loss from an investment portfolio over a fixed period of time. A negative value for X indicates gains. Given a probability level α, α-VaR of the random variable X is given by the following relation:

$$\text{VaR}_\alpha(X) := \min\{\gamma : P(X \geq \gamma) \leq 1 - \alpha\}. \tag{17.1}$$

When the loss distribution is continuous, $\text{VaR}_\alpha(X)$ is simply the loss such that

$$P(X \leq \text{VaR}_\alpha(X)) = \alpha.$$

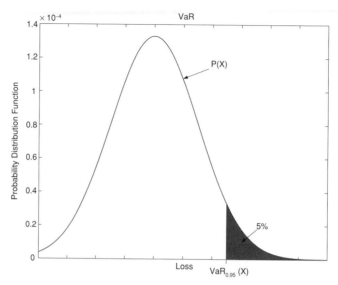

Figure 17.1 The 0.95-VaR on a portfolio loss distribution plot

Figure 17.1 illustrates the 0.95-VaR on a portfolio loss distribution plot. VaR is widely used by people in the financial industry and VaR calculators are common features in most financial software. Despite this popularity, VaR has one important undesirable property – it lacks subadditivity. Risk measures should respect the maxim "diversification reduces risk" and, therefore, satisfy the following property: "the total risk of two different investment portfolios does not exceed the sum of the individual risks." This is precisely what we mean by saying that a risk measure should be a subadditive function, i.e., for a risk measure f, we should have

$$f(x_1 + x_2) \leq f(x_1) + f(x_2), \quad \forall x_1, x_2.$$

Consider the following simple example that illustrates that diversification can actually increase the risk measured by VaR:

Example 17.1 *Consider two independent investment opportunities each returning a $1 gain with probability 0.96 and $2 loss with probability 0.04. Then, 0.95-VaR for both investments are* -1*. Now consider the sum of these two investment opportunities. Because of independence, this sum has the following loss distribution: $4 with probability* $0.04 \times 0.04 = 0.0016$*, $1 with probability* $2 \times 0.96 \times 0.04 = 0.0768$*, and* $-$*$2 with probability* $0.96 \times 0.96 = 0.9216$*. Therefore, the 0.95-VaR of the sum of the two investments is 1, which exceeds* -2*, the sum of the 0.95-VaR values for individual investments.*

An additional difficulty with VaR is in its computation and optimization. When VaR is computed by generating scenarios, it turns out to be a nonsmooth and nonconvex function of the positions in the investment portfolio. Therefore, when one tries to optimize VaR computed in this manner, multiple local optimizers are encountered, hindering the global optimization process.

Another criticism of VaR is that it pays no attention to the magnitude of losses beyond the VaR value. This and other undesirable features of VaR led to the development of alternative risk measures. One well-known modification of VaR is obtained by computing the *expected* loss *given* that the loss exceeds VaR. This quantity is often called *Conditional Value-at-Risk* or CVaR. There are several alternative names for this measure in the finance literature including mean expected loss, mean shortfall, and tail VaR. We now describe this risk measure in more detail and discuss how it can be optimized using linear programming techniques when the loss function is linear in the portfolio positions. Our discussion follows parts of articles by Rockafellar and Uryasev [68, 82].

We consider a portfolio of assets with random returns. We denote the portfolio choice vector by x and the random events by the vector y. Let $f(x, y)$ denote the loss function when we choose the portfolio x from a set X of feasible portfolios and y is the realization of the random events. We assume that the random vector y has a probability density function denoted by $p(y)$.

For a fixed decision vector x, we compute the cumulative distribution function of the loss associated with that vector x:

$$\Psi(x, \gamma) := \int_{f(x,y)<\gamma} p(y)\mathrm{d}y. \tag{17.2}$$

Then, for a given confidence level α, the α-VaR associated with portfolio x is given by

$$\mathrm{VaR}_\alpha(x) := \min\{\gamma \in \mathit{I\!R} : \Psi(x, \gamma) \geq \alpha\}. \tag{17.3}$$

We define the α-CVaR associated with portfolio x as:

$$\mathrm{CVaR}_\alpha(x) := \frac{1}{1-\alpha} \int_{f(x,y)\geq \mathrm{VaR}_\alpha(x)} f(x, y)p(y)\mathrm{d}y. \tag{17.4}$$

Note that

$$\mathrm{CVaR}_\alpha(x) = \frac{1}{1-\alpha} \int_{f(x,y)\geq \mathrm{VaR}_\alpha(x)} f(x, y)p(y)\mathrm{d}y$$

$$\geq \frac{1}{1-\alpha} \int_{f(x,y)\geq \mathrm{VaR}_\alpha(x)} \mathrm{VaR}_\alpha(x)p(y)\mathrm{d}y$$

$$= \frac{\mathrm{VaR}_\alpha(x)}{1-\alpha} \int_{f(x,y)\geq \mathrm{VaR}_\alpha(x)} p(y)\mathrm{d}y$$

$$\geq \mathrm{VaR}_\alpha(x),$$

i.e., the CVaR of a portfolio is always at least as big as its VaR. Consequently, portfolios with small CVaR also have small VaR. In general however, minimizing CVaR and VaR are not equivalent.

For a discrete probability distribution (where event y_j occurs with probability p_j, for $j = 1, \ldots, n$), the above definition of CVaR becomes

$$\text{CVaR}_\alpha(x) = \frac{1}{1 - \alpha} \sum_{j:f(x,y_j) \geq \text{VaR}_\alpha(x)} p_j f(x, y_j).$$

Example 17.2 *Suppose we are given the loss function $f(x, y)$ for a given decision x as $f(x, y) = -y$ where $y = 75 - j$ with probability 1% for $j = 0, \ldots, 99$. We would like to determine the Value-at-Risk $VaR_\alpha(x)$ for $\alpha = 95\%$. We have $VaR_{95\%}(x) = 20$ since the loss is 20 or more with probability 5%.*

To compute the Conditional Value-at-Risk, we use the above formula: $CVaR_{95\%}(x) = \frac{1}{0.05}(20 + 21 + 22 + 23 + 24) \times 1\% = 22.$

Exercise 17.1

(i) Compute the 0.90-VaR and 0.90-CVaR for the rates of return of stocks between 1961 and 2003 (see Section 8.1.1 for the data).
(ii) Compute the 0.90-VaR and 0.90-CVaR for the rates of return of bonds and a money market account. Again use the data of Section 8.1.1.

17.2 Minimizing CVaR

Since the definition of CVaR involves the VaR function explicitly, it is difficult to work with and optimize this function. Instead, we consider the following simpler auxiliary function:

$$F_\alpha(x, \gamma) := \gamma + \frac{1}{1 - \alpha} \int_{f(x,y) \geq \gamma} (f(x, y) - \gamma) \, p(y) dy. \qquad (17.5)$$

Alternatively, we can write $F_{\alpha,x}(\gamma)$ as follows:

$$F_\alpha(x, \gamma) = \gamma + \frac{1}{1 - \alpha} \int (f(x, y) - \gamma)^+ p(y) dy, \qquad (17.6)$$

where $a^+ = \max\{a, 0\}$. This function, viewed as a function of γ, has the following important properties that make it useful for the computation of VaR and CVaR:

1. $F_\alpha(x, \gamma)$ is a convex function of γ.
2. $\text{VaR}_\alpha(x)$ is a minimizer over γ of $F_\alpha(x, \gamma)$.
3. The minimum value over γ of the function $F_\alpha(x, \gamma)$ is $\text{CVaR}_\alpha(x)$.

Exercise 17.2 Prove the properties of $F_{\alpha,x}(\gamma)$ stated above.

As a consequence of the listed properties, we immediately deduce that, in order to minimize $\text{CVaR}_\alpha(x)$ over x, we need to minimize the function $F_\alpha(x, \gamma)$ with respect to x and γ simultaneously:

$$\min_{x \in X} \text{CVaR}_\alpha(x) = \min_{x \in X, \gamma} F_\alpha(x, \gamma). \tag{17.7}$$

Consequently, we can optimize CVaR directly, without needing to compute VaR first. If the loss function $f(x, y)$ is a convex (linear) function of the portfolio variables x, then $F_\alpha(x, \gamma)$ is also a convex (linear) function of x. In this case, provided the feasible portfolio set X is also convex, the optimization problems in (17.7) are smooth convex optimization problems that can be solved using well-known optimization techniques for such problems (see Chapter 5).

Often it is not possible or desirable to compute/determine the joint density function $p(y)$ of the random events in our formulation. Instead, we may have a number of scenarios, say y_s for $s = 1, \ldots, S$, which may represent some historical values of the random events or some values obtained via computer simulation. We will assume that all scenarios have the same probability. In this case, we obtain the following approximation to the function $F_\alpha(x, \gamma)$ by using the empirical distribution of the random events based on the available scenarios:

$$\tilde{F}_\alpha(x, \gamma) := \gamma + \frac{1}{(1 - \alpha)S} \sum_{s=1}^{S} (f(x, y_s) - \gamma)^+. \tag{17.8}$$

Compare this definition to (17.6). Now, the problem $\min_{x \in X} \text{CVaR}_\alpha(x)$ can be approximated by replacing $F_\alpha(x, \gamma)$ with $\tilde{F}_\alpha(x, \gamma)$ in (17.7):

$$\min_{x \in X, \gamma} \gamma + \frac{1}{(1 - \alpha)S} \sum_{s=1}^{S} (f(x, y_s) - \gamma)^+. \tag{17.9}$$

To solve this optimization problem, we introduce artificial variables z_s to replace $(f(x, y_s) - \gamma)^+$. This is achieved by imposing the constraints $z_s \geq f(x, y_s) - \gamma$ and $z_s \geq 0$:

$$\begin{aligned} \min_{x, z, \gamma} \quad & \gamma + \frac{1}{(1-\alpha)S} \sum_{s=1}^{S} z_s \\ \text{s.t.} \quad & z_s \geq 0, \ s = 1, \ldots, S, \\ & z_s \geq f(x, y_s) - \gamma, \ s = 1, \ldots, S, \\ & x \in X. \end{aligned} \tag{17.10}$$

Note that the constraints $z_s \geq f(x, y_s) - \gamma$ and $z_s \geq 0$ alone cannot ensure that $z_s = (f(x, y_s) - \gamma)^+ = \max\{f(x, y_s) - \gamma, 0\}$ since z_s can be larger than both right-hand sides and be still feasible. However, since we are minimizing the objective function, which involves a positive multiple of z_s, it will never be optimal to assign

z_s a value larger than the maximum of the two quantities $f(x, y_s) - \gamma$ and 0, and, therefore, in an optimal solution z_s will be precisely $(f(x, y_s) - \gamma)^+$, justifying our substitution.

In the case that $f(x, y)$ is linear in x, all the expressions $z_s \geq f(x, y_s) - \gamma$ represent linear constraints and therefore the problem (17.10) is a linear programming problem that can be solved using the simplex method or alternative LP algorithms.

Other optimization problems arise naturally within the context of risk management. For example, risk managers often try to optimize a performance measure (e.g., expected return) while making sure that certain risk measures do not exceed a threshold value. When the risk measure is CVaR, the resulting optimization problem is:

$$\max_x \quad \mu^T x$$
$$\text{s.t.} \quad \text{CVaR}_{\alpha^j}(x) \leq U_{\alpha^j}, \quad j = 1, \ldots, J, \tag{17.11}$$
$$x \in X.$$

Above, J is an index set for different confidence levels used for CVaR computations and U_{α^j} represents the maximum tolerable CVaR value at the confidence level α^j. As above, we can replace the CVaR functions in the constraints of this problem with the function $F_\alpha(x, \gamma)$ and then approximate this function using the scenarios for random events. This approach results in the following approximation of the CVaR-constrained problem (17.11):

$$\max_{x,z,\gamma} \quad \mu^T x$$
$$\text{s.t.} \quad \gamma + \frac{1}{(1-\alpha^j)S} \sum_{s=1}^{S} z_s \leq U_{\alpha^j}, \qquad j = 1, \ldots, J,$$
$$z_s \geq 0, \qquad s = 1, \ldots, S, \tag{17.12}$$
$$z_s \geq f(x, y_s) - \gamma, \quad s = 1, \ldots, S,$$
$$x \in X.$$

17.3 Example: bond portfolio optimization

A portfolio of risky bonds might be characterized by a large likelihood of small earnings, coupled with a small chance of loosing a large amount of the investment. The loss distribution is heavily skewed and, in this case, standard mean-variance analysis to characterize market risk is inadequate. VaR and CVaR are more appropriate criteria for minimizing *portfolio credit risk*. Credit risk is the risk of a trading partner not fulfilling their obligation in full on the due date or at any time thereafter.

Losses can result both from default and from a decline in market value stemming from downgrades in credit ratings. A good reference is the paper of Anderson *et al.* [3].

Anderson *et al.* consider a portfolio of 197 bonds from 29 different countries with a market value of $8.8 billion and duration of approximately five years. Their goal is to rebalance the portfolio in order to minimize credit risk. That is they want to minimize losses resulting from default and from a decline in market value stemming from downgrades in credit ratings (credit migration). The loss due to credit migration is simply

$$f(x, y) = (b - y)^{\mathrm{T}} x,$$

where b are the future values of each bond with no credit migration and y are the future values with credit migration (so y is a random vector). The one-year portfolio credit loss was generated using a Monte Carlo simulation: 20 000 scenarios of joint credit states of obligators and related losses. The distribution of portfolio losses has a long fat tail. The authors rebalanced the portfolio by minimizing CVaR. The set X of feasible porfolios was described by the following constraints. Let x_i denote the weight of asset i in the portfolio. Upper and lower bounds were set on each x_i:

$$l_i \leq x_i \leq u_i \ i = 1, \ldots, n,$$
$$\sum_i x_i = 1.$$

To calculate the efficient frontier, the expected portfolio return was set to at least R:

$$\sum_i \mu_i x_i \geq R.$$

To summarize, the linear program (17.10) to be solved was as follows:

$$\min_{x,z,\gamma} \gamma + \frac{1}{(1-\alpha)S} \sum_{s=1}^{S} z_s$$

$$\text{subject to } z_s \geq \sum_i (b_i - y_{is}) x_i - \gamma \quad \text{for } s = 1, \ldots, S,$$
$$z_s \geq 0 \qquad\qquad\qquad\qquad \text{for } s = 1, \ldots, S,$$
$$l_i \leq x_i \leq u_i \qquad\qquad\qquad i = 1, \ldots, n,$$
$$\sum_i x_i = 1,$$
$$\sum_i \mu_i x_i \geq R.$$

Consider $\alpha = 99\%$. The original bond portfolio had an expected portfolio return of 7.26%. The expected loss was 95 million dollars with a standard deviation of 232 million. The VaR was 1.03 billion dollars and the CVaR was 1.32 billion.

After optimizing the portfolio (with expected return of 7.26%), the expected loss was only $5000, with a standard deviation of 152 million. The VaR was reduced to $210 million and the CVaR to $263 million. So all around, the characteristics of the portfolio were much improved. Positions were reduced in bonds from Brazil, Russia and, Venezuela, whereas positions were increased in bonds from Thailand, Malaysia, and Chile. Positions in bonds from Colombia, Poland, and Mexico remained high and each accounted for about 5% of the optimized CVaR.

18

Stochastic programming models: asset/liability management

18.1 Asset/liability management

The financial health of any company, and in particular those of financial institutions, is reflected in the balance sheets of the company. Proper management of the company requires attention to both sides of the balance sheet – assets and liabilities. Asset/liability management (ALM) offers sophisticated mathematical tools for an integrated management of assets and liabilities and is the focus of many studies in financial mathematics.

ALM recognizes that static, one-period investment planning models (such as mean-variance optimization) fail to incorporate the multi-period nature of the liabilities faced by the company. A multi-period model that emphasizes the need to meet liabilities in each period for a finite (or possibly infinite) horizon is often required. Since liabilities and asset returns usually have random components, their optimal management requires tools of "optimization under uncertainty" and, most notably, stochastic programming approaches.

We recall the ALM setting we introduced in Section 1.3.4: let L_t be the liability of the company in year t for $t = 1, \ldots, T$. The L_t's are random variables. Given these liabilities, which assets (and in which quantities) should the company hold each year to maximize its expected wealth in year T? The assets may be domestic stocks, foreign stocks, real estate, bonds, etc. Let R_{it} denote the return on asset i in year t. The R_{it}'s are random variables. The decision variables are:

$$x_{it} = \text{market value invested in asset } i \text{ in year } t.$$

The decisions x_{it} in year t are made after the random variables L_t and R_{it} are realized. That is, the decision problem is multistage, stochastic, with recourse. The stochastic program can be written as follows:

$$\max \ E\left[\sum_i x_{iT} \right]$$

subject to

asset accumulation: $\sum_i (1 + R_{it}) x_{i,t-1} - \sum_i x_{it} = L_t$ for $t = 1, \ldots, T$,

$$x_{it} \geq 0.$$

The constraint says that the surplus left after liability L_t is covered will be invested as follows: x_{it} invested in asset i. In this formulation, $x_{0,t}$ are the fixed, and possibly nonzero initial positions in different asset classes. The objective selected in the model above is to maximize the expected wealth at the end of the planning horizon. In practice, one might have a different objective. For example, in some cases, minimizing Value-at-Risk (VaR) might be more appropriate. Other priorities may dictate other objective functions.

To address the issue of the most appropriate objective function, one must understand the role of liabilities. Pension funds and insurance companies are among the most typical arenas for the integrated management of assets and liabilities through ALM. We consider the case of a Japanese insurance company, the Yasuda Fire and Marine Insurance Co., Ltd, following the work of Cariño *et al.* [19]. In this case, the liabilities are mainly savings-oriented policies issued by the company. Each new policy sold represents a deposit, or inflow of funds. Interest is periodically credited to the policy until maturity, typically three to five years, at which time the principal amount plus credited interest is refunded to the policyholder. The crediting rate is typically adjusted each year in relation to a market index like the prime rate. Therefore, we cannot say with certainty what future liabilities will be. Insurance business regulations stipulate that interest credited to some policies be earned from investment income, not capital gains. So, in addition to ensuring that the maturity cash flows are met, the firm must seek to avoid interim shortfalls in income earned versus interest credited. In fact, it is the risk of not earning adequate income quarter by quarter that the decision-makers view as the primary component of risk at Yasuda.

The problem is to determine the optimal allocation of the deposited funds into several asset categories: cash, fixed- and floating-rate loans, bonds, equities, real estate, and other assets. Since we can revise the portfolio allocations over time, the decision we make is not just among allocations today but among allocation strategies over time. A realistic dynamic asset/liability model must also account for the payment of taxes. This is made possible by distinguishing between interest income and price return.

A stochastic linear program is used to model the problem. The linear program has uncertainty in many coefficients. This uncertainty is modeled through a finite number of scenarios. In this fashion, the problem is transformed into a very large scale linear program of the form (16.6). The random elements include price return and interest income for each asset class, as well as policy crediting rates.

We now present a multistage stochastic program that was developed for the Yasuda Fire and Marine Insurance Co., Ltd. Our presentation follows the description of the model as stated in [19].

Stages are indexed by $t = 0, 1, \ldots, T$.

Decision variables of the stochastic program:

$$x_{it} = \text{market value in asset } i \text{ at } t,$$
$$w_t = \text{interest income shortfall at } t \geq 1,$$
$$v_t = \text{interest income surplus at } t \geq 1.$$

Random variables appearing in the stochastic linear program, for $t \geq 1$:

$$RP_{it} = \text{price return of asset } i \text{ from } t - 1 \text{ to } t,$$
$$RI_{it} = \text{interest income of asset } i \text{ from } t - 1 \text{ to } t,$$
$$F_t = \text{deposit inflow from } t - 1 \text{ to } t,$$
$$P_t = \text{principal payout from } t - 1 \text{ to } t,$$
$$I_t = \text{interest payout from } t - 1 \text{ to } t,$$
$$g_t = \text{rate at which interest is credited to policies from } t - 1 \text{ to } t,$$
$$L_t = \text{liability valuation at } t.$$

Parameterized function appearing in the objective:

$$c_t = \text{piecewise linear convex cost function.}$$

The objective of the model is to allocate funds among available assets to maximize expected wealth at the end of the planning horizon T less expected penalized shortfalls accumulated through the planning horizon:

$$\max \quad E\Big[\sum_i x_{iT} - \sum_{t=1}^{T} c_t(w_t) \Big]$$

subject to

asset accumulation:
$$\sum_i x_{it} - \sum_i (1 + RP_{it} + RI_{it}) x_{i,t-1}$$
$$= F_t - P_t - I_t \qquad \text{for } t = 1, \ldots, T,$$

interest income shortfall:
$$\sum_i RI_{it} x_{i,t-1} + w_t - v_t = g_t L_{t-1} \quad \text{for } t = 1, \ldots, T,$$
$$x_{it} \geq 0, \quad w_t \geq 0, \quad v_t \geq 0.$$

$$(18.1)$$

Liability balances and cash flows are computed so as to satisfy the liability accumulation relations:

$$L_t = (1 + g_t) L_{t-1} + F_t - P_t - I_t \quad \text{for } t \geq 1.$$

The stochastic linear program (18.1) is converted into a large linear program using a finite number of scenarios to deal with the random elements in the data. Creation of scenario inputs is made in stages using a tree. The tree structure can be described by the number of branches at each stage. For example, a 1-8-4-4-2-1 tree has 256 scenarios. Stage $t = 0$ is the initial stage. Stage $t = 1$ may be chosen to be the end of quarter 1 and has eight different scenarios in this example. Stage $t = 2$ may be chosen to be the end of year 1, with each of the previous scenarios giving rise to four new scenarios, and so on. For the Yasuda Fire and Marine Insurance Co., Ltd, a problem with seven asset classes and six stages gives rise to a stochastic linear program (18.1) with 12 constraints (other than nonnegativity) and 54 variables. Using 256 scenarios, this stochastic program is converted into a linear program with several thousand constraints and over 10 000 variables. Solving this model yielded extra income estimated to be about US$80 million per year for the company.

Exercise 18.1   Discuss the relevance of the techniques from Chapter 16 in the solution of the Yasuda Fire and Marine Insurance Co., Ltd, such as scenario-generation (correlation of the random parameters over time, variance reduction techniques in constructing the scenario tree), decomposition techniques to solve the large-scale linear programs.

18.1.1 Corporate debt management

A closely related problem to the asset/liability management (ALM) problem in corporate financial planning is the problem of debt management. Here the focus is on retiring (paying back) outstanding debt at minimum cost. More specifically, corporate debt managers must make financial decisions to minimize the costs and risks of borrowing to meet debt financing requirements. These requirements are often determined by the firm's investment decisions. Our discussion in this subsection is based on the article [26].

Debt managers need to choose the sources of borrowing, types of debts to be used, timing and terms of debts, whether the debts will be callable,[1] etc., in a multi-period framework where the difficulty of the problem is compounded by the fact that the interest rates that determine the cost of debt are uncertain. Since interest rate movements can be modeled by random variables this problem presents an attractive setting for the use of stochastic programming techniques. Below, we discuss a deterministic linear programming equivalent of stochastic LP model for the debt management problem.

---

[1] A *callable debt* is a debt security whose issuer has the right to redeem the security prior to its stated maturity date at a price established at the time of issuance, on or after a specified date.

We consider a multi-period framework with T time periods. We will use the indices s and t ranging between 0 (now) and T (termination date, or horizon) to denote different time periods in the model. We consider K types of debt that are distinguished by market of issue, term, and the presence (or absence) of a call option available to the borrower. In our notation, the superscript k ranging between 1 and K will denote the different types of debt being considered.

The evolution of the interest rates are described using a scenario tree. We denote by $e_j = e_{j1}, e_{j2}, \ldots, e_{jT}$, $j = 1, \ldots, J$, a sample path of this scenario tree which corresponds to a sequence of interest rate events. When a parameter or variable is contingent on the event sequence e_j we use the notation (e_j) (see below).

The decision variables in this model are the following:

- $B_t^k(e_j)$: dollar amount at par[2] of debt type k **B**orrowed at the beginning of period t.
- $O_{s,t}^k(e_j)$: dollar amount at par of debt type k borrowed in period s and **O**utstanding at the beginning of period t.
- $R_{s,t}^k(e_j)$: dollar amount at par of debt type k borrowed in period s and **R**etired (paid back) at the beginning of period t.
- $S_t(e_j)$: dollar value of **S**urplus cash held at the beginning of period t.

Next, we list the input parameters to the problem:

- $r_{s,t}^k(e_j)$: interest payment in period t per dollar outstanding of debt type k issued in period s.
- f_t^k: issue costs (excluding premium or discount) per dollar borrowed of debt type k issued in period t.
- $g_{s,t}^k(e_j)$: retirement premium or discount per dollar for debt type k issued in period s, if retired in period t.[3]
- $i_t(e_j)$: interest earned per dollar on surplus cash in period t.
- $p(e_j)$: probability of the event sequence e_j. Note that $p(e_j) \geq 0$, $\forall j$ and $\sum_{j=1}^{J} p(e_j) = 1$.
- C_t: cash requirements for period t, which can be negative to indicate an operating surplus.
- M_t: maximum allowable cost of debt service in period t.
- $q_t^k(Q_t^k)$: minimum (maximum) borrowing of debt type k in period t.
- $L_t(e_j)(U_t(e_j))$: minimum (maximum) dollar amount of debt (at par) retired in period t.

The objective function of this problem is expressed as follows:

$$\min \sum_{j=1}^{J} p(e_j) \left(\sum_{k=1}^{K} \sum_{t=1}^{T} \left(1 + g_{t,T}^k(e_j)\right) \left[O_{t,T}^k(e_j) - R_{t,T}^k(e_j)\right] + \left(1 - f_T^k\right) B_T^k(e_j) \right).$$
(18.2)

This function expresses the expected retirement cost of the total debt outstanding at the end of period T.

---

[2] At a price equal to the par (face) value of the security; the original issue price of a security.
[3] These parameters are used to define call options and to value the debt portfolio at the end of the planning period.

We complete the description of the deterministic equivalent of the stochastic LP by listing the constraints of the problem:

- **Cash requirements:** For each time period $t = 1, \ldots, T$ and scenario path $j = 1, \ldots, J$:

$$C_t + S_t(e_j) = \sum_{k=1}^{K} \left\{ \left(1 - f_t^k\right) B_t^k(e_j) + (1 + i_{t-1}(e_j))S_{t-1}(e_j) \right.$$
$$\left. - \sum_{s=0}^{t-1} \left[r_{s,t}^k(e_j)O_{s,t}^k(e_j) - \left(1 + g_{s,t}^k(e_j)\right) R_{s,t}^k(e_j) \right] \right\}.$$

This balance equation indicates that the difference between cash available (new net borrowing, surplus cash from previous period, and the interest earned on this cash) and the debt payments (interest on outstanding debt and cash outflows on repayment) should equal the cash requirements plus the surplus cash left for this period.

- **Debt balance constraints:** For $j = 1, \ldots, J$, $t = 1, \ldots, T$, $s = 0, \ldots, t-2$, and $k = 1, \ldots K$:

$$O_{s,t}^k(e_j) - O_{s,t-1}^k(e_j) + R_{s,t-1}^k(e_j) = 0,$$
$$O_{t-1,t}^k(e_j) - B_{t-1}^k(e_j) - R_{t-1,t}^k(e_j) = 0.$$

- **Maximum cost of debt:** For $j = 1, \ldots, J$, $t = 1, \ldots, T$, and $k = 1, \ldots K$:

$$\sum_{s=1}^{t-1} \left(r_{s,t}^k(e_j)O_{s,t}^k(e_j) - i_{t-1}(e_j)S_{t-1}(e_j) \right) \leq M_t.$$

- **Borrowing limits:** For $j = 1, \ldots, J$, $t = 1, \ldots, T$, and $k = 1, \ldots K$:

$$q_t^k \leq B_t^k(e_j) \leq Q_t^k.$$

- **Payoff limits:** For $j = 1, \ldots, J$ and $t = 1, \ldots, T$:

$$L_t(e_j) \leq \sum_{k=1}^{K} \sum_{s=0}^{t-1} R_{s,t}^k(e_j) \leq U_t(e_j).$$

- **Nonnegativity:** For $j = 1, \ldots, J$, $t = 1, \ldots, T$, $s = 0, \ldots, t-2$, and $k = 1, \ldots K$:

$$B_t^k(e_j) \geq 0, \quad O_{s,t}^k(e_j) \geq 0, \quad R_{s,t}^k(e_j) \geq 0, \quad S_t(e_j) \geq 0.$$

In the formulation above, we used the notation of the article [26]. However, since the parameters and variables dependent on e_j can only depend on the portion of the sequence that is revealed by a certain time, a more precise notation can be obtained using the following ideas. First, let $e_j^t = e_{j1}, e_{j2}, \ldots, e_{jt}$, $j = 1, \ldots, J$, $t = 1, \ldots, T$, i.e., e_j^t represents the portion of e_j observed by time period t. Then, one replaces the expressions such as $S_t(e_j)$ with $S_t(e_j^t)$, etc.

18.2 Synthetic options

An important issue in portfolio selection is the potential decline of the portfolio value below some critical limit. How can we control the risk of downside losses? A possible answer is to create a payoff structure similar to a European call option.

While one may be able to construct a diversified portfolio well suited for a corporate investor, there may be no option market available on this portfolio. One solution may be to use index options. However, exchange-traded options with sufficient liquidity are limited to maturities of about three months. This makes the cost of long-term protection expensive, requiring the purchase of a series of high priced short-term options. For large institutional or corporate investors, a cheaper solution is to artificially produce the desired payoff structure using available resources. This is called a "synthetic option strategy."

18.2.1 The model

The model is based on the following data:

W_0 = investor's initial wealth,
T = planning horizon,
R = riskless return for one period,
R_t^i = return for asset i at time t,
θ_t^i = transaction cost for purchases and sales of asset i at time t.

The R_t^i's are random, but we know their distributions.
The variables used in the model are the following:

x_t^i = amount allocated to asset i at time t,
A_t^i = amount of asset i bought at time t,
D_t^i = amount of asset i sold at time t,
α_t = amount allocated to riskless asset at time t.

We formulate a stochastic program that produces the desired payoff at the end of the planning horizon T, much in the flavor of the stochastic programs developed in the previous two sections. Let us first discuss the constraints.

The initial portfolio is

$$\alpha_0 + x_0^1 + \cdots + x_0^n = W_0.$$

The portfolio at time t is

$$x_t^i = R_t^i x_{t-1}^i + A_t^i - D_t^i \quad \text{for } t = 1, \ldots, T,$$

$$\alpha_t = R\alpha_{t-1} - \sum_{i=1}^n \left(1 + \theta_t^i\right) A_t^i + \sum_{i=1}^n \left(1 - \theta_t^i\right) D_t^i \quad \text{for } t = 1, \ldots, T.$$

One can also impose upper bounds on the proportion of any risky asset in the portfolio:

$$0 \le x_t^i \le m_t \left(\alpha_t + \sum_{j=1}^{n} x_t^j \right),$$

where m_t is chosen by the investor.

The value of the portfolio at the end of the planning horizon is

$$v = R\alpha_{T-1} + \sum_{i=1}^{n} \left(1 - \theta_T^i \right) R_T^i x_{T-1}^i,$$

where the summation term is the value of the risky assets at time T.

To construct the desired synthetic option, we split v into the riskless value of the portfolio Z and a surplus $z \ge 0$ which depends on random events. Using a scenario approach to the stochastic program, Z is the worst-case payoff over all the scenarios. The surplus z is a random variable that depends on the scenario. Thus

$$v = Z + z,$$
$$z \ge 0.$$

We consider Z and z as variables of the problem, and we optimize them together with the asset allocations x and other variables described earlier. The objective function of the stochastic program is

$$\max E(z) + \mu Z,$$

where $\mu \ge 1$ is the risk aversion of the investor. The risk aversion μ is given data.

When $\mu = 1$, the objective is to maximize expected return.

When μ is very large, the objective is to maximize "riskless profit" as we defined it in Chapter 4 (Exercise 4.10).

As an example, consider an investor with initial wealth $W_0 = 1$ who wants to construct a portfolio comprising one risky asset and one riskless asset using the "synthetic option" model described above. We write the model for a two-period planning horizon, i.e., $T = 2$. The return on the riskless asset is R per period. For the risky asset, the return is R_1^+ with probability 0.5 and R_1^- with the same probability at time $t = 1$. Similarly, the return of the risky asset is R_2^+ with probability 0.5 and R_2^- with the same probability at time $t = 2$. The transaction cost for purchases and sales of the risky asset is θ.

There are four scenarios in this example, each occurring with probability 0.25, which we can represent by a binary tree. The initial node will be denoted by 0, the up node from it by 1 and the down node by 2. Similarly the up node from node 1 will be denoted by 3, the down node by 4, and the successors of 2 by 5 and 6 respectively. Let x_i, α_i denote the amount of risky asset and of riskless asset respectively in the portfolio at node i of this binary tree. Z is the riskless value of the portfolio and z_i is the surplus at node i. The linear program is:

$$\max \quad 0.25z_3 + 0.25z_4 + 0.25z_5 + 0.25z_6 + \mu Z$$

$$\text{subject to}$$

$$\text{initial portfolio:} \quad \alpha_0 + x_0 = 1$$

$$\text{rebalancing constraints:} \quad x_1 = R_1^+ x_0 + A_1 - D_1$$

$$\alpha_1 = R\alpha_0 - (1+\theta)A_1 + (1-\theta)D_1$$

$$x_2 = R_1^- x_0 + A_2 - D_2$$

$$\alpha_2 = R\alpha_0 - (1+\theta)A_2 + (1-\theta)D_2$$

$$\text{payoff:} \quad z_3 + Z = R\alpha_1 + (1-\theta)R_2^+ x_1$$

$$z_4 + Z = R\alpha_1 + (1-\theta)R_2^- x_1$$

$$z_5 + Z = R\alpha_2 + (1-\theta)R_2^+ x_2$$

$$z_6 + Z = R\alpha_2 + (1-\theta)R_2^- x_2$$

$$\text{nonnegativity:} \quad \alpha_i, x_i, z_i, A_i, D_i \geq 0.$$

18.2.2 An example

An interesting paper discussing synthetic options is the paper of Zhao and Ziemba [85]. Zhao and Ziemba apply the synthetic option model to an example with three assets (cash, bonds and stocks) and four periods (a one-year horizon with quarterly portfolio reviews). The quarterly return on cash is constant at $\rho = 0.0095$. For stocks and bonds, the expected logarithmic rates of returns are $s = 0.04$ and $b = 0.019$ respectively. Transaction costs are 0.5% for stocks and 0.1% for bonds. The scenarios needed in the stochastic program are generated using an auto regression model which is constructed based on historical data (quarterly returns from 1985 to 1998; the Salomon Brothers bond index and S&P 500 index respectively). Specifically, the auto regression model is

$$\begin{cases} s_t = 0.037 - 0.193s_{t-1} + 0.418b_{t-1} - 0.172s_{t-2} + 0.517b_{t-2} + \epsilon_t, \\ b_t = 0.007 - 0.140s_{t-1} + 0.175b_{t-1} - 0.023s_{t-2} + 0.122b_{t-2} + \eta_t, \end{cases}$$

where the pair (ϵ_t, η_t) characterizes uncertainty. The scenarios are generated by selecting 20 pairs of (ϵ_t, η_t) to estimate the empirical distribution of one period uncertainty. In this way, a scenario tree with $160\,000 (= 20 \times 20 \times 20 \times 20)$ paths describing possible outcomes of asset returns is generated for the four periods.

The resulting large-scale linear program is solved. When this linear program is solved for a risk aversion of $\mu = 2.5$, the value of the terminal portfolio is always

Table 18.1 *A typical portfolio*

	Cash	Stocks	Bonds	Portfolio value at end of period
				100
Period 1	12%	18%	70 %	103
2		41%	59%	107
3		70%	30%	112
4	30%		70%	114

at least 4.6% more than the initial portfolio wealth and the distribution of terminal portfolio values is skewed to larger values because of dynamic downside risk control. The expected return is 16.33% and the volatility is 7.2%. It is interesting to compare these values with those obtained from a static Markowitz model, which gives an expected return of 15.4% for the same volatility but no guaranteed minimum return! In fact, in some scenarios, the value of the Markowitz portfolio is 5% *less* at the end of the one-year horizon than it was at the beginning.

It is also interesting to look at an example of a typical portfolio (one of the 160 000 paths) generated by the synthetic option model (the linear program was set up with an upper bound of 70% placed on the fraction of stocks or bonds in the portfolio); see Table 18.1.

Exercise 18.2   Computational exercise: Develop a synthetic option model in the spirit of that used by Zhao and Ziemba, adapted to the size limitation of your linear programming solver. Compare with a static model.

18.3 Case study: option pricing with transaction costs

A European call option on a stock with maturity T and strike price X gives the right to buy the stock at price X at time T. The holder of the option will not exercise this option if the stock has a price S lower than X at time T. Therefore, the value of a European call option is $\max(S - X, 0)$. Since S is random, the question of pricing the option correctly is of interest. The Black–Scholes–Merton option-pricing model relates the price of an option to the volatility of the stock return. The assumptions are that the market is efficient and that the returns are lognormal. From the volatility σ of the stock return, one can compute the option price for any strike price X. Conversely, from option prices one can compute the implied volatility σ. For a given stock, options with different strike prices should lead to the same σ (if the assumptions of the Black–Scholes–Merton model are correct).

The aim of the model developed in this section is to examine the extent to which market imperfections can explain the deviation of observed option prices from the Black–Scholes–Merton option-pricing model. One way to measure the deviation of the Black–Scholes–Merton model from observed option prices is through the "volatility smile": for a given maturity date, the implied volatility of a stock computed by the Black–Scholes–Merton model from observed option prices at different strike prices is typically not constant, but instead often exhibits a convex shape as the strike price increases (the "smile"). One explanation for the deviation is that the smile occurs because the Black–Scholes–Merton model assumes the ability to rebalance portfolios without costs imposed either by the inability to borrow or due to a bid–ask spread or other trading costs. Here we will look at the effect of transaction costs on option prices.

The derivation of the Black–Scholes–Merton formula is through a replicating portfolio containing the stock and a riskless bond. If the market is efficient, we should be able to replicate the option payoff at time T by rebalancing the portfolio between now and time T, as the stock price evolves. Rather than work with a continuous-time model, we discretize this process. This discretization is called the binomial approximation to the Black–Scholes–Merton option-pricing model. In this model, we specify a time period Δ between trading opportunities and postulate the behavior of stock and bond prices along successive time periods. The binomial model assumes that in between trading periods, only two possible stock price movements are possible.

1. There are N stages in the tree, indexed $0, 1 \ldots, N$, where stage 0 is the root of the tree and stage N is the last stage. If we divide the maturity date T of an option by N, we get that the length of a stage is $\Delta = T/N$.
2. Label the initial node k_0.
3. For a node $k \neq k_0$, let k^- be the node that is the immediate predecessor of k.
4. Let $S(k)$ be the stock price at node k and let $B(k)$ be the bond price at node k.
5. We assume that the interest rate is fixed at the annualized rate r so that $B(k) = B(k^-)e^{r\Delta}$.
6. Letting σ denote the volatility of the stock return, we use the standard parametrization $u = e^{\sigma\sqrt{\Delta}}$ and $d = 1/u$. So $S(k) = S(k^-)e^{\sigma\sqrt{\Delta}}$ if an uptick occurs from k^- to k and $S(k) = S(k^-)e^{-\sigma\sqrt{\Delta}}$ if a downtick occurs.
7. Let $n(k)$ be the quantity of stocks at node k and let $m(k)$ be the quantity of bonds at k.

18.3.1 The standard problem

In the binomial model, we have dynamically complete markets. This means that by trading the stock and the bond dynamically, we can replicate the payoffs (and values) from a call option. The option value is simply the cost of the replicating portfolio, and the replicating portfolio is self-financing after the first stage. This means that after we initially buy the stock and the bond, all subsequent trades do not require any additional money and, at the last stage, we reproduce the payoffs from the call option.

Therefore, we can represent the option-pricing problem as the following linear program. Choose quantities $n(k)$ of the stock, quantities $m(k)$ of the bond at each nonterminal node k to

$$\min \quad n(k_0)S(k_0) + m(k_0)B(k_0)$$

$$\text{subject to}$$

$$\text{rebalancing constraints:} \quad n(k^-)S(k) + m(k^-)B(k) \geq n(k)S(k) + m(k)B(k)$$

$$\text{for every node } k \neq k_0,$$

$$\text{replication constraints:} \quad n(k^-)S(k) + m(k^-)B(k) \geq \max(S(k) - X, 0)$$

$$\text{for every terminal node } k,$$

$$(18.3)$$

where k^- denotes the predecessor of k.

Note that we do not impose nonnegativity constraints since we will typically have a short position in the stock or bond.

Exercise 18.3   For a nondividend paying stock, collect data on four or five call options for the nearest maturity (but at least one month). Calculate the implied volatility for each option. Solve the standard problem (18.3) when the number of stages is seven using the implied volatility of the at-the-money option to construct the tree.

18.3.2 Transaction costs

To model transaction costs, we consider the simplest case where there are no costs of trading at the initial and terminal nodes, but there is a bid–ask spread on stocks at other nodes. So assume that if you buy a stock at node k, you pay $S(k)(1 + \theta)$, while if you sell a stock, you receive $S(k)(1 - \theta)$. This means that the rebalancing constraint becomes

$$n(k^-)S(k) + m(k^-)B(k) \geq n(k)S(k) + m(k)B(k) + |n(k) - n(k^-)|\theta S(k).$$

As there is an absolute value in this constraint, it is not linear. However, it can be linearized as follows. Define two nonnegative variables:

$$x(k) = \text{number of stocks } \textit{bought} \text{ at node } k, \text{ and}$$

$$y(k) = \text{number of stocks } \textit{sold} \text{ at node } k.$$

The rebalancing constraint now becomes

$$n(k^-)S(k) + m(k^-)B(k) \geq n(k)S(k) + m(k)B(k) + (x(k) + y(k))\theta S(k),$$

$$n(k) - n(k^-) = x(k) - y(k),$$

$$x(k) \geq 0, \quad y(k) \geq 0.$$

Note that this constraint leaves the possibility of simultaneously buying and selling stocks at the same node. But obviously this cannot improve the objective function that we minimize in (18.3), so we do not need to impose a constraint to prevent it.

The modified formulation is

$$\min \quad n(k_0)S(k_0) + m(k_0)B(k_0)$$

subject to

rebalancing constraints: $n(k^-)S(k) + m(k^-)B(k) \geq n(k)S(k) + m(k)B(k)$

$$+ (x(k) + y(k))\theta S(k) \quad \text{for every node } k \neq k_0,$$

$$n(k) - n(k^-) = x(k) - y(k) \quad \text{for every node } k \neq k_0,$$

replication constraints: $n(k^-)S(k) + m(k^-)B(k) \geq \max(S(k) - X, 0)$

for every terminal node k,

nonnegativity: $x(k) \geq 0, \quad y(k) \geq 0 \quad \text{for every node } k \neq k_0.$

$$(18.4)$$

Exercise 18.4 Repeat Exercise 18.3 allowing for transaction costs, with different values of θ, to see if the volatility smile can be explained by transaction costs. Specifically, given a value for σ and for θ, calculate option prices and see how they match up to observed prices. Try $\theta = 0.001, 0.005, 0.01, 0.02, 0.05$.

19

Robust optimization: theory and tools

19.1 Introduction to robust optimization

In many optimization models the inputs to the problem are not known at the time the problem must be solved, are computed inaccurately, or are otherwise uncertain. Since the solutions obtained can be quite sensitive to these inputs, one serious concern is that we are solving the wrong problem, and that the solution we find is far from optimal for the correct problem.

Robust optimization refers to the modeling of optimization problems with data uncertainty to obtain a solution that is guaranteed to be "good" for all or most possible realizations of the uncertain parameters. Uncertainty in the parameters is described through *uncertainty sets* that contain many possible values that may be realized for the uncertain parameters. The size of the uncertainty set is determined by the level of desired robustness.

Robust optimization can be seen as a complementary alternative to sensitivity analysis and stochastic programming. Robust optimization models can be especially useful in the following situations:

- Some of the problem parameters are estimates and carry estimation risk.
- There are constraints with uncertain parameters that *must* be satisfied regardless of the values of these parameters.
- The objective function or the optimal solutions are particularly sensitive to perturbations.
- The decision-maker cannot afford to take low-probability but high-magnitude risks.

Recall from Chapter 1 that there are different definitions and interpretations of robustness; the resulting models and formulations differ accordingly. In particular, we can distinguish between *constraint robustness* and *objective robustness*. In the first case, data uncertainty puts the feasibility of potential solutions at risk. In the second, feasibility constraints are fixed and the uncertainty of the objective function affects the proximity of the generated solutions to optimality.

Both the constraint and objective robustness models we considered in the introduction have a worst-case orientation. That is, we try to optimize the behavior of the solutions under the most adverse conditions. Following Kouvelis and Yu [47], we call solutions that optimize the worst-case behavior under uncertainty *absolute-robust* solutions. While such conservatism is necessary in some optimization settings, it may not be desirable in others. Absolute robustness is not always consistent with a decision-theoretic approach and with common utility functions. An alternative is to seek robustness in a *relative* sense.

In uncertain decision environments, people whose performance is judged relative to their peers will want to make decisions that avoid falling severely behind their competitors under all scenarios rather than protecting themselves against the worst-case scenarios. For example, a portfolio manager will be considered successful in a down market as long as she loses less than her peers or a benchmark. These considerations motivate the concept of *relative robustness*, which we discuss in Section 19.3.3.

Another variant of the robust optimization models called *adjustable-robust optimization* is attractive in multi-period models. To motivate these models one can consider a multi-period uncertain optimization problem where uncertainty is resolved progressively through periods. We assume that a subset of the decision variables can be chosen after these parameters are observed in a way to correct the sub-optimality of the decisions made with less information in earlier stages. In spirit, these models are closely related to two- (or multi-)stage stochastic programming problems with recourse. They were introduced by Guslitzer and co-authors [6, 37] and we summarize this approach in Section 19.3.4.

Each different interpretation of robustness and each different description of uncertainty leads to a different robust optimization formulation. These robust optimization problems often are or at least appear to be more difficult than their nonrobust counterparts. Fortunately, many of them can be reformulated in a tractable manner. While it is difficult to expect a single approach to handle each one of the different variations in a unified manner, a close study of the existing robust optimization formulations reveals many common threads. In particular, methods of conic optimization appear frequently in the solution of robust optimization problems. We review some of the most commonly used reformulation techniques used in robust optimization at the end of the chapter.

19.2 Uncertainty sets

In robust optimization, the description of the uncertainty of the parameters is formalized via *uncertainty sets*. Uncertainty sets can represent or may be formed by differences of opinions on future values of certain parameters, alternative estimates of parameters generated via statistical techniques from historical data and/or Bayesian techniques, among other things.

Common types of uncertainty sets encountered in robust optimization models include the following:

- Uncertainty sets representing a finite number of scenarios generated for the possible values of the parameters:

$$\mathcal{U} = \{p_1, p_2, \ldots, p_k\}.$$

- Uncertainty sets representing the convex hull of a finite number of scenarios generated for the possible values of the parameters (these are sometimes called polytopic uncertainty sets):

$$\mathcal{U} = \text{conv}(p_1, p_2, \ldots, p_k).$$

- Uncertainty sets representing an interval description for each uncertain parameter:

$$\mathcal{U} = \{p : l \leq p \leq u\}.$$

Confidence intervals encountered frequently in statistics can be the source of such uncertainty sets.

- Ellipsoidal uncertainty sets:

$$\mathcal{U} = \{p : p = p_0 + Mu, \|u\| \leq 1\}$$

These uncertainty sets can also arise from statistical estimation in the form of *confidence regions*, see [30]. In addition to their mathematically compact description, ellipsoidal uncertainty sets have the nice property that they smoothen the optimal value function [73].

It is a non-trivial task to determine the uncertainty set that is appropriate for a particular model as well as the type of uncertainty sets that lead to tractable problems. As a general guideline, the shape of the uncertainty set will often depend on the sources of uncertainty as well as the sensitivity of the solutions to these uncertainties. The size of the uncertainty set, on the other hand, will often be chosen based on the desired level of robustness.

When uncertain parameters reflect the "true" values of moments of random variables, as is the case in mean-variance portfolio optimization, we simply have no way of knowing these unobservable true values exactly. In such cases, after making some assumptions about the stationarity of these random processes we can generate estimates of these true parameters using statistical procedures. Goldfarb and Iyengar, for example, show that if we use a linear factor model for the multivariate returns of several assets and estimate the factor loading matrices via linear regression, the confidence regions generated for these parameters are ellipsoidal sets, and they advocate their use in robust portfolio selection as uncertainty sets [30].

To generate interval-type uncertainty sets, Tütüncü and Koenig use bootstrapping strategies as well as moving averages of returns from historical data [80]. The shape and the size of the uncertainty set can significantly affect the robust solutions generated. However, with few guidelines backed by theoretical and empirical studies, their choice remains an art form at the moment.

19.3 Different flavors of robustness

In this section we discuss each one of the robust optimization models we mentioned above in more detail. We start with constraint robustness.

19.3.1 Constraint robustness

One of the most important concepts in robust optimization is *constraint robustness*. This refers to situations where the uncertainty is in the constraints and we seek solutions that remain *feasible* for all possible values of the uncertain inputs. This type of solution is required in many engineering applications. Typical instances include multi-stage problems where the uncertain outcomes of earlier stages have an effect on the decisions of the later stages and the decision variables must be chosen to satisfy certain balance constraints (e.g., inputs to a particular stage can not exceed the outputs of the previous stage) no matter what happens with the uncertain parameters of the problem. Therefore, our solution must be constraint robust with respect to the uncertainties of the problem. We present a mathematical model for finding constraint-robust solutions. Consider an optimization problem of the form:

$$\min_x \quad f(x)$$
$$G(x, p) \in K, \tag{19.1}$$

where x are the decision variables, f is the (certain) objective function, G and K are the structural elements of the constraints that are assumed to be certain, and p are the possibly uncertain parameters of the problem. Consider an uncertainty set $\mathcal{U}$ that contains all possible values of the uncertain parameters p. Then, a constraint-robust optimal solution can be found by solving the following problem:

$$\min_x \quad f(x)$$
$$G(x, p) \in K, \quad \forall p \in \mathcal{U}. \tag{19.2}$$

As (19.2) indicates, the robust feasible set is the intersection of the feasible sets $S(p) = \{x : G(x, p) \in K\}$ indexed by the uncertainty set $\mathcal{U}$. We illustrate this in Figure 19.1 for an ellipsoidal feasible set with $\mathcal{U} = \{p_1, p_2, p_3, p_4\}$, where p_i correspond to the uncertain center of the ellipse.

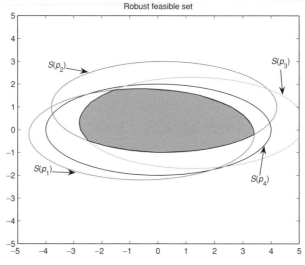

Figure 19.1 Constraint robustness

There are no uncertain parameters in the objective function of the problem (19.2). This, however, is not a restrictive assumption. An optimization problem with uncertain parameters in both the objective function and constraints can be easily reformulated to fit the form in (19.2). In fact,

$$\begin{aligned} \min_x \quad & f(x, p) \\ & G(x, p) \in K \end{aligned} \tag{19.3}$$

is equivalent to the problem:

$$\begin{aligned} \min_{t,x} \quad & t \\ & t - f(x, p) \geq 0, \\ & G(x, p) \in K. \end{aligned} \tag{19.4}$$

This last problem has all its uncertainties in its constraints.

Exercise 19.1 Show that if $S(p) = \{x : G(x, p) \in K\}$ is convex for all p, then the robust feasible set $S := \bigcap_{p \in \mathcal{U}} S(p)$ is also convex. If $S(p)$ is polyhedral for all p, is S necessarily polyhedral?

19.3.2 Objective robustness

Another important robustness concept is *objective robustness*. This refers to solutions that will remain close to optimal for all possible realizations of the uncertain problem parameters. Since such solutions may be difficult to obtain, especially when uncertainty sets are relatively large, an alternative goal for objective robustness is to

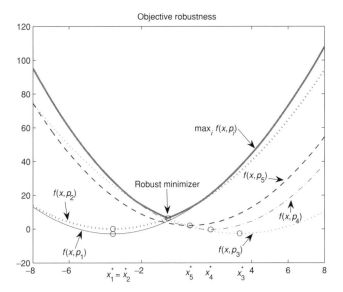

Figure 19.2 Objective robustness

find solutions whose worst-case behavior is optimized. The worst-case behavior of a solution corresponds to the value of the objective function for the worst possible realization of the uncertain data for that particular solution.

We now develop a mathematical model that addresses objective robustness. Consider an optimization problem of the form:

$$\min_x f(x, p) \\ x \in S,$$ (19.5)

where S is the (certain) feasible set and f is the objective function that depends on uncertain parameters p. As before, $\mathcal{U}$ denotes the uncertainty set that contains all possible values of the uncertain parameters p. Then, an objective-robust solution can be obtained by solving:

$$\min_{x \in S} \ \max_{p \in \mathcal{U}} f(x, p).$$ (19.6)

We illustrate the objective robustness problem (19.6) in Figure 19.2. In this example, the feasible set S is the real line, the uncertainty set is $\mathcal{U} = \{p_1, p_2, p_3, p_4, p_5\}$, and the objective function $f(x, p_i)$ is a convex quadratic function whose parameters p_i determine its shape. Note that the robust minimizer is different from the minimizers of each $f(x, p_i)$, which are denoted by x_i^* in the figure. In fact, none of the x_i^*'s is particularly close to the robust minimizer.

As the argument at the end of the previous subsection shows, objective robustness can be seen as a special case of constraint robustness via a reformulation. However, it is important to distinguish between these two problem variants as their "natural" robust formulations lead to two different classes of optimization formulations, namely semi-infinite and min-max optimization problems respectively. This way, different methodologies available for these two problem classes can be readily used for respective problems.

Exercise 19.2　　In Chapter 8, for a given constant λ, expected return vector μ, and a positive definite covariance matrix Σ we considered the following mean-variance optimization problem:

$$\max_{x \in \mathcal{X}} \mu^{\mathrm{T}} x - \lambda x^{\mathrm{T}} \Sigma x, \tag{19.7}$$

where $\mathcal{X} = \{x : e^{\mathrm{T}} x = 1\}$ with $e = [1\ 1\ \ldots\ 1]^{\mathrm{T}}$. Here, we consider the situation where we assume Σ to be certain and given but μ is assumed to be uncertain. For a fixed μ let $z(\mu)$ represent the optimal value of this problem. Determine $z(\mu)$ as an explicit function of μ. Verify that $z(\mu)$ is a quadratic function. Is it convex? Let $\mathcal{U}$ represent the uncertainty set for μ and formulate the objective robustness problem.

19.3.3 Relative robustness

The focus of constraint and objective robustness models on an *absolute* measure of worst-case performance is not consistent with the risk tolerances of many decision-makers. Instead, we may prefer to measure the worst case in a relative manner, relative to the best possible solution under each scenario. This leads us to the notion of relative robustness.

Consider the following optimization problem:

$$\begin{aligned} \min_x\ & f(x, p) \\ & x \in S, \end{aligned} \tag{19.8}$$

where p is uncertain with uncertainty set $\mathcal{U}$. To simplify the description, we restrict our attention to the case with objective uncertainty and assume that the constraints are certain.

Given a fixed $p \in \mathcal{U}$, let $z^*(p)$ denote the optimal-value function, i.e.,

$$z^*(p) = \min_x\ f(x, p)\ \text{s.t.}\ x \in S.$$

Furthermore, we define the optimal-solution map:

$$x^*(p) = \arg\min_x\ f(x, p)\ \text{s.t.}\ x \in S.$$

Note that $z^*(p)$ can be extended-valued and $x^*(p)$ can be set-valued.

To motivate the notion of relative robustness we first define a measure of *regret* associated with a decision after the uncertainty is resolved. If we choose x as our vector and p is the realized value of the uncertain parameter, the regret associated with choosing x instead of an element of $x^*(p)$ is defined as:

$$r(x, p) = f(x, p) - z^*(p) = f(x, p) - f(x^*(p), p). \qquad (19.9)$$

Note that the regret function is always nonnegative and can also be regarded as a measure of the "benefit of hindsight."

Now, for a given x in the feasible set we consider the maximum-regret function:

$$R(x) := \max_{p \in \mathcal{U}} r(x, p) = \max_{p \in \mathcal{U}} f(x, p) - f(x^*(p), p). \qquad (19.10)$$

A *relative-robust* solution to problem (19.8) is a vector x that minimizes the maximum regret:

$$\min_{x \in S} \max_{p \in \mathcal{U}} f(x, p) - z^*(p). \qquad (19.11)$$

While they are intuitively attractive, relative-robust formulations can also be significantly more difficult than the standard absolute-robust formulations. Indeed, since $z^*(p)$ is the optimal-value function and involves an optimization problem itself, problem (19.11) is a three-level optimization problem as opposed to the two-level problems in absolute-robust formulations. Furthermore, the optimal-value function $z^*(p)$ is rarely available in analytic form, is typically nonsmooth and is often hard to analyze. Another difficulty is that, if f is linear in p as is often the case, then $z^*(p)$ is a concave function. Therefore, the inner maximization problem in (19.11) is a convex maximization problem and is difficult for most $\mathcal{U}$.

A simpler variant of (19.11) can be constructed by deciding on the maximum level of regret to be tolerated beforehand and by solving a feasibility problem instead with this level imposed as a constraint. For example, if we decide to limit the maximum regret to R, the problem to solve becomes the following: find an x satisfying $G(x) \in K$ such that

$$f(x, p) - z^*(p) \le R, \quad \forall p \in \mathcal{U}.$$

If desired, one can then use bi-section on R to find its optimal value.

Another variant of relative robustness models arises when we measure the regret in terms of the proximity of our chosen solution to the optimal solution set rather than in terms of the optimal objective values. For this model, consider the following distance function for a given x and p:

$$d(x, p) = \inf_{x^* \in x^*(p)} \|x - x^*\|. \qquad (19.12)$$

When the solution set is a singleton, there is no optimization involved in the definition. As above, we then consider the maximum-distance function:

$$D(x) := \max_{p \in \mathcal{U}} d(x, p) = \max_{p \in \mathcal{U}} \inf_{x^* \in x^*(p)} \|x - x^*\|. \qquad (19.13)$$

For relative robustness in this new sense, we seek x such that

$$\min_{x \in S} \max_{p \in \mathcal{U}} d(x, p). \tag{19.14}$$

This variant is an attractive model for cases where we have time to revise our decision variables x, perhaps only slightly, once p is revealed. In such cases, we will want to choose an x that will not need much perturbation under any scenario, i.e., we seek the solution to (19.14). This model can also be useful for multi-period problems where revisions of decisions between periods can be costly. Portfolio rebalancing problems with transaction costs are examples of such settings.

Exercise 19.3 Formulate the relative robustness formulation for the optimization problem discussed in Exercise 19.2 . Comment on the consequences of the convexity of the function $z^*(\mu)$. Show that the relative robustness problems for $\mathcal{U} = \{p_1, p_2, \ldots, p_k\}$ and $\mathcal{U} = \mathrm{conv}(p_1, p_2, \ldots, p_k)$ are equivalent.

Exercise 19.4 Recall the setting in Exercise 19.2 . Let $x^*(\mu)$ denote the unique optimal solution of the problem for a given μ and obtain an explicit expression for $x^*(\mu)$. Using this expression, formulate the variant of the relative robustness problem given in (19.14).

19.3.4 Adjustable robust optimization

Robust optimization formulations we saw above assume that the uncertain parameters will not be observed until all the decision variables are determined and therefore do not allow for recourse actions that may be based on realized values of some of these parameters. This is not always the appropriate model for uncertain optimization problems. In particular, multi-period decision models involve uncertain parameters, some of which are revealed during the decision process. After observing these parameters, later stage decisions can respond to this new information and can correct any sub-optimality resulting from less desirable outcomes in the uncertain parameters. *Adjustable robust optimization* (ARO) formulations model these decision environment and allow recourse action. These models are closely related to, and in fact, partly inspired by the multi-stage stochastic programming formulations with recourse.

ARO models were recently introduced in [6, 37] for uncertain linear programming problems. Consider, for example, the two-stage linear optimization problem given below whose first-stage decision variables x^1 need to be determined now, while the second-stage decision variables x^2 can be chosen after the uncertain parameters of the problem A^1, A^2, and b are realized:

$$\min_{x^1, x^2}\{c^Tx^1 : A^1x^1 + A^2x^2 \leq b\}. \tag{19.15}$$

Note that the second-stage variables x^2 do not appear in the objective function – this is what Ben-Tal *et al.* call the "normalized" form of the problem [6]. Problems with objective functions involving variables x^2 can be reformulated as in (19.15) after introducing an artificial variable; see Exercise 19.5 . Therefore, we can focus on this simpler and convenient form without loss of generality.

Let $\mathcal{U}$ denote the *uncertainty set* for parameters A^1, A^2, and b. The standard constraint-robust optimization formulation for this problem seeks to find vectors x^1 and x^2 that optimize the objective function and satisfy the constraints of the problem for all possible realizations of the constraint coefficients. In this formulation, both sets of variables must be chosen before the uncertain parameters can be observed and therefore cannot depend on these parameters. Consequently, the standard robust counterpart of this problem can be written as follows:

$$\min_{x^1}\{c^\mathsf{T}x^1 : \exists x^2 \ \forall (A^1, A^2, b) \in \mathcal{U} : A^1 x^1 + A^2 x^2 \le b\}. \qquad (19.16)$$

Note that this formulation is equivalent to the formulation we saw before, i.e.,

$$\min_{x^1, x^2}\{c^\mathsf{T}x^1 : A^1 x^1 + A^2 x^2 \le b, \forall (A^1, A^2, b) \in \mathcal{U}\}. \qquad (19.17)$$

We prefer (19.16) since it illustrates the difference between this formulation and the adjustable version more clearly.

In contrast, the adjustable robust optimization formulation allows the choice of the second-period variables x^2 to depend on the realized values of the uncertain parameters. As a result, the adjustable robust counterpart problem is given as follows:

$$\min_{x^1}\{c^\mathsf{T}x^1 : \forall (A^1, A^2, b) \in \mathcal{U}, \ \exists x^2 \equiv x^2(A^1, A^2, b) : A^1 x^1 + A^2 x^2 \le b\}.$$
$$(19.18)$$

The feasible set of the second problem is larger than that of the first problem in general and, therefore, the model is more flexible. ARO models can be especially useful when robust counterparts are unnecessarily conservative. The price to pay for this additional modeling flexibility appears to be the increased difficulty of the resulting ARO formulations. Even for problems where the robust counterpart is tractable, it can happen that the ARO formulation leads to an NP-hard problem. One of the factors that contribute to the added difficulty in ARO models is the fact that the feasible set of the recourse actions (second-period decisions) depends not only on the realization of the uncertain parameters but also the first-period decisions. One way to overcome this difficulty is to consider simplifying assumptions either on the uncertainty set, or on the dependence structure of recourse actions to uncertain parameters. For example, if the recourse actions are restricted to be affine functions of the uncertain parameters. While this restriction will likely give us suboptimal solutions, it may be the only strategy to obtain tractable formulations.

Exercise 19.5   Consider the following adjustable robust optimization problem:

$$\min_{x^1} \left\{ c_1^{\mathsf{T}} x^1 + c_2^{\mathsf{T}} x_2 : \forall (A^1, A^2, b) \in \mathcal{U}, \ \exists x^2 \equiv x^2(A^1, A^2, b) : A^1 x^1 + A^2 x^2 \leq b \right\}.$$

Show how this problem can be expressed in the "normalized" form (19.18) after introducing an artificial variable.

19.4 Tools and strategies for robust optimization

In this section we review a few of the commonly used techniques for the solution of robust optimization problems. The tools we discuss are essentially reformulation strategies for robust optimization problems so that they can be rewritten as a deterministic optimization problem with no uncertainty. In these reformulations, we look for *economy* so that the new formulation is not much bigger than the original, "uncertain" problem and *tractability* so that the new problem can be solved efficiently using standard optimization methods.

The variety of the robustness models and the types of uncertainty sets rule out a unified approach. However, there are some common threads and the material in this section can be seen as a guide to the available tools, which can be combined or appended with other techniques to solve a given problem in the robust optimization setting.

19.4.1 Sampling

One of the simplest strategies for achieving robustness under uncertainty is to sample several *scenarios* for the uncertain parameters from a set that contains possible values of these parameters. This sampling can be done with or without using distributional assumptions on the parameters and produces a robust optimization formulation with a finite uncertainty set.

If uncertain parameters appear in the constraints, we create a copy of each such constraint corresponding to each scenario. Uncertainty in the objective function can be handled in a similar manner. Recall, for example, the generic uncertain optimization problem given in (19.3):

$$\begin{aligned} \min_x \quad & f(x, p) \\ & G(x, p) \in K \end{aligned} \tag{19.19}$$

If the uncertainty set $\mathcal{U}$ is a finite set, i.e., $\mathcal{U} = \{p_1, p_2, \ldots, p_k\}$, the robust formulation is obtained as follows:

$$\begin{aligned} \min_{t,x} \quad & t \\ & t - f(x, p_i) \geq 0, \ i = 1, \ldots, k, \\ & G(x, p_i) \in K, \ i = 1, \ldots, k. \end{aligned} \tag{19.20}$$

Note that no reformulation is necessary in this case and the duplicated constraints preserve the structural properties (linearity, convexity, etc.) of the original constraints. Consequently, when the uncertainty set is a finite set the resulting robust optimization problem is larger but theoretically no more difficult than the non-robust version of the problem. The situation is somewhat similar to stochastic programming formulations. Examples of robust optimization formulations with finite uncertainty sets can be found, e.g., in the recent book by Rustem and Howe [70].

19.4.2 Conic optimization

Moving from finite uncertainty sets to continuous sets such as intervals or ellipsoids presents a theoretical challenge. The robust version of an uncertain constraint that has to be satisfied for all values of the uncertain parameters in a continuous set results in a semi-infinite optimization formulation. These problems are called semi-infinite since there are infinitely many constraints – indexed by the uncertainty set – but only finitely many variables.

Fortunately, it is possible to reformulate certain semi-infinite optimization problems using a finite set of conic constraints. Such reformulations were already introduced in Chapter 9. We recall two constraint robustness examples from that chapter:

- The robust formulation for the linear programming problem

$$\begin{aligned} \min \quad & c^\mathrm{T} x \\ \text{s.t.} \quad & a^\mathrm{T} x + b \geq 0, \end{aligned} \tag{19.21}$$

where the uncertain parameters $[a; b]$ belong to the ellipsoidal uncertainty set

$$\mathcal{U} = \{[a; b] = [a^0; b^0] + \sum_{j=1}^{k} u_j [a^j; b^j], \|u\| \leq 1\},$$

is equivalent to the following second-order cone program:

$$\begin{aligned} \min_{x,z} \quad & c^\mathrm{T} x \\ \text{s.t.} \quad & a_j^\mathrm{T} x + b_j = z_j, \quad j = 0, \dots, k, \\ & (z_0, z_1, \dots, z_k) \in C_q \end{aligned}$$

where C_q is the second-order cone defined in (9.2).

- The robust formulation for the quadratically constrained optimization problem

$$\begin{aligned} \min \quad & c^\mathrm{T} x \\ \text{s.t.} \quad & -x^\mathrm{T} (A^\mathrm{T} A) x + 2b^\mathrm{T} x + \gamma \geq 0, \end{aligned} \tag{19.22}$$

where the uncertain parameters $[A; b; \gamma]$ belong to the ellipsoidal uncertainty set

$$\mathcal{U} = \left\{ [A; b; \gamma] = [A^0; b^0; \gamma^0] + \sum_{j=1}^{k} u_j [A^j; b^j; \gamma^j], \|u\| \leq 1 \right\},$$

is equivalent to the following semidefinite program:

$$\min_{x, z^0, \ldots, z^k, y, \lambda} \quad c^T x$$

$$\text{s.t. } A^j x = z^j, \ j = 0, \ldots, k,$$

$$(b^j)^T x = y^j, \ j = 0, \ldots, k,$$

$$\lambda \geq 0,$$

$$\begin{bmatrix} \gamma^0 + 2y^0 - \lambda & \left[y^1 + \frac{1}{2}\gamma^1 \cdots y^k + \frac{1}{2}\gamma^k \right] & (z^0)^T \\ \begin{bmatrix} y^1 + \frac{1}{2}\gamma^1 \\ \vdots \\ y^k + \frac{1}{2}\gamma^k \end{bmatrix} & \lambda I & \begin{bmatrix} (z^1)^T \\ \vdots \\ (z^k)^T \end{bmatrix} \\ z^0 & [z^1 \cdots z^k] & I \end{bmatrix} \succeq 0.$$

Exercise 19.6   Consider a simple, two-variable LP with nonnegative variables and a single uncertain constraint $a_1 x_1 + a_2 x_2 + b \geq 0$, where $[a_1, a_2, b]$ belongs to the following uncertainty set:

$$\mathcal{U} = \left\{ [a_1, a_2, b]: \ [a_1, a_2, b] = [1, 1, 1] + u_1 \left[\frac{1}{2}, 0, 0 \right] + u_2 \left[0, \frac{1}{3}, 0 \right], \|u\| \leq 1 \right\}.$$

Determine the robust formulation of this constraint and the projection of the robust feasible set to the (x_1, x_2) space. Try to approximate this set using the sampling strategy outlined above. Comment on the number of samples required until the approximate robust feasible set is a relatively good approximation of the true robust feasible set.

Exercise 19.7   When $A = 0$ for the quadratically constrained problem (19.22) above, the problem reduces to a linearly constrained problem. Verify that when $A^j = 0$ for all $j = 0, 1, \ldots, k$ in the uncertainty set $\mathcal{U}$, the robust formulation of this problem reduces to the robust formulation of the linearly constrained problem.

Exercise 19.8   Note that the quadratically constrained optimization problem given above can alternatively be parameterized as follows:

$$\min \quad c^T x$$

$$\text{s.t.} \quad -x^T \Sigma x + 2b^T x + \gamma \geq 0,$$

where we used a positive semidefinite matrix Σ instead of $A^T A$ in the constraint definition. How can we define an ellipsoidal uncertainty set for this parameterization of the problem? What are the potential advantages and potential problems with using this parameterization?

19.4.3 Saddle-point characterizations

For the solution of problems arising from objective uncertainty, the robust solution can be characterized using saddle-point conditions when the original problem satisfies certain convexity assumptions. The benefit of this characterization is that we can then use algorithms such as interior-point methods already developed and available for saddle-point problems.

As an example of this strategy consider the problem (19.5) from Section 19.3.2 and its robust formulation reproduced below:

$$\min_{x \in S} \max_{p \in \mathcal{U}} f(x, p). \qquad (19.23)$$

We note that the *dual* of this robust optimization problem is obtained by changing the order of the minimization and maximization problems:

$$\max_{p \in \mathcal{U}} \min_{x \in S} f(x, p). \qquad (19.24)$$

From standard results in convex analysis we have the following conclusion:

Lemma 19.1 *If $f(x, p)$ is a convex function of x and a concave function of p, and if S and $\mathcal{U}$ are nonempty and at least one of them is bounded, the optimal values of the problems (19.23) and (19.24) coincide and there exists a saddle point (x^*, p^*) such that*

$$f(x^*, p) \le f(x^*, p^*) \le f(x, p^*), \forall x \in S, p \in \mathcal{U}.$$

This characterization is the basis of the robust optimization algorithms given in [38, 80].

20

Robust optimization models in finance

As we discussed in the previous chapter, robust optimization formulations address problems with input uncertainty. Since many financial optimization problems involve future values of security prices, interest rates, exchange rates, etc. which are not known in advance but can only be forecasted or estimated, such problems fit perfectly into the framework of robust optimization. In this chapter, we give examples of robust optimization formulations for a variety of financial optimization problems including portfolio selection, risk management, and derivatives pricing/hedging.

We start with the application of constraint-robust optimization approach to a multi-period portfolio selection problem:

20.1 Robust multi-period portfolio selection

This section is adapted from an article by Ben-Tal *et al.* [7]. We consider an investor who currently holds the portfolio $x^0 = (x_1^0, \ldots, x_n^0)$, where x_i^0 denotes the number of shares of asset i in the portfolio, for $i = 1, \ldots, n$. Also, let x_0^0 denote her cash holdings. She wants to determine how to adjust her portfolio in the next L investment periods to maximize her total wealth at the end of period L.

We use the following decision variables to model this multi-period portfolio selection problem: b_i^l denotes the number of additional shares of asset i bought at the beginning of period l and s_i^l denotes the number of asset i shares sold at the beginning of period l, for $i = 1, \ldots, n$ and $l = 1, \ldots, L$. Then, the number of shares of asset i in the portfolio at the beginning of period l, denoted x_i^l, is given by the following simple equation:

$$x_i^l = x_i^{l-1} - s_i^l + b_i^l, \quad i = 1, \ldots, n, \ l = 1, \ldots, L. \tag{20.1}$$

Let P_i^l denote the price of a share of asset i in period l. We make the assumption that the cash account earns no interest so that $P_0^l = 1, \forall l$. This is not a restrictive assumption – we can always reformulate the problem in this way after a change of numeraire.

We assume that proportional transaction costs are paid on asset purchases and sales and denote them with α_i^l and β_i^l for sales and purchases, respectively, for asset i and period l. We assume that α_i^l's and β_i^l's are all known at the beginning of period 0, although they can vary from period to period and from asset to asset. Transaction costs are paid from the investor's cash account and, therefore, we have the following balance equation for the cash account:

$$x_0^l = x_0^{l-1} + \sum_{i=1}^{n}(1 - \alpha_i)P_i^l s_i^l - \sum_{i=1}^{n}(1 + \beta_i)P_i^l b_i^l, l = 1, \ldots, L.$$

This balance condition indicates that the cash available at the beginning of period l is the sum of last period's cash holdings and the proceeds from sales (discounted by transaction costs) minus the cost of new purchases. For technical reasons, we will replace the equation above with an inequality, effectively allowing the investor to "burn" some of her cash if she wishes to:

$$x_0^l \leq x_0^{l-1} + \sum_{i=1}^{n}(1 - \alpha_i)P_i^l s_i^l - \sum_{i=1}^{n}(1 + \beta_i)P_i^l b_i^l, l = 1, \ldots, L.$$

The objective of the investor is to maximize her total wealth at the end of period L. This objective can be represented as follows:

$$\max \sum_{i=0}^{n} P_i^L x_i^L.$$

If we assume that all the future prices P_i^l are known at the time this investment problem is to be solved, we obtain the following deterministic optimization problem:

$$
\begin{aligned}
&\max_{x,s,b} \sum_{i=0}^{n} P_i^L x_i^L \\
&\left.
\begin{array}{ll}
x_0^l \leq x_0^{l-1} + \sum_{i=1}^{n}(1 - \alpha_i)P_i^l s_i^l - \sum_{i=1}^{n}(1 + \beta_i)P_i^l b_i^l, & \\
x_i^l = x_i^{l-1} - s_i^l + b_i^l, & i = 1, \ldots, n, \\
s_i^l \geq 0, & i = 1, \ldots, n, \\
b_i^l \geq 0, & i = 1, \ldots, n, \\
x_i^l \geq 0, & i = 0, \ldots, n,
\end{array}
\right\} \quad l = 1, \ldots, L.
\end{aligned}
$$

$$(20.2)$$

This is a linear programming problem that can be solved easily using the simplex method or interior-point methods. The nonnegativity constraints imposed by Ben-Tal *et al.* [7] on x_i^l's disallow short positions and borrowing. We note that these constraints are not essential to the model and some or all of them can be removed to allow short sales on a subset of the assets or to allow borrowing. Observe that the investor would, of course, never choose to burn money if she is trying to maximize her final wealth. Therefore, the cash balance inequalities will always be satisfied with equality in any optimal solution of this problem.

In a realistic setting, we do not know the P_i^l's in advance and therefore cannot solve the optimal portfolio allocation problem as the linear program we developed above. Instead, we will develop a robust optimization model that incorporates the uncertainty in the P_i^l's in (20.2). This is an alternative approach to the stochastic programming models discussed in Chapter 18. Since the objective function involves uncertain parameters P_i^L, we first reformulate the problem as in (19.4) to move all the uncertainty to the constraints:

$$\max_{x,s,b,t} \ t$$

$$\left.\begin{array}{ll} t \ \leq \sum_{i=0}^n P_i^L x_i^L \\ x_0^l \leq x_0^{l-1} + \sum_{i=1}^n (1-\alpha_i) P_i^l s_i^l - \sum_{i=1}^n (1+\beta_i) P_i^l b_i^l, \\ x_i^l = x_i^{l-1} - s_i^l + b_i^l, \quad i = 1, \ldots, n, \\ s_i^l \geq 0, \qquad\qquad\quad i = 1, \ldots, n, \\ b_i^l \geq 0, \qquad\qquad\quad i = 1, \ldots, n, \\ x_i^l \geq 0, \qquad\qquad\quad i = 0, \ldots, n, \end{array}\right\} \quad l = 1, \ldots, L. \qquad (20.3)$$

The first two constraints of this reformulation are the constraints that are affected by uncertainty and we would like to find a solution that satisfies these constraints for most possible realizations of the uncertain parameters P_i^l. To determine the robust version of these constraints, we need to choose an appropriate uncertainty set for these uncertain parameters. For this purpose, we follow a 3-σ approach common in engineering and statistical applications.

Future prices can be assumed to be random quantities. Let us denote the expected value of the vector

$$P^l = \begin{bmatrix} P_1^l \\ \vdots \\ P_n^l \end{bmatrix}$$

with

$$\mu^l = \begin{bmatrix} \mu_1^l \\ \vdots \\ \mu_n^l \end{bmatrix}$$

and its covariance matrix with V^l. First, consider the constraint:

$$t \leq \sum_{i=0}^n P_i^L x_i^L.$$

Letting $x^L = (x_1^L, \ldots, x_n^L)$, the expected value and the standard deviation of the right-hand-side expression are given by $x_0^L + (\mu^L)^\mathrm{T} x^L = x_0^L + \sum_{i=1}^n \mu_i^L x_i^L$ and $\sqrt{(x^L)^\mathrm{T} V^L x^L}$. If the P_i^L quantities are normally distributed, by requiring

$$t \leq \mathrm{E}(\mathrm{RHS}) - 3\,\mathrm{STD}(\mathrm{RHS}) = x_0^L + (\mu^L)^\mathrm{T} x^L - 3\sqrt{(x^L)^\mathrm{T} V^L x^L}, \qquad (20.4)$$

we would guarantee that the (random) inequality $t \leq \sum_{i=0}^n P_i^L x_i^L$ would be satisfied more than 99% of the time. Therefore, we regard (20.4) as the "robust" version of $t \leq \sum_{i=0}^n P_i^L x_i^L$.

We can apply a similar logic to other constraints affected by uncertainty:

$$x_0^l - x_0^{l-1} \leq \sum_{i=1}^n (1 - \alpha_i) P_i^l s_i^l - \sum_{i=1}^n (1 + \beta_i) P_i^l b_i^l, \quad l = 1, \ldots, L,$$

where we moved x_0^{l-1} to the left-hand side to isolate the uncertain terms on the right-hand side of the inequality. In this case, the expected value and variance of the right-hand-side expression are given by the following formulas:

$$\mathrm{E}\left[\sum_{i=1}^n (1 - \alpha_i) P_i^l s_i^l - \sum_{i=1}^n (1 + \beta_i) P_i^l b_i^l \right] = (\mu^l)^\mathrm{T} D_\alpha^l s^l - (\mu^l)^\mathrm{T} D_\beta^l b^l$$

$$= (\mu^l)^\mathrm{T} \begin{bmatrix} D_\alpha^l & -D_\beta^l \end{bmatrix} \begin{bmatrix} s^l \\ b^l \end{bmatrix},$$

and

$$\mathrm{Var}\left[\sum_{i=1}^n (1 - \alpha_i) P_i^l s_i^l - \sum_{i=1}^n (1 + \beta_i) P_i^l b_i^l \right] = \begin{bmatrix} s^l \\ b^l \end{bmatrix}^\mathrm{T} \begin{bmatrix} D_\alpha^l \\ -D_\beta^l \end{bmatrix} V^l \begin{bmatrix} D_\alpha^l & -D_\beta^l \end{bmatrix} \begin{bmatrix} s^l \\ b^l \end{bmatrix}.$$

Above, D_α^ℓ and D_β^ℓ are the diagonal matrices

$$D_\alpha^\ell := \begin{bmatrix} (1 - \alpha_1^l) & & \\ & \ddots & \\ & & (1 - \alpha_n^l) \end{bmatrix}, \text{ and } D_\beta^\ell := \begin{bmatrix} (1 + \beta_1^l) & & \\ & \ddots & \\ & & (1 + \beta_n^l) \end{bmatrix}.$$

Also, $s^l = (s_1^l, \ldots, s_n^l)^\mathrm{T}$, and $b^l = (b_1^l, \ldots, b_n^l)^\mathrm{T}$. Replacing

$$x_0^l - x_0^{l-1} \leq \sum_{i=1}^n (1 - \alpha_i) P_i^l s_i^l - \sum_{i=1}^n (1 + \beta_i) P_i^l b_i^l, l = 1, \ldots, L$$

with

$$x_0^l - x_0^{l-1} \leq (\mu^l)^\mathrm{T} \begin{bmatrix} D_\alpha^l & -D_\beta^l \end{bmatrix} \begin{bmatrix} s^l \\ b^l \end{bmatrix} - 3\sqrt{ \begin{bmatrix} s^l \\ b^l \end{bmatrix}^\mathrm{T} \begin{bmatrix} D_\alpha^l \\ -D_\beta^l \end{bmatrix} V^l \begin{bmatrix} D_\alpha^l & -D_\beta^l \end{bmatrix} \begin{bmatrix} s^l \\ b^l \end{bmatrix} },$$

we obtain a "robust" version of the constraint. Once again, assuming normality in the distribution of the uncertain parameters, by satisfying this robust constraint we can guarantee that the original constraint will be satisfied with probability more than 0.99.

The approach above corresponds to choosing the uncertainty sets for the uncertain parameter vectors P^l in the following manner:

$$\mathcal{U}^l := \left\{ P^l : \sqrt{(P^l - \mu^l)^{\mathrm{T}}(V^l)^{-1}(P^l - \mu^l)} \leq 3 \right\}, \quad l = 1, \ldots, L. \qquad (20.5)$$

The complete uncertainty set $\mathcal{U}$ for all the uncertain parameters is the Cartesian product of the sets $\mathcal{U}^l$ defined as $\mathcal{U} = \mathcal{U}^1 \times \ldots \times \mathcal{U}^L$.

Exercise 20.1 Let $\mathcal{U}^L$ be as in (20.5). Show that

$$t \leq \sum_{i=0}^{n} P_i^L x_i^L, \quad \forall P^L \in \mathcal{U}^L$$

if and only if

$$t \leq (\mu^L)^{\mathrm{T}} x^L - 3\sqrt{(x^L)^{\mathrm{T}} V^L x^L}.$$

Thus, our 3-σ approach is equivalent to the robust formulation of this constraint using an appropriate uncertainty set. Hint: You may first want to show that

$$\mathcal{U}^L = \left\{ \mu^L + (V^L)^{1/2} u : \|u\| \leq 3 \right\}.$$

The resulting problem has nonlinear constraints, because of the square-roots and quadratic terms within the square roots as indicated in Exercise 20.1. Fortunately, however, these constraints can be written as *second-order cone* constraints and result in a *second-order cone optimization* problem.

Exercise 20.2 A vector $(y^0, y^1) \in \mathbb{R} \times \mathbb{R}^k$ belongs to the $k+1$ dimensional second-order cone if it satisfies the following inequality:

$$y^0 \geq \|y^1\|_2.$$

Constraints of the form above are called second-order cone constraints. Show that the constraint

$$t \leq (\mu^L)^{\mathrm{T}} x^L - 3\sqrt{(x^L)^{\mathrm{T}} V^L x^L}$$

can be represented as a second-order cone constraint using an appropriate change of variables. You can assume that V^L is a given positive definite matrix.

20.2 Robust profit opportunities in risky portfolios

Consider an investment environment with n financial securities whose future price vector $r \in I\!R^n$ is a random variable. Let $p \in I\!R^n$ represent the current prices of these securities. Consider an investor who chooses a portfolio $x = (x_1, \ldots, x_n)$, where x_i denote the number of shares of security i in the portfolio. If x satisfies

$$p^T x < 0,$$

meaning that the portfolio is formed with negative cash flow (by pocketing money), and if the realization $\tilde{r}$ at the end of the investment period of the random variable r satisfies

$$\tilde{r}^T x \geq 0,$$

meaning that the portfolio has a nonnegative value at the end, then the investor would get to keep the money pocketed initially, and perhaps even more. A type-A arbitrage opportunity would correspond to the situation when the ending portfolio value is guaranteed to be nonnegative, i.e., when the investor can choose a portfolio x such that $p^T x < 0$ and

$$\Pr[r^T x \geq 0] = 1. \tag{20.6}$$

Since arbitrage opportunities generally do not persist in financial markets, one might be interested in the alternative and weaker profitability notion where the nonnegativity of the final portfolio is not guaranteed but is highly likely. Consider, for example, the following relaxation of (20.6):

$$\Pr[r^T x \geq 0] \geq 0.99. \tag{20.7}$$

This approach can be formalized using a similar construction to what we have seen in Section 20.1. Let μ and Σ represent the expected future price vector and covariance matrix of the random vector r. Then, $E(r^T x) = \mu^T x$ and $STD(r^T x) = \sqrt{x^T \Sigma x}$.

Exercise 20.3 If r is a Gaussian random vector with mean μ and covariance matrix Σ, then show that

$$\Pr[r^T x \geq 0] \geq 0.99 \iff \mu^T x - \theta \sqrt{x^T \Sigma x} \geq 0,$$

where $\theta = \Phi^{-1}(0.99)$ and $\Phi^{-1}(\cdot)$ is the inverse map of standard normal cumulative distribution function.

As Exercise 20.3 indicates, the inequality (20.6) can be relaxed as

$$\mu^T x - \theta \sqrt{x^T \Sigma x} \geq 0,$$

where θ determines the likelihood of the inequality being satisfied. Therefore, if we find an x satisfying

$$\mu^\mathsf{T}x - \theta\sqrt{x^\mathsf{T}\Sigma x} \geq 0, \quad p^\mathsf{T}x < 0$$

for a large enough positive value of θ we have an approximation of an arbitrage opportunity called a *robust profit opportunity* in [64]. Note that, by relaxing the constraint $p^\mathsf{T}x < 0$ as $p^\mathsf{T}x \leq 0$ or using $p^\mathsf{T}x \leq -\varepsilon$ for some $\varepsilon > 0$, we obtain a conic feasibility system. Therefore, the resulting system can be solved using the conic optimization approaches. These ideas are explored in detail in [63, 64].

Exercise 20.4 Consider the robust profit opportunity formulation for a given θ:

$$\mu^\mathsf{T}x - \theta\sqrt{x^\mathsf{T}\Sigma x} \geq 0, \quad p^\mathsf{T}x \leq 0. \tag{20.8}$$

In this exercise, we investigate the problem of finding the largest θ for which (20.8) has a solution other than the zero vector. Namely, we want to solve

$$\begin{aligned}
\max_{\theta,x} \quad & \theta \\
\text{s.t.} \quad & \mu^\mathsf{T}x - \theta\sqrt{x^\mathsf{T}\Sigma x} \geq 0, \\
& p^\mathsf{T}x \leq 0.
\end{aligned} \tag{20.9}$$

This problem is no longer a convex optimization problem (why?). However, we can rewrite the first constraint as

$$\frac{\mu^\mathsf{T}x}{\sqrt{x^\mathsf{T}\Sigma x}} \geq \theta.$$

Using the strategy we employed in Section 8.2, we can take advantage of the homogeneity of the constraints in x and impose the normalizing constraint $x^\mathsf{T}\Sigma x = 1$ to obtain the following equivalent problem:

$$\begin{aligned}
\max_{\theta,x} \quad & \theta \\
\text{s.t.} \quad & \mu^\mathsf{T}x - \theta \geq 0, \\
& p^\mathsf{T}x \leq 0, \\
& x^\mathsf{T}\Sigma x = 1.
\end{aligned} \tag{20.10}$$

While we got rid of the fractional terms, we now have a nonlinear equality constraint that creates nonconvexity for the optimization. We can now relax the constraint $x^\mathsf{T}\Sigma x = 1$ as $x^\mathsf{T}\Sigma x \leq 1$ and obtain a convex optimization problem.

$$\begin{aligned}
\max_{\theta,x} \quad & \theta \\
\text{s.t.} \quad & \mu^\mathsf{T}x - \theta \geq 0, \\
& p^\mathsf{T}x \leq 0, \\
& x^\mathsf{T}\Sigma x \leq 1.
\end{aligned} \tag{20.11}$$

This relaxation can be expressed in conic form and solved using the methods discussed in Chapter 9. However, (20.11) is not equivalent to (20.9) and its solution need not be a solution to that problem in general. Find sufficient conditions under which the optimal solution of (20.11) satisfies $x^T \Sigma x \le 1$ with equality and therefore the relaxation is equivalent to the original problem.

Exercise 20.5 Note that the fraction $\mu^T x / \sqrt{x^T \Sigma x}$ in the θ-maximization exercise above resembles the Sharpe ratio. Assume that one of the assets in consideration is a riskless asset with a return of r_f. Show that the θ-maximization problem is equivalent to maximizing the Sharpe ratio in this case.

20.3 Robust portfolio selection

This section is adapted from an article by Tütüncü and Koenig [80]. Recall that Markowitz' mean-variance optimization problem can be stated in the following form that combines the reward and risk in the objective function:

$$\max_{x \in \mathcal{X}} \mu^T x - \lambda x^T \Sigma x, \tag{20.12}$$

where μ_i is an estimate of the expected return of security i, σ_{ii} is the variance of this return, σ_{ij} is the covariance between the returns of securities i and j, and λ is a risk-aversion constant used to trade-off the reward (expected return) and risk (portfolio variance). The set $\mathcal{X}$ is the set of feasible portfolios that may carry information on short-sale restrictions, sector distribution requirements, etc. Since such restrictions are typically predetermined, we can assume that the set $\mathcal{X}$ is known without any uncertainty at the time the problem is solved.

Recall that solving the problem above for different values of λ we can obtain the *efficient frontier* of the set of feasible portfolios. The optimal portfolio will be different for individuals with different risk-taking tendencies, but it will always be on the efficient frontier.

One of the limitations of this model is its need to estimate accurately the expected returns and covariances. In [5], Bawa *et al.* argue that using estimates of the unknown expected returns and covariances leads to an *estimation risk* in portfolio choice, and that methods for optimal selection of portfolios must take this risk into account. Furthermore, the optimal solution is sensitive to perturbations in these input parameters – a small change in the estimate of the return or the variance may lead to a large change in the corresponding solution, see, for example, [56, 57]. This property of the solutions is undesirable for many reasons. Most importantly, results can be unintuitive and the performance often suffers as the inaccuracies in the inputs lead to severely inefficient portfolios. If the modeler wants periodically to rebalance the portfolio based on new data, he/she may incur significant transaction costs, as small changes in inputs may dictate large changes in positions. Furthermore, using point estimates of the expected return and covariance parameters do not respond to the needs of a conservative investor who does not necessarily trust these estimates and would be more comfortable choosing a portfolio that will perform well under a number of different scenarios. Of course, such an investor cannot expect to get better performance on some of the more likely scenarios, but may prefer to accept it in exchange for insurance against more extreme cases. All these arguments point to the need of a portfolio optimization formulation that incorporates robustness and tries to find a solution that is relatively insensitive to inaccuracies in the input data. Since all the uncertainty is in the objective function coefficients, we seek an objective robust portfolio, as outlined in the previous chapter.

For *robust portfolio optimization* we consider a model that allows return and covariance matrix information to be given in the form of intervals. For example, this information may take the form "the expected return on security j is between 8% and 10%" rather than claiming that it is, say, 9%. Mathematically, we will represent this information as membership in the following set:

$$\mathcal{U} = \{(\mu, \Sigma) : \mu^L \le \mu \le \mu^U, \ \Sigma^L \le \Sigma \le \Sigma^U, \ \Sigma \succeq 0\}, \qquad (20.13)$$

where $\mu^L, \mu^U, \Sigma^L, \Sigma^U$ are the extreme values of the intervals we just mentioned. Recall that the notation $\Sigma \succeq 0$ indicates that the matrix Σ is a symmetric and positive semidefinite matrix. This restriction is necessary for Σ to be a valid covariance matrix.

The uncertainty intervals in (20.13) may be generated in different ways. An extremely cautious modeler may want to use historical lows and highs of certain input parameters as the range of their values. In a linear factor model of returns, one may generate different scenarios for factor return distributions and combine these scenarios to generate the uncertainty set. Different analysts may produce different estimates for these parameters and one may choose the extreme estimates as the endpoints of the intervals. One may choose a confidence level and then generate estimates of covariance and return parameters in the form of prediction intervals.

Using the objective robustness model in (19.6), we want to find a portfolio that maximizes the objective function in (20.12) in the worst-case realization of the input parameters μ and Σ from their uncertainty set $\mathcal{U}$ in (20.13). Given these considerations, the robust optimization problem takes the following form:

$$\max_{x \in \mathcal{X}} \{ \min_{(\mu, \Sigma) \in \mathcal{U}} \mu^\mathrm{T} x - \lambda x^\mathrm{T} \Sigma x \}. \qquad (20.14)$$

Since $\mathcal{U}$ is bounded, using classical results of convex analysis [67], it is easy to show that (20.14) is equivalent to its dual where the order of the min and the max is reversed:

$$\min_{(\mu,\Sigma)\in\mathcal{U}} \{\max_{x\in\mathcal{X}} -\mu^\mathrm{T}x + \lambda x^\mathrm{T}\Sigma x\}.$$

Furthermore, the solution to (20.14) is a saddle-point of the function $f(x, \mu, \Sigma) = \mu^\mathrm{T}x - \lambda x^\mathrm{T}\Sigma x$ and can be determined using the technique outlined in [38].

Exercise 20.6 Consider a special case of problem (20.14) where we make the following assumptions:

- $x \geq 0$, $\forall x \in \mathcal{X}$ (i.e., $\mathcal{X}$ includes no-shorting constraints);
- Σ^U is positive semidefinite.

Under these assumptions, show that (20.14) reduces to the following single-level maximization problem:

$$\max_{x\in\mathcal{X}}(\mu^L)^\mathrm{T}x - \lambda x^\mathrm{T}\Sigma^U x. \tag{20.15}$$

Observe that this new problem is a simple concave quadratic maximization problem and can be solved easily using, for example, interior-point methods. (Hint: Note that the objective function of (20.14) is separable in μ and Σ and that $x^\mathrm{T}\Sigma x = \sum_{i,j} \sigma_{ij} x_{ij}$ with $x_{ij} = x_i x_j \geq 0$ when $x \geq 0$.)

20.4 Relative robustness in portfolio selection

We consider the following simple portfolio optimization example derived from an example in [20].

Example 20.1

$$\begin{aligned}
\max\ &\mu_1 x_1 + \mu_2 x_2 + \mu_3 x_3 \\
&\mathrm{TE}(x_1, x_2, x_3) \leq 0.10 \\
&x_1 + x_2 + x_3 = 1 \\
&x_1 \geq 0, x_2 \geq 0, x_3 \geq 0,
\end{aligned} \tag{20.16}$$

where

$$TE(x_1, x_2, x_3) = \sqrt{\begin{bmatrix} x_1 - 0.5 \\ x_2 - 0.5 \\ x_3 \end{bmatrix}^\mathrm{T} \begin{bmatrix} 0.1764 & 0.09702 & 0 \\ 0.09702 & 0.1089 & 0 \\ 0 & 0 & 0 \end{bmatrix} \begin{bmatrix} x_1 - 0.5 \\ x_2 - 0.5 \\ x_3 \end{bmatrix}}.$$

This is essentially a two-asset portfolio optimization problem where the third asset (x_3) represents proportion of the funds that are not invested. The first two assets have standard deviations of 42% and 33% respectively and a correlation coefficient

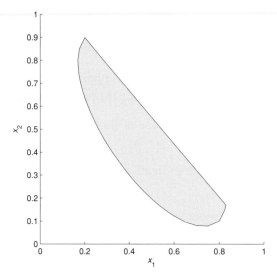

Figure 20.1 The feasible set of the MVO problem in (20.16)

of 0.7. The "benchmark" is the portfolio that invests funds half-and-half in the two assets. The function TE(x) *represents the tracking error of the portfolio with respect to the half-and-half benchmark and the first constraint indicates that this tracking error should not exceed 10%. The second constraint is the budget constraint, the third enforces no shorting. We depict the projection of the feasible set of this problem onto the space spanned by variables x_1 and x_2 in Figure 20.1.*

We now build a relative robustness model for this portfolio problem. Assume that the covariance matrix estimate is certain and consider a simple uncertainty set for expected return estimates consisting of three scenarios represented with arrows in Figure 20.2. These three scenarios correspond to the following values for (μ_1, μ_2, μ_3): $(6, 4, 0)$, $(5, 5, 0)$, and $(4, 6, 0)$. The optimal solution when $(\mu_1, \mu_2, \mu_3) = (6, 4, 0)$ is $(0.831, 0.169, 0)$ with an objective value of 5.662. Similarly, when $(\mu_1, \mu_2, \mu_3) = (4, 6, 0)$ the optimal solution is $(0.169, 0.831, 0)$ with an objective value of 5.662. When $(\mu_1, \mu_2, \mu_3) = (5, 5, 0)$ all points between the previous two optimal solutions are optimal with a shared objective value of 5.0. Therefore, the relative robust formulation for this problem can be written as follows:

$$\min_{x,t} \quad t$$
$$5.662 - (6x_1 + 4x_2) \leq t$$
$$5.662 - (4x_1 + 6x_2) \leq t$$
$$5.0 - (5x_1 + 5x_2) \leq t \qquad\qquad (20.17)$$
$$\text{TE}(x_1, x_2, x_3) \leq 0.10$$
$$x_1 + x_2 + x_3 = 1$$
$$x_1 \geq 0, x_2 \geq 0, x_3 \geq 0.$$

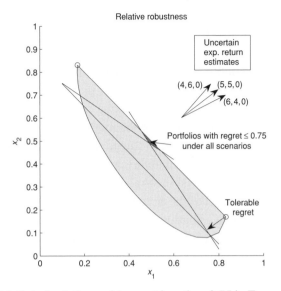

Figure 20.2 Set of solutions with regret less than 0.75 in Example 20.1

Instead of solving the problem where the optimal regret level is a variable (t in the formulation), an easier strategy is to choose a level of regret that can be tolerated and find portfolios that do not exceed this level of regret in any scenario. For example, choosing a maximum tolerable regret level of 0.75 we get the following feasibility problem:

$$
\begin{aligned}
\text{Find } & x \\
\text{s.t.} \quad & 5.662 - (6x_1 + 4x_2) \leq 0.75 \\
& 5.662 - (4x_1 + 6x_2) \leq 0.75 \\
& 5.0 - (5x_1 + 5x_2) \leq 0.75 \\
& \text{TE}(x_1, x_2, x_3) \leq 0.10 \\
& x_1 + x_2 + x_3 = 1 \\
& x_1 \geq 0, x_2 \geq 0, x_3 \geq 0.
\end{aligned}
\tag{20.18}
$$

This problem and its feasible set of solutions is illustrated in Figure 20.2. The small shaded triangle represents the portfolios that have a regret level of 0.75 or less under all three scenarios.

Exercise 20.7 Interpret the objective function of (20.17) geometrically in Figure 20.2. Verify that the vector $x^* = (0.5, 0.5, 0)$ solves (20.17) with the maximum regret level of $t^* = 0.662$.

20.5 Moment bounds for option prices

To price derivative securities, a common strategy is to first assume a stochastic process for the future values of the underlying process and then derive a differential equation satisfied by the price function of the derivative security that can be solved analytically or numerically. For example, this is the strategy used in the derivation of the Black–Scholes–Merton (BSM) formula for European options.

The prices obtained in this manner are sensitive to the model assumptions made to determine them. For example, the removal of the constant volatility assumption used in the BSM derivation deems the resulting pricing formulas incorrect. Since there is uncertainty in the correctness of the models or model parameters used for pricing derivatives, robust optimization can be used as an alternative approach.

One variation considered in the literature assumes that we have reliable estimates of the first few moments of the risk-neutral density of the underlying asset price but have uncertainty with respect to the actual shape of this density. Then, one asks the following question: what distribution for the risk neutral density with pre-specified moments produces the highest/lowest price estimate for the derivative security? This is the approach considered in [11] where the authors argue that the convex optimization models provide a natural framework for addressing the relationship between option and stock prices in the absence of distributional information for the underlying price dynamics.

Another strategy, often called *arbitrage pricing*, or *robust pricing*, makes no model assumptions at all and tries to produce lower and upper price bounds by examining the known prices of related securities such as other options on the same underlying, etc. This is the strategy we employed for pricing forward start options in Section 10.4. Other examples of this strategy include the work of Laurence and Wang [51].

Each one of these considerations leads to optimization problems. Some of these problems are easy. For example, one can find an arbitrage bound for a (possibly exotic) derivative security from a static super- or sub-replicating portfolio by solving a linear optimization problem. Other robust pricing and hedging problems can appear quite intractable. Fortunately, modern optimization models and methods continue to provide efficient solution techniques for an expanding array of financial optimization problems including pricing and hedging problems.

20.6 Additional exercises

Exercise 20.8 Recall that we considered the following *two-stage stochastic linear program with recourse* in Section 16.2:

$$
\begin{array}{ll}
\max & (c^1)^{\mathrm{T}} x^1 + E[\max c^2(\omega)^{\mathrm{T}} x^2(\omega)] \\
& A^1 x^1 = b^1 \\
& B^2(\omega) x^1 + A^2(\omega) x^2(\omega) = b^2(\omega) \\
& x^1 \geq 0, \qquad\qquad x^2(\omega) \geq 0.
\end{array}
\tag{20.19}
$$

In this problem, it was assumed the uncertainty in ω was of "random" nature, and therefore, the stochastic programming approach was appropriate. Now consider the case where ω is not a random variable but is known to belong to an uncertainty set $\mathcal{U}$. Formulate a *two-stage robust linear program with recourse* using the ideas developed in Section 20.1. Next, assume that B^2 and A^2 are certain (they do not depend on ω), but b^2 and c^2 are uncertain and depend affinely on ω: $b^2(\omega) = b^2 + P\omega$ and $c^2(\omega) = c^2 + R\omega$, where b^2, c^2, P, R are (certain) vectors/matrices of appropriate dimension. Also, assume that $\mathcal{U} = \{\omega : \sum_i d_i w_i^2 \leq 1\}$ for some positive constants d_i. Can you simplify the two-stage robust linear program with recourse under these assumptions?

Exercise 20.9 For a given constant λ, expected return vector μ, and a positive definite covariance matrix Σ consider the following MVO problem:

$$\max_{x \in \mathcal{X}} \mu^{\mathrm{T}} x - \lambda x^{\mathrm{T}} \Sigma x, \qquad (20.20)$$

where $\mathcal{X} = \{x : e^{\mathrm{T}} x = 1\}$ with $e = [1\ 1\ \ldots\ 1]^{\mathrm{T}}$. Let $z(\mu, \Sigma)$ represent the optimal value of this problem. Determine $z(\mu, \Sigma)$ as an explicit function of μ and Σ. Next, assume that μ and Σ are uncertain and belong to the uncertainty set $\mathcal{U} := \{(\mu_i, \Sigma_i) : i = 1, \ldots, m\}$, i.e., we have a finite number of scenarios for μ and Σ. Assume also that $z(\mu_i, \Sigma_i) > 0 \ \forall i$. Now formulate the following robust optimization problem: find a feasible portfolio vector x such that the objective value with this portfolio under each scenario is within 10% of the optimal objective value corresponding to that scenario. Discuss how this problem can be solved. What would be a good objective function for this problem?

Appendix A

Convexity

Convexity is an important concept in mathematics, and especially in optimization, that is used to describe certain sets and certain functions. Convex sets and convex functions are related but separate mathematical entities.

Let x and y be given points in some vector space. Then, for any $\lambda \in [0, 1]$, the point $\lambda x + (1 - \lambda)y$ is called a *convex combination* of x and y. The set of all convex combinations of x and y is the line segment joining these two points.

A subset S of a given vector space X is called a *convex set* if $x \in S$, $y \in S$, and $\lambda \in [0, 1]$ always imply that $\lambda x + (1 - \lambda)y \in S$. In other words, a convex set is characterized by the following property: for any two points in the set, the line segment connecting these two points lies entirely in the set.

Polyhedral sets (or *polyhedra*) are sets defined by linear equalities and inequalities. So, for example, the feasible region of a linear optimization problem is a polyhedral set. It is a straightforward exercise to show that polyhedral sets are convex.

Given a convex set S, a function $f : S \to I\!R$ is called a *convex function* if $\forall x \in S, y \in S$ and $\lambda \in [0, 1]$ the following inequality holds:

$$f(\lambda x + (1 - \lambda)y) \leq \lambda f(x) + (1 - \lambda)f(y).$$

We say that f is a *strictly convex function* if $x \in S, y \in S$ and $\lambda \in (0, 1)$ implies the following strict inequality:

$$f(\lambda x + (1 - \lambda)y) < \lambda f(x) + (1 - \lambda)f(y).$$

A function f is *concave* if $-f$ is convex. Equivalently, f is concave for all if $x \in S, y \in S$ and $\lambda \in [0, 1]$ the following inequality holds:

$$f(\lambda x + (1 - \lambda)y) \geq \lambda f(x) + (1 - \lambda)f(y).$$

A function f is *strictly concave* if $-f$ is strictly convex.

Given $f : S \to I\!R$ with $S \subset X$, epi(f) – the epigraph of f – is the following subset of $X \times I\!R$:

$$\text{epi}(f) := \{(x, r) : x \in S, \; f(x) \leq r\}.$$

f is a convex *function* if and only if epi(f) is a convex *set*.

For a twice-continuously differentiable function $f : S \to I\!R$ with $S \subset I\!R$, we have a simple characterization of convexity: f is convex on S if and only if $f''(x) \geq 0$, $\forall x \in S$. For multivariate functions, we have the following generalization: if $f : S \to I\!R$ with $S \subset I\!R^n$ is twice-continuously differentiable, then f is convex on S if and only if $\nabla^2 f(x)$ is positive semidefinite for all $x \in S$. Here, $\nabla^2 f(x)$ denotes the (symmetric) Hessian matrix of f; namely,

$$[\nabla^2 f(x)]_{ij} = \frac{\partial^2 f(x)}{\partial x_i \partial x_j}, \; \forall i, j.$$

Recall that a symmetric matrix $H \in I\!R^{n \times n}$ is positive semidefinite (positive definite) if $y^T H y \geq 0$, $\forall y \in I\!R^n$ ($y^T H y > 0$, $\forall \, y \in I\!R^n$, $y \neq 0$).

The following theorem is one of the many reasons for the importance of convex functions and convex sets for optimization:

Theorem A.1   *Consider the following optimization problem:*

$$\begin{aligned} \min_x \quad & f(x) \\ s.t. \quad & x \in S. \end{aligned} \tag{A.1}$$

If S is a convex set and if f is a convex function of x on S, then all local optimal solutions of (A.1) are also global optimal solutions.

Appendix B

Cones

A *cone* is a set that is closed under positive scalar multiplication. In other words, a set C is a cone if $\lambda x \in C$ for all $\lambda \geq 0$ and $x \in C$. A cone is called *pointed* if it does not include any lines. We will generally be dealing with closed, convex, and pointed cones. Here are a few important examples:

- $C_l := \{x \in I\!R^n : x \geq 0\}$, the nonnegative orthant. In general, any set of the form $C := \{x \in I\!R^n : Ax \geq 0\}$ for some matrix $A \in I\!R^{m \times n}$ is called a *polyhedral cone*. The subscript l is used to indicate that this cone is defined by linear inequalities.
- $C_q := \{x = (x_0, x_1, \ldots, x_n) \in I\!R^{n+1} : x_0 \geq \|(x_1, \ldots, x_n)\|\}$, the second-order cone. This cone is also called the quadratic cone (hence the subscript q), Lorentz cone, and the ice-cream cone.
- $C_s := \left\{ X = \begin{bmatrix} x_{11} & \cdots & x_{1n} \\ \vdots & \ddots & \vdots \\ x_{n1} & \cdots & x_{nn} \end{bmatrix} \in I\!R^{n \times n} : X = X^{\mathrm{T}}, X \text{ is positive semidefinite} \right\}$,

  the cone of symmetric positive semidefinite matrices.

If C is a cone in a vector space X with an inner product denoted by $\langle \cdot, \cdot \rangle$, then its *dual cone* is defined as follows:

$$C^* := \{x \in X : \langle x, y \rangle \geq 0, \forall y \in C\}.$$

It is easy to see that the nonnegative orthant in $I\!R^n$ (with the usual inner product) is equal to its dual cone. The same holds for the second-order cone and the cone of symmetric positive semidefinite matrices, but not for general cones.

The *polar cone* is the negative of the dual cone, i.e.,

$$C^P := \{x \in X : \langle x, y \rangle \leq 0, \forall y \in C\}.$$

Appendix C

A probability primer

One of the most basic concepts in probability theory is a *random experiment*, which is an experiment whose outcome can not be determined in advance. In most cases, however, one has a (possibly infinite) set of all possible outcomes of the event; we call this set the *sample space* of the random experiment. For example, flipping a coin is a random experiment, so is the score of the next soccer game between Japan and Korea. The set $\Omega = \{\texttt{heads}, \texttt{tails}\}$ is the sample space of the first experiment, $\Omega = I\!N \times I\!N$ with $I\!N = \{0, 1, 2, \ldots\}$ is the sample space for the second experiment.

Another important concept is an *event*: a subset of the sample space. It is customary to say that an event *occurs* if the outcome of the random experiment is in the corresponding subset. So, "Japan beats Korea" is an event for the second random experiment of the previous paragraph. A class $\mathcal{F}$ of subsets of a sample space Ω is called a *field* if it satisfies the following conditions:

(i) $\Omega \in \mathcal{F}$;
(ii) $A \in \mathcal{F}$ implies that $A^c \in \mathcal{F}$, where A^c is the complement of A;
(iii) $A, B \in \mathcal{F}$ implies $A \cup B \in \mathcal{F}$.

The second and third conditions are known as *closure under complements and (finite) unions*. If, in addition, $\mathcal{F}$ satisfies

(iv) $A_1, A_2, \ldots \in \mathcal{F}$ implies $\cup_{i=1}^{\infty} A_i \in \mathcal{F}$,

then $\mathcal{F}$ is called a σ-field. The condition (iv) is *closure under countable unions*. Note that, for subtle reasons, Condition (iii) does not necessarily imply Condition (iv).

A *probability measure* or *distribution* Pr is a real-valued function defined on a field $\mathcal{F}$ (whose elements are subsets of the sample space Ω), and satisfies the following conditions:

(i) $0 \leq \Pr(A) \leq 1$, for $\forall A \in \mathcal{F}$;
(ii) $\Pr(\emptyset) = 0$, and $\Pr(\Omega) = 1$;

(iii) If $A_1, A_2, \ldots$ is a sequence of disjoint sets in $\mathcal{F}$ and if $\cup_{i=1}^{\infty} A_i \in \mathcal{F}$, then

$$\Pr\left(\cup_{i=1}^{\infty} A_i\right) = \sum_{i=1}^{\infty} \Pr(A_i).$$

The last condition above is called *countable additivity*.

A probability measure is said to be *discrete* if Ω has countably many (and possibly finite) number of elements. A *density function* f is a nonnegative valued integrable function that satisfies

$$\int_{\Omega} f(x) dx = 1.$$

A continuous probability distribution is a probability defined by the following relation:

$$\Pr[X \in A] = \int_A f(x) dx,$$

for a density function f.

The collection Ω, $\mathcal{F}$ (a σ-field in Ω), and Pr (a probability measure on $\mathcal{F}$) is called a *probability space*.

Now we are ready to define a *random variable*. A random variable X is a real-valued function defined on the set Ω.[1] Continuing with the soccer example, the difference between the goals scored by the two teams is a random variable, and so is the "winner", a function which is equal to, say, 1 if the number of goals scored by Japan is higher, 2 if the number of goals scored by Korea is higher, and 0 if they are equal. A random variable is said to be discrete (respectively, continuous) if the underlying probability space is discrete (respectively, continuous).

The *probability distribution* of a random variable X is, by definition, the probability measure $\Pr_X$ in the probability space $(\Omega, \mathcal{F}, \Pr)$:

$$\Pr_X(B) = \Pr[X \in B].$$

The *distribution function* F of the random variable X is defined as:

$$F(x) = \Pr[X \le x] = \Pr[X \in (-\infty, x]].$$

For a continuous random variable X with the density function f,

$$F(x) = \int_{-\infty}^{x} f(x) dx$$

and therefore $f(x) = dF(x)/dx$.

---

[1] Technically speaking, for X to be a random variable, it has to satisfy the condition that for each $B \in \mathcal{B}$, the Euclidean Borel field on $\mathbb{R}$, the set $\{\omega : X(\omega) \in B\} =: X^{-1}(B) \in \mathcal{F}$. This is a purely technical requirement which is met for discrete probability spaces (Ω is finite or countably infinite) and by any function that we will be interested in.

A *random vector* $X = (X_1, X_2, \ldots, X_k)$ is a k-tuple of random variables, or equivalently, a function from Ω to IR^k that satisfies a technical condition similar to the one mentioned in the footnote. The *joint distribution function F* of random variables $X_1, \ldots, X_k$ is defined by

$$F(x_1, \ldots, x_k) = \Pr_X[X_1 \le x_1, \ldots, X_k \le x_k].$$

In the special case of $k = 2$ we have

$$F(x_1, x_2) = \Pr_X[X_1 \le x_1, X_2 \le x_2].$$

Given the joint distribution function of random variables X_1 and X_2, their *marginal distribution functions* are given by the following formulas:

$$F_{X_1}(x_1) = \lim_{x_2 \to \infty} F(x_1, x_2)$$

and

$$F_{X_2}(x_2) = \lim_{x_1 \to \infty} F(x_1, x_2).$$

We say that random variables X_1 and X_2 are *independent* if

$$F(x_1, x_2) = F_{X_1}(x_1) F_{X_2}(x_2)$$

for every x_1 and x_2.

The *expected value (expectation, mean)* of the random variable X is defined by

$$E[X] = \int_\Omega x \, dF(x)$$
$$= \begin{cases} \sum_{x \in \Omega} x \Pr[X = x] & \text{if } X \text{ is discrete} \\ \int_\Omega x f(x) dx & \text{if } X \text{ is continuous} \end{cases}$$

(provided that the integrals exist) and is denoted by $E[X]$. For a function $g(X)$ of a random variable, the expected value of $g(X)$ (which is itself a random variable) is given by

$$E[g(X)] = \int_\Omega x \, dF_g(x) = \int_\Omega g(x) dF(x).$$

The *variance* of a random variable X is defined by

$$\text{Var}[X] = E[(X - E[X])^2]$$
$$= E[X^2] - (E[X])^2.$$

The *standard deviation* of a random variable is the square root of its variance.

For two jointly distributed random variables X_1 and X_2, their *covariance* is defined to be

$$\text{Cov}(X_1, X_2) = \text{E}\left[(X_1 - \text{E}[X_1])(X_2 - \text{E}[X_2])\right]$$
$$= \text{E}[X_1 X_2] - \text{E}[X_1]\text{E}[X_2].$$

The *correlation coefficient* of two random variables is the ratio of their covariance to the product of their standard deviations.

For a collection of random variables $X_1, \ldots, X_n$, the expected value of the sum of these random variables is equal to the sum of their expected values:

$$\text{E}\left[\sum_{i=1}^{n} X_i\right] = \sum_{i=1}^{n} \text{E}[X_i].$$

The formula for the variance of the sum of the random variables $X_1, \ldots, X_n$ is a bit more complicated:

$$\text{Var}\left[\sum_{i=1}^{n} X_i\right] = \sum_{i=1}^{n} \text{Var}[X_i] + 2 \sum_{1 \le i < j \le n} \text{Cov}(X_i, X_j).$$

Appendix D

The revised simplex method

As we discussed in Chapter 2, in each iteration of the simplex method, we first choose an entering variable looking at the objective row of the current tableau, and then identify a leaving variable by comparing the ratios of the numbers on the right-hand side and the column for the entering variable. Once these two variables are identified we update the tableau. Clearly, the most time-consuming job among these steps of the method is the tableau update. If we can save some time on this bottleneck step then we can make the simplex method much faster. The *revised simplex method* is a variant of the simplex method developed with precisely that intention.

The crucial question here is whether it is necessary to update the *whole* tableau in *every* iteration. To answer this question, let us try to identify what parts of the tableau are absolutely necessary to run the simplex algorithm. As we mentioned before, the first task in each iteration is to find an entering variable. Let us recall how we do that. In a maximization problem, we look for a nonbasic variable with a *positive* rate of improvement. In terms of the tableau notation, this translates into having a *negative* coefficient in the objective row, where Z is the basic variable.

To facilitate the discussion below let us represent a simplex tableau in an algebraic form, using the notation from Section 2.4.1. As before, we consider a linear programming problem of the form:

$$\max \; c\,x$$
$$Ax \leq b$$
$$x \geq 0.$$

After adding the slack variables and choosing them as the initial set of basic variables we get the following "initial" or "original" tableau:

Current basic variables	Z	Coefficient of		RHS
		Original nonbasics	Original basics	
Z	1	$-c$	0	0
x_B	0	A	I	b

Note that we wrote the objective function equation $Z = cx$ as $Z - cx = 0$ to keep variables on the left-hand side and the constants on the right. In matrix form this can be written as:

$$\begin{bmatrix} 1 & -c & 0 \\ 0 & A & I \end{bmatrix} \begin{bmatrix} Z \\ x \\ x_s \end{bmatrix} = \begin{bmatrix} 0 \\ b \end{bmatrix}.$$

Pivoting, which refers to the algebraic operations performed by the simplex method in each iteration to get a representation of the problem in a particular form, can be expressed in matrix form as a premultiplication of the original matrix representation of the problem with an appropriate matrix. If the current basis matrix is B, the premultiplying matrix happens to be the following:

$$\begin{bmatrix} 1 & c_B B^{-1} \\ 0 & B^{-1} \end{bmatrix}.$$

Multiplying this matrix with the matrices in the matrix form of the equations above we get:

$$\begin{bmatrix} 1 & c_B B^{-1} \\ 0 & B^{-1} \end{bmatrix} \begin{bmatrix} 1 & -c & 0 \\ 0 & A & I \end{bmatrix} = \begin{bmatrix} 1 & c_B B^{-1} A - c & c_B B^{-1} \\ 0 & B^{-1} A & B^{-1} \end{bmatrix},$$

and

$$\begin{bmatrix} 1 & c_B B^{-1} \\ 0 & B^{-1} \end{bmatrix} \begin{bmatrix} 0 \\ b \end{bmatrix} = \begin{bmatrix} c_B B^{-1} b \\ B^{-1} b \end{bmatrix},$$

which gives us the matrix form of the *set of equations in each iteration represented with respect to the current set of basic variables*:

$$\begin{bmatrix} 1 & c_B B^{-1} A - c & c_B B^{-1} \\ 0 & B^{-1} A & B^{-1} \end{bmatrix} \begin{bmatrix} Z \\ x \\ x_s \end{bmatrix} = \begin{bmatrix} c_B B^{-1} b \\ B^{-1} b \end{bmatrix}.$$

This is observed in the following tableau:

Current basic variables	Z	Coefficient of			RHS
		Original nonbasics	Original basics		
Z	1	$c_B B^{-1} A - c$	$c_B B^{-1}$		$c_B B^{-1} b$
x_B	0	$B^{-1} A$	B^{-1}		$B^{-1} b$

Equipped with this algebraic representation of the simplex tableau, we continue our discussion of the revised simplex method. Recall that, for a maximization problem, an entering variable must have a negative objective row coefficient. Using the tableau above, we can look for entering variables by checking whether:

1. $c_B B^{-1} \geq 0$;
2. $c_B B^{-1} A - c \geq 0$.

Furthermore, we only need to compute the parts of these vectors corresponding to *nonbasic* variables, since the parts corresponding to basic variables will be zero. Now, if both inequalities above are satisfied, we stop concluding that we found an optimal solution. If not, we pick a nonbasic variable, say x_k, for which the updated objective row coefficient is negative, to enter the basis. So in this step we use the updated objective function row.

Next step is to find the leaving variable. For that, we use the updated column k for the variable x_k and the updated right-hand-side vector. If the column that corresponds to x_k in the original tableau is A_k, then the updated column is $\bar{A}_k = B^{-1} A_k$ and the updated RHS vector is $\bar{b} = B^{-1} b$.

Next, we make a crucial observation: for the steps above, we do not need to calculate the updated columns for the nonbasic variables that are not selected to enter the basis. Notice that, if there are a lot of nonbasic variables (which would happen if there were many more variables than constraints) this would translate into substantial savings in terms of computation time. However, we need to be able to compute $\bar{A}_k = B^{-1} A_k$, which requires the matrix B^{-1}. So, how do we find B^{-1} in each iteration? Taking the inverse from scratch in every iteration would be too expensive, instead we can keep track of B^{-1} in the tableau as we iterate the simplex method. We will also keep track of the updated RHS $\bar{b} = B^{-1} b$. Finally, we will keep track of the expression $\pi = c_B B^{-1}$. Looking at the tableau in the previous page, we see that the components of π are just the updated objective function coefficients of the initial basic variables. The components of the vectors π are often called the *shadow prices*, or *dual prices*.

Now we are ready to give an outline of the revised simplex method:

Step 0 *Find an initial feasible basis B and compute B^{-1}, $\bar{b} = B^{-1}b$, and $\pi = c_B B^{-1}$.*

Now assuming that we are given the current basis B and we know $B^{-1}, \bar{b} = B^{-1}b$, and $\pi = c_B B^{-1}$ let us try to describe the iterative steps of the revised simplex method:

Step 1 *For each nonbasic variable x_i calculate $\bar{c}_i = c_i - c_B B^{-1}A_i = c_i - \pi A_i$. If $\bar{c}_i \leq 0$ for all nonbasic variables x_i, then STOP, the current basis is optimal. Otherwise choose a variable x_k such that $\bar{c}_k > 0$.*

Step 2 *Compute the updated column $\bar{A}_k = B^{-1}A_k$ and perform the ratio test, i.e., find*

$$\min_{\bar{a}_{ik} > 0} \left\{ \frac{\bar{b}_i}{\bar{a}_{ik}} \right\}.$$

Here $\bar{a}_{ik}$ and $\bar{b}_i$ denote the ith entry of the vectors $\bar{A}_k$ and $\bar{b}$, respectively. If $\bar{a}_{ik} \leq 0$ for every row i, then STOP, the problem is unbounded. Otherwise, choose the basic variable of the row that gives the minimum ratio in the ratio test (say row r) as the leaving variable.

The pivoting step is where we achieve the computational savings:

Step 3 *Pivot on the entry $\bar{a}_{rk}$ in the following truncated tableau:*

Current basic variables	Coefficient of		RHS
	x_k	Original basics	
Z	$-\bar{c}_k$	$\pi = c_B B^{-1}$	$c_B B^{-1}b$
$\vdots$ x_{B_r} $\vdots$	$\vdots$ $\bar{a}_{rk}$ $\vdots$	B^{-1}	$B^{-1}b$

Replace the current values of B^{-1}, $\bar{b}$, and π with the matrices and vectors that appear in their respective positions after pivoting. Go back to Step 1.

Once again, notice that when we use the revised simplex method, we work with a truncated tableau. This tableau has $m + 2$ columns; m columns corresponding to the initial basic variables, one for the entering variable, and one for the right-hand side. In the standard simplex method, we work with $n + 1$ columns, n of them for *all* variables, and one for the RHS vector. For a problem that has many more variables (say, $n = 50\,000$) than constraints (say, $m = 10\,000$) the savings are very significant.

An example

Now we apply the revised simplex method described above to a linear programming problem. We will consider the following problem:

Maximize $Z = \quad x_1 + 2x_2 + x_3 - 2x_4$

subject to:

$$
\begin{array}{rcr}
-2x_1 + x_2 + x_3 + 2x_4 \quad\quad + x_6 \quad\quad\quad\quad\quad &=& 2 \\
-x_1 + 2x_2 + x_3 \quad\quad + x_5 \quad\quad + x_7 \quad\quad &=& 7 \\
x_1 \quad\quad + x_3 + x_4 + x_5 \quad\quad\quad\quad + x_8 &=& 3
\end{array}
$$

$$x_1 \geq 0, x_2 \geq 0, x_3 \geq 0, x_4 \geq 0, x_5 \geq 0, x_6 \geq 0, x_7 \geq 0, x_8 \geq 0.$$

The variables x_6, x_7, and x_8 form a feasible basis and we will start the algorithm with this basis. Then the initial simplex tableau is as follows:

Basic var.	x_1	x_2	x_3	x_4	x_5	x_6	x_7	x_8	RHS
Z	−1	−2	−1	2	0	0	0	0	0
x_6	−2	1	1	2	0	1	0	0	2
x_7	−1	2	1	0	1	0	1	0	7
x_8	1	0	1	1	1	0	0	1	3

Once a feasible basis B is determined, the first thing to do in the revised simplex method is to calculate the quantities B^{-1}, $\bar{b} = B^{-1}b$, and $\pi = c_B B^{-1}$. Since the basis matrix B for the basis above is the identity, we calculate these quantities easily:

$$B^{-1} = I,$$

$$\bar{b} = B^{-1}b = \begin{bmatrix} 2 \\ 7 \\ 3 \end{bmatrix},$$

$$\pi = c_B B^{-1} = [0\ 0\ 0]\,I = [0\ 0\ 0].$$

Above, I denotes the identity matrix of size 3. Note that, c_B, i.e., the sub-vector of the objective function vector $c = [1\ 2\ 1\ -2\ 0\ 0\ 0\ 0]^{\mathrm{T}}$ that corresponds to the current basic variables, consists of all zeroes.

Now we calculate $\bar{c}_i$ values for nonbasic variables using the formula $\bar{c}_i = c_i - \pi A_i$, where A_i refers to the ith column of the initial tableau. So,

$$\bar{c}_1 = c_1 - \pi A_1 = 1 - [0 \ 0 \ 0] \begin{bmatrix} -2 \\ -1 \\ 1 \end{bmatrix} = 1,$$

$$\bar{c}_2 = c_2 - \pi A_2 = 2 - [0 \ 0 \ 0] \begin{bmatrix} 1 \\ 2 \\ 0 \end{bmatrix} = 2,$$

and similarly,

$$\bar{c}_3 = 1, \ \bar{c}_4 = -1, \ \bar{c}_5 = 0.$$

The quantity $\bar{c}_i$ is often called the *reduced cost* of the variable x_i and it tells us the rate of improvement in the objective function when x_i is introduced into the basis. Since $\bar{c}_2$ is the largest of all $\bar{c}_i$ values we choose x_2 as the entering variable.

To determine the leaving variable, we need to compute the updated column $\bar{A}_2 = B^{-1}A_2$:

$$\bar{A}_2 = B^{-1}A_2 = I \begin{bmatrix} 1 \\ 2 \\ 0 \end{bmatrix} = \begin{bmatrix} 1 \\ 2 \\ 0 \end{bmatrix}.$$

Now using the updated right-hand-side vector $\bar{b} = [2 \ 7 \ 3]^T$ we perform the ratio test and find that x_6, the basic variable in the row that gives the minimum ratio has to leave the basis. (Remember that we only use the *positive* entries of $\bar{A}_2$ in the ratio test, so the last entry, which is a zero, does not participate in the ratio test.)

Up to here, what we have done was exactly the same as in regular simplex, only the language was different. The next step, the pivoting step, is going to be significantly different. Instead of updating the whole tableau, we will only update a reduced tableau which has one column for the entering variable, three columns for the initial basic variables, and one more column for the RHS. So, we will use the following tableau for pivoting:

Basic var.	x_2	Init. basics			RHS
		x_6	x_7	x_8	
Z	-2	0	0	0	0
x_6	1^*	1	0	0	2
x_7	2	0	1	0	7
x_8	0	0	0	1	3

As usual we pivot in the column of the entering variable and try to get a 1 in the position of the pivot element, and zeros elsewhere in the column. After pivoting we get:

Basic		Init. basics			
var.	x_2	x_6	x_7	x_8	RHS
Z	0	2	0	0	4
x_2	1	1	0	0	2
x_7	0	−2	1	0	3
x_8	0	0	0	1	3

Now we can read the basis inverse B^{-1}, updated RHS vector $\bar{b}$, and the shadow prices π for the new basis from this new tableau. Recalling the algebraic form of the simplex tableau we discussed above, we see that the new basis inverse lies in the columns corresponding to the initial basic variables, so

$$B^{-1} = \begin{bmatrix} 1 & 0 & 0 \\ -2 & 1 & 0 \\ 0 & 0 & 1 \end{bmatrix}.$$

Updated values of the objective function coefficients of initial basic variables and the updated RHS vector give us the π and $\bar{b}$ vectors we will use in the next iteration:

$$\bar{b} = \begin{bmatrix} 2 \\ 3 \\ 3 \end{bmatrix}, \quad \pi = [2\,0\,0].$$

Above, we only updated five columns and did not worry about the four columns that correspond to x_1, x_3, x_4, and x_5. These are the variables that are neither in the initial basis, nor are selected to enter the basis in this iteration.

Now, we repeat the steps above. To determine the new entering variable, we need to calculate the reduced costs $\bar{c}_i$ for nonbasic variables:

$$\bar{c}_1 = c_1 - \pi A_1 = 1 - [2\,0\,0] \begin{bmatrix} -2 \\ -1 \\ 1 \end{bmatrix} = 5,$$

$$\bar{c}_3 = c_3 - \pi A_3 = 1 - [2\,0\,0] \begin{bmatrix} 1 \\ 1 \\ 1 \end{bmatrix} = -1,$$

and similarly,

$$\bar{c}_4 = -6, \ \bar{c}_5 = 0, \ \text{and} \ \bar{c}_6 = -2.$$

When we look at the $-\bar{c}_i$ values we find that only x_1 is eligible to enter. So, we generate the updated column $\bar{A}_1 = B^{-1}A_1$:

$$\bar{A}_1 = B^{-1}A_1 = \begin{bmatrix} 1 & 0 & 0 \\ -2 & 1 & 0 \\ 0 & 0 & 1 \end{bmatrix} \begin{bmatrix} -2 \\ -1 \\ 0 \end{bmatrix} = \begin{bmatrix} -2 \\ 3 \\ 1 \end{bmatrix}.$$

The ratio test indicates that x_7 is the leaving variable:

$$\min \left\{ \frac{3}{3}, \frac{3}{1} \right\} = 1.$$

Next, we pivot on the following tableau:

Basic var.	x_1	Init. basics x_6	Init. basics x_7	Init. basics x_8	RHS
Z	-5	2	0	0	4
x_2	-2	1	0	0	2
x_7	3^*	-2	1	0	3
x_8	1	0	0	1	3

And we obtain:

Basic var.	x_1	Init. basics x_6	Init. basics x_7	Init. basics x_8	RHS
Z	0	$-\frac{4}{3}$	$\frac{5}{3}$	0	9
x_2	0	$-\frac{1}{3}$	$\frac{2}{3}$	0	4
x_1	1	$-\frac{2}{3}$	$\frac{1}{3}$	0	1
x_8	0	$\frac{2}{3}$	$-\frac{1}{3}$	1	2

Once again, we read new values of B^{-1}, $\bar{b}$, and π from this tableau:

$$B^{-1} = \begin{bmatrix} -\frac{1}{3} & \frac{2}{3} & 0 \\ -\frac{2}{3} & \frac{1}{3} & 0 \\ \frac{2}{3} & -\frac{1}{3} & 1 \end{bmatrix}, \ \bar{b} = \begin{bmatrix} 4 \\ 1 \\ 2 \end{bmatrix}, \ \pi = \begin{bmatrix} -\frac{4}{3} & \frac{5}{3} & 0 \end{bmatrix}.$$

We start the third iteration by calculating the reduced costs:

$$\bar{c}_3 = c_3 - \pi A_3 = 1 - \begin{bmatrix} -\frac{4}{3} & \frac{5}{3} & 0 \end{bmatrix} \begin{bmatrix} 1 \\ 1 \\ 1 \end{bmatrix} = \frac{2}{3},$$

$$\bar{c}_4 = c_4 - \pi A_4 = -2 - \begin{bmatrix} -\frac{4}{3} & \frac{5}{3} & 0 \end{bmatrix} \begin{bmatrix} 2 \\ 0 \\ 1 \end{bmatrix} = \frac{2}{3},$$

and similarly,

$$\bar{c}_5 = -\frac{2}{3}, \quad \bar{c}_6 = \frac{4}{3}, \quad \text{and } \bar{c}_7 = -\frac{5}{3}.$$

So, x_6 is chosen as the next entering variable. Once again, we calculate the updated column $\bar{A}_6$:

$$\bar{A}_6 = B^{-1} A_6 = \begin{bmatrix} -\frac{1}{3} & \frac{2}{3} & 0 \\ -\frac{2}{3} & \frac{1}{3} & 0 \\ \frac{2}{3} & -\frac{1}{3} & 1 \end{bmatrix} \begin{bmatrix} 1 \\ 0 \\ 0 \end{bmatrix} = \begin{bmatrix} -\frac{1}{3} \\ -\frac{2}{3} \\ \frac{2}{3} \end{bmatrix}.$$

The ratio test indicates that x_8 is the leaving variable, since it is the basic variable in the only row where $\bar{A}_6$ has a positive coefficient. Now we pivot on the following tableau:

| Basic | | Init. basics | | | |
var.	x_6	x_6	x_7	x_8	RHS
Z	$-\frac{4}{3}$	$-\frac{4}{3}$	$\frac{5}{3}$	0	9
x_2	$-\frac{1}{3}$	$-\frac{1}{3}$	$\frac{2}{3}$	0	4
x_1	$-\frac{2}{3}$	$-\frac{2}{3}$	$\frac{1}{3}$	0	1
x_8	$\frac{2}{3}^{*}$	$\frac{2}{3}$	$-\frac{1}{3}$	1	2

Pivoting yields:

| Basic | | Init. basics | | | |
var.	x_6	x_6	x_7	x_8	RHS
Z	0	-0	1	2	13
x_2	0	0	$\frac{1}{2}$	$\frac{1}{2}$	5
x_1	0	0	0	1	3
x_6	1	1	$-\frac{1}{2}$	$\frac{3}{2}$	3

The new value of the vector π is given by:

$$\pi = [0\ 1\ 2].$$

Using π we compute:

$$\bar{c}_3 = c_3 - \pi A_3 = 1 - [0\ 1\ 2]\begin{bmatrix} 1 \\ 1 \\ 1 \end{bmatrix} = -2,$$

$$\bar{c}_4 = c_4 - \pi A_4 = -2 - [0\ 1\ 2]\begin{bmatrix} 2 \\ 0 \\ 1 \end{bmatrix} = -4,$$

$$\bar{c}_5 = c_5 - \pi A_5 = 0 - [0\ 1\ 2]\begin{bmatrix} 0 \\ 1 \\ 1 \end{bmatrix} = -3,$$

$$\bar{c}_7 = c_7 - \pi A_7 = 0 - [0\ 1\ 2]\begin{bmatrix} 0 \\ 1 \\ 0 \end{bmatrix} = -1,$$

$$\bar{c}_8 = c_8 - \pi A_8 = 0 - [0\ 1\ 2]\begin{bmatrix} 0 \\ 0 \\ 1 \end{bmatrix} = -2.$$

Since all the $\bar{c}_i$ values are negative we conclude that the last basis is optimal. The optimal solution is:

$$x_1 = 3,\ x_2 = 5,\ x_6 = 3,\ x_3 = x_4 = x_5 = x_7 = x_8 = 0,\ \text{and } z = 13.$$

Exercise D.1   Consider the following linear programming problem:

$$\max Z = 20x_1 + 10x_2$$

$$x_1 - x_2 + x_3 = 1$$
$$3x_1 + x_2 + x_4 = 7$$
$$x_1 \geq 0,\ x_2 \geq 0,\ x_3 \geq 0,\ x_4 \geq 0.$$

The initial simplex tableau for this problem is given below:

Basic var.	Z	Coefficient of				RHS
		x_1	x_2	x_3	x_4	
Z	1	-20	-10	0	0	0
x_3	0	1	-1	1	0	1
x_4	0	3	1	0	1	7

Optimal set of basic variables for this problem happen to be $\{x_2, x_3\}$. Write the basis matrix B for this set of basic variables and determine its inverse. Then, using the algebraic representation of the simplex tableau given in Appendix D, determine the optimal tableau corresponding to this basis.

Exercise D.2   One of the insights of the algebraic representation of the simplex tableau we considered in Appendix D is that the simplex tableau at any iteration can be computed from the initial tableau and the matrix B^{-1}, the inverse of the current basis matrix. Using this insight, one can easily answer many types of "what if" questions. As an example, consider the LP problem given in the previous exercise. What would happen if the right-hand-side coefficients in the initial representation of the example above were 2 and 5 instead of 1 and 7? Would the optimal basis $\{x_2, x_3\}$ still be optimal? If yes, what would the new optimal solution and new optimal objective value be?

References

1. F. Alizadeh and D. Goldfarb, Second-order cone programming. *Mathematical Programming*, **95**:1 (2003), 3–51.
2. A. Altay-Salih, M. Ç. Pınar, and S. Leyffer, Constrained nonlinear programming for volatility estimation with GARCH models. *SIAM Review*, **45**:3 (2003), 485–503.
3. F. Anderson, H. Mausser, D. Rosen, and S. Uryasev, Credit risk optimization with conditional value-at-risk criterion. *Mathematical Programming B*, **89** (2001), 273–91.
4. Y. Baba, R. F. Engle, D. Kraft, and K. F. Kroner, *Multivariate Simultaneous Generalized ARCH*. Technical report, Department of Economics, University of California San Diego (1989).
5. V. S. Bawa, S. J. Brown, and R. W. Klein, *Estimation Risk and Optimal Portfolio Choice* (Amsterdam: North-Holland, 1979).
6. A. Ben-Tal, A. Goyashko, E. Guslitzer, and A. Nemirovski, Adjustable robust solutions of uncertain linear programs. *Mathematical Programming*, **99**:2 (2004), 351–76.
7. A. Ben-Tal, T. Margalit, and A. N. Nemirovski, Robust modeling of multi-stage portfolio problems. In H. Frenk, K. Roos, T. Terlaky, and S. Zhang, editors, *High Performance Optimization* (Dordrecht: Kluwer, 2002), pp. 303–28.
8. A. Ben-Tal and A. N. Nemirovski, Robust convex optimization. *Mathematics of Operations Research*, **23**:4 (1998), 769–805.
9. A. Ben-Tal and A. N. Nemirovski, Robust solutions of uncertain linear programs. *Operations Research Letters*, **25**:1 (1999), 1–13.
10. M. Bénichou, J. M. Gauthier, P. Girodet, G. Hentges, G. Ribière, and O. Vincent, Experiments in mixed-integer linear programming. *Mathematical Programming*, **1** (1971), 76–94.
11. D. Bertsimas and I. Popescu, On the relation between option and stock prices: a convex programming approach. *Operations Research*, **50** (2002), 358–74.
12. D. Bienstock, Computational study of a family of mixed-integer quadratic programming problems. *Mathematical Programming A*, **74** (1996), 121–40.
13. J. R. Birge and F. Louveaux, *Introduction to Stochastic Programming* (New York: Springer, 1997).
14. F. Black and R. Litterman, Global portfolio optimization. *Financial Analysts Journal*, **48**:5 (1992), 28–43.
15. F. Black and M. Scholes, The pricing of options and corporate liabilities. *Journal of Political Economy*, **81** (1973), 673–59.

16. P. T. Boggs and J. W. Tolle, Sequential quadratic programming. *Acta Numerica*, **4** (1996), 1–51.

17. T. Bollerslev, Generalized autoregressive conditional heteroskedasticity. *Journal of Econometrics*, **31** (1986), 307–27.

18. T. Bollerslev, R. F. Engle, and D. B. Nelson, GARCH models. In R. F. Engle and D. L. McFadden, editors, *Handbook of Econometrics*, vol. 4 (Amsterdam: Elsevier, 1994), pp. 2961–3038.

19. D. R. Cariño, T. Kent, D. H. Myers, C. Stacy, M. Sylvanus, A. L. Turner, K. Watanabe, and W. Ziemba, The Russell–Yasuda Kasai model: an asset/liability model for a Japanese insurance company using multistage stochastic programming. *Interfaces*, **24** (1994), 29–49.

20. S. Ceria and R. Stubbs, Incorporating estimation errors into portfolio selection. Robust portfolio selection. To appear in *Journal of Asset Management* (2006).

21. V. Chvátal, *Linear Programming* (New York: W. H. Freeman and Company, 1983).

22. T. F. Coleman, Y. Kim, Y. Li, and A. Verma, *Dynamic Hedging in a Volatile Market*. Technical report, Cornell Theory Center (1999).

23. T. F. Coleman, Y. Li, and A. Verma, Reconstructing the unknown volatility function. *Journal of Computational Finance*, **2**:3 (1999), 77–102.

24. G. Cornuéjols, M. L. Fisher, and G. L. Nemhauser, Location of bank accounts to optimize float: An analytic study of exact and approximate algorithms. *Management Science*, **23** (1977), 789–810.

25. J. Cox, S. Ross, and M. Rubinstein, Option pricing: a simplified approach. *Journal of Financial Economics*, **7**:3 (1979), 229–63.

26. M. A. H. Dempster and A. M. Ireland, A financial expert decision support system. In G. Mitra, editor, *Mathematical Models for Decision Support*, vol. F48 of *NATO ASI Series*, pp. 415–40 (1988).

27. R. F. Engle, Autoregressive conditional heteroskedasticity with estimates of the variance of the U.K. inflation. *Econometrica*, **50** (1982), 987–1008.

28. M. Fischetti and A. Lodi, Local branching. *Mathematical Programming B*, **98** (2003), 23–47.

29. R. Fletcher and S. Leyffer, *User manual for FILTER/SQP* (Dundee, Scotland: University of Dundee, 1998).

30. D. Goldfarb and G. Iyengar, Robust portfolio selection problems. *Mathematics of Operations Research*, **28** (2003), 1–38.

31. A. J. Goldman and A. W. Tucker, *Linear Equalities and Related Systems* (Princeton, NJ: Princeton University Press, 1956) pp. 53–97.

32. R. Gomory, Outline of an algorithm for integer solutions to linear programs. *Bulletin of the American Mathematical Society*, **64** (1958), 275–8.

33. R. Gomory, *An Algorithm for the Mixed Integer Problem*. Technical Report RM-2597, The Rand Corporation (1960).

34. J. Gondzio and R. Kouwenberg, High performance computing for asset liability management. *Operations Research*, **49** (2001), 879–91.

35. C. Gourieroux, *ARCH Models and Financial Applications* (New York: Springer-Verlag, 1997).

36. R. Green and B. Hollifield, When will mean-variance efficient portfolios be well-diversified. *Journal of Finance*, **47** (1992), 1785–810.

37. E. Guslitser, Uncertainty-immunized solutions in linear programming. Master's thesis, The Technion, Haifa (2002).

38. B. Halldorsson and R. H. Tütüncü, An interior-point method for a class of saddle point problems. *Journal of Optimization Theory and Applications*, **116**:3 (2003), 559–90.

39. R. Hauser and D. Zuev, Robust portfolio optimisation using maximisation of min eigenvalue methodology. Presentation at the Workshop on Optimization in Finance, Coimbra, Portugal, July 2005.

40. S. Herzel, *Arbitrage Opportunities on Derivatives: A Linear Programming Approach.* Dynamics of Continuous, Discrete and Impulsive Systems, Series B: Applications and Algorithms (2005), 589–606.

41. K. Hoyland and S. W. Wallace, Generating scenario trees for multistage decision problems. *Management Science*, **47** (2001), 295–307.

42. P. Jorion, Portfolio optimization with tracking error constraints. *Financial Analysts Journal*, **59**:5 (2003), 70–82.

43. N. K. Karmarkar, A new polynomial-time algorithm for linear programming. *Combinatorica*, **4** (1984), 373–95.

44. L. G. Khachiyan, A polynomial algorithm in linear programming. *Soviet Mathematics Doklady*, **20** (1979), 191–4.

45. P. Klaassen, Comment on 'generating scenario trees for multistage decision problems'. *Management Science*, **48** (2002), 1512–16.

46. H. Konno and H. Yamazaki, Mean-absolute deviation portfolio optimization model and its applications to Tokyo stock market. *Management Science*, **37** (1991), 519–31.

47. P. Kouvelis and G. Yu, *Robust Discrete Optimization and its Applications* (Amsterdam: Kluwer, 1997).

48. R. Kouwenberg, Scenario generation and stochastic programming models for asset liability management. *European Journal of Operational Research*, **134** (2001), 279–92.

49. R. Lagnado and S. Osher, Reconciling differences. *Risk*, **10** (1997), 79–83.

50. A. H. Land and A. G. Doig, An automatic method for solving discrete programming problems. *Econometrica*, **28** (1960), 497–520.

51. P. Laurence and T. H. Wang, What's a basket worth? *Risk*, **17**:2 (2004), 73–8.

52. R. Litterman and Quantitative Resources Group, *Modern Investment Management: An Equilibrium Approach* (Hoboken, NJ: John Wiley and Sons, 2003).

53. M. S. Lobo, L. Vandenberghe, S. Boyd, and H. Lebret, Applications of second-order cone programming. *Linear Algebra and Its Applications*, **284** (1998), 193–228.

54. H. Markowitz, Portfolio selection. *Journal of Finance*, **7** (1952), 77–91.

55. R. C. Merton, Theory of rational option pricing. *Bell Journal of Economics and Management Science*, **4**:1 (1973), 141–83.

56. R. O. Michaud, The Markowitz optimization enigma: is optimized optimal? *Financial Analysts Journal*, **45** (1989), 31–42.

57. R. O. Michaud, *Efficient Asset Management* (Boston, MA: Harvard Business School Press, 1998).

58. J. J. Moré and S. J. Wright, *Optimization Software Guide* (Philadelphia, PA: SIAM, 1993).

59. J. M. Mulvey, Generating scenarios for the Towers Perrin investment system. *Interfaces*, **26** (1996), 1–15.

60. Yu. Nesterov and A. Nemirovski, *Interior-Point Polynomial Algorithms in Convex Programming* (Philadelphia, PA: SIAM, 1994).

61. J. Nocedal and S. J. Wright, *Numerical Optimization* (New York: Springer-Verlag, 1999).

62. M. Padberg and G. Rinaldi, Optimization of a 532-city traveling salesman problem by branch and cut, *Operations Research Letters*, **6** (1987), 1–8.

63. M. Pınar, *Minimum Risk Arbitrage with Risky Financial Contracts*. Technical report, Bilkent University, Ankara, Turkey (2001).
64. M. Pınar and R. H. Tütüncü, Robust profit opportunities in risky financial portfolios. *Operations Research Letters*, **33**:4 (2005), 331–40.
65. I. Pólik and T. Terlaky, *S-lemma: A Survey*. Technical Report 2004/14, AdvOL, McMaster University, Department of Computing and Software (2004).
66. C.R. Rao, *Linear Stastistical Inference and its Applications* (New York: Wiley, 1965).
67. R. T. Rockafellar, *Convex Analysis* (Princeton, NJ: Princeton University Press, 1970).
68. R. T. Rockafellar and S. Uryasev, Optimization of conditional value-at-risk. *The Journal of Risk*, **2** (2000), 21–41.
69. E. I. Ronn, A new linear programming approach to bond portfolio management. *Journal of Financial and Quantitative Analysis*, **22** (1987), 439–66.
70. B. Rustem and M. Howe, *Algorithms for Worst-Case Design and Applications to Risk Management* (Princeton, NJ: Princeton University Press, 2002).
71. A. Ruszczynski and R. J. Vanderbei, Frontiers of stochastically nondominated portfolios. *Econometrica*, **71**:4 (2003), 1287–97.
72. S. M. Schaefer, Tax induced clientele effects in the market for British government securities. *Journal of Financial Economics*, **10** (1982), 121–59.
73. K. Schöttle and R. Werner, *Benefits and Costs of Robust Conic Optimization*. Technical report, T. U. München (2006).
74. W. F. Sharpe, Determining a fund's effective asset mix. *Investment Management Review*, December (1988), 59–69.
75. W. F. Sharpe, Asset allocation: Management style and performance measurement. *Journal of Portfolio Management*, Winter (1992), 7–19.
76. W. F. Sharpe, The Sharpe ratio. *Journal of Portfolio Management*, Fall (1994), 49–58.
77. M. C. Steinbach, Markowitz revisited: mean-variance models in financial portfolio analysis. *SIAM Review*, **43**:1 (2001), 31–85.
78. J. F. Sturm, Using SeDuMi 1.02, a matlab toolbox for optimization over symmetric cones. *Optimization Methods and Software*, **11–12** (1999), 625–53.
79. M. J. Todd, Semidefinite optimization. *Acta Numerica*, **10** (2001), 515–60.
80. R. H. Tütüncü and M. Koenig, Robust asset allocation. *Annals of Operations Research*, **132** (2004), 157–87.
81. R. H. Tütüncü, K. C. Toh, and M. J. Todd, Solving semidefinite-quadratic-linear programs using SDPT3. *Mathematical Programming*, **95** (2003), 189–217.
82. S. Uryasev, Conditional value-at-risk: Optimization algorithms and applications. *Financial Engineering News*, **14** (2000), 1–6.
83. L. A. Wolsey, *Integer Programming*. John Wiley and Sons, New York, 1988.
84. D. B. Yudin and A. S. Nemirovski, Informational complexity and efficient methods for the solution of convex extremal problems (in Russian). *Ekonomika i Matematicheskie metody*, 12 (1976), 357–69. (English translation: *Matekon* **13**(2), 3–25.)
85. Y. Zhao and W. T. Ziemba, The Russell–Yasuda Kasai model: a stochastic programming model using a endogenously determined worst case risk measure for dynamic asset allocation. *Mathematical Programming B*, **89** (2001), 293–309.

Index